Knowledge Regulation and National Security in Postwar America

Knowledge Regulation and National Security in Postwar America

MARIO DANIELS AND JOHN KRIGE

THE UNIVERSITY OF CHICAGO PRESS   CHICAGO AND LONDON

The University of Chicago Press, Chicago 60637
The University of Chicago Press, Ltd., London
© 2022 by The University of Chicago
All rights reserved. No part of this book may be used or reproduced in any manner whatsoever without written permission, except in the case of brief quotations in critical articles and reviews. For more information, contact the University of Chicago Press, 1427 E. 60th St., Chicago, IL 60637.
Published 2022
Printed in the United States of America

31 30 29 28 27 26 25 24 23 22   1 2 3 4 5

ISBN-13: 978-0-226-81748-4 (cloth)
ISBN-13: 978-0-226-81753-8 (paper)
ISBN-13: 978-0-226-81752-1 (e-book)
DOI: https://doi.org/10.7208/chicago/9780226817521.001.0001

Library of Congress Cataloging-in-Publication Data

Names: Daniels, Mario, author. | Krige, John, author.
Title: Knowledge regulation and national security in postwar America / Mario Daniels and John Krige.
Description: Chicago ; London : The University of Chicago Press, 2022. | Includes bibliographical references and index.
Identifiers: LCCN 2021040769 | ISBN 9780226817484 (cloth) | ISBN 9780226817538 (paperback) | ISBN 9780226817521 (ebook)
Subjects: LCSH: Export controls—United States—History—20th century. | Technology transfer—Government policy—United States. | Technology and international relations—United States. | National security—United States. | United States—Foreign relations. | United States—Commercial policy. | United States—Politics and government—1945–1989. | United States—Politics and government—1989–
Classification: LCC HF1414.55.U6 D36 2022 | DDC 382/.640973—dc23
LC record available at https://lccn.loc.gov/2021040769

Contents

List of Abbreviations   vii

CHAPTER 1. Introduction: What Are Export Controls, and Why Do They Matter?   1

PART 1

CHAPTER 2. The Invention of Export Controls over Unclassified Technological Data and Know-How (1917–45)   37

CHAPTER 3. The Cold War National Security State and the Export Control Regime   61

PART 2

CHAPTER 4. The Recalibration of American Power, the Bucy Report, and the Reshaping of Export Controls in the 1970s   101

CHAPTER 5. The Reagan Administration's Attempts to Control Soviet Knowledge Acquisition in Academia   135

CHAPTER 6. Academia Fights Back: The Corson Panel and the Fundamental Research Exclusion   166

PART 3

CHAPTER 7. "Economic Security" and the Politics of Export Controls over Technology Transfers to Japan in the 1980s   193

CHAPTER 8. Paradigm Shifts in Export Control Policies by Reagan, Bush, and Clinton and the Evolving US-China Relations   231

CHAPTER 9. The Conflict over Technology Sharing in Clinton's Second Term: The Cox Report and the Use of Chinese Launchers   258

PART 4

CHAPTER 10. Epilogue: Export Controls, US Academia, and the Chinese-American Clash during the Trump Administration   299

Notes   337

Index   429

Abbreviations

AAAS	American Association for the Advancement of Science
ACDA	Arms Control and Disarmament Agency
ACEP	Advisory Committee on Export Policy
AI	Artificial Intelligence
AVS	American Vacuum Society
CASC	China Aerospace Corporation
CATIC	China National Aero-Technology Import and Export Corporation
CCL	Commodity Control List
CFIUS	Committee on Foreign Investment in the United States
CGWIC	China Great Wall Industry Corporation
CIA	Central Intelligence Agency
Cocom	Coordinating Committee for Multilateral Export Controls
CORSI	Committee on Release of Scientific Information
CREST	CIA Records Search Tool
CSIS	Center for Strategic and International Studies
DARPA	Defense Advanced Research Projects Agency
DEAC	Deemed Export Advisory Committee
DoC	Department of Commerce
DoD	Department of Defense
DoE	Department of Energy
DoS	Department of State
DSB	Defense Science Board
DTSA	Defense Technology Security Administration

EAA	Export Administration Act
EAR	Export Administration Regulations
ECA	Export Control Act
ECRA	Export Control Reform Act
EO	Executive Order
EPCI	Enhanced Proliferation Control Initiative
FBI	Federal Bureau of Investigation
FDI	Foreign direct investment
FNUS	Foreign nationals in the United States
FRE	Fundamental Research Exclusion
GAO	General Accounting Office
HPC	High-performance computer
HRI	Hydrocarbon Research Incorporated
IAEA	International Atomic Energy Agency
ICBM	Intercontinental ballistic missile
IEEE	Institute of Electrical and Electronic Engineers
IP	Intellectual property
IPR	Intellectual property rights
IRC	Independent Review Committee
ITAR	International Traffic in Arms Regulations
KGB	(Soviet) Committee for State Security
MCTL	Militarily Critical Technologies List
MFN	Most-favored nation
MIT	Massachusetts Institute of Technology
MTCR	Missile Technology Control Regime
MTOPS	Million theoretical operations per second
NARA	National Archives and Records Administration
NASA	National Aeronautics and Space Administration
NATO	North Atlantic Treaty Organization
NGO	Nongovernmental organization
NIH	National Institutes of Health
NSA	National Security Agency
NSC	National Security Council
NSDD	National Security Decision Directive

NSF	National Science Foundation
OECD	Organization for Economic Cooperation and Development
OPEC	Organization of the Petroleum Exporting Countries
OSRD	Office of Scientific Research and Development
OSTP	Office of Science and Technology Policy
OTA	Office of Technology Assessment
PDR	Processing Data Rate
PI	Principal investigator
PLA	People's Liberation Army
PRC	People's Republic of China
RMA	Revolution in Military Affairs
SBU	Sensitive but unclassified
SCI	Sensitive Compartmentalized Information
SDI	Strategic Defense Initiative
STEM	Science, technology, engineering, and mathematics
TPPC	Trade Promotion Coordinating Committee
TTCP	Technology Transfer Control Plan
TTP	Thousand Talents Plan
USML	United States Munitions List
UTI	Unclassified Technological Information Committee
VHSIC	Very high speed integrated circuits
WMDs	Weapons of mass destruction
ZTE	Zhongxing Telecommunication Equipment Company Limited

CHAPTER ONE

Introduction

*What Are Export Controls, and
Why Do They Matter?*

In December 1962, the New York–based company Hydrocarbon Research Incorporated (HRI) was the first American firm punished for exporting unclassified intangible scientific-technological knowledge. The case set a legal precedent that had far-reaching implications. It was the first official interpretation of the meaning of "technical data" in the US export control regulations specified in the *Federal Register*. The case also summed up some of the most important results of two decades of reflection within the US government on how to control knowledge and information in an age of intensifying global communication. A discussion of the issues and misunderstandings that it raised can help orient our readers in the intricacies of knowledge controls before we plunge into our detailed account of the history of export controls in the US national security state, whose roots stretch back to World War I.

In 1959 HRI signed a $17 million contract with the Rumanian government to build an oil-refining complex in Ploesti. In the following years, HRI built facilities "of the latest design" to turn low-grade petroleum into high-grade motor and aviation fuels, plastics, synthetic rubber, and synthetic fibers.[1] For the work in Ploesti, HRI drew on the support of two European partners, Hydrocarbon Engineering SARL in Paris, wholly owned by HRI, and the German Hydrocarbon Mineralöl GmbH in Düsseldorf, an HRI affiliate. The three companies had cooperated before in the design, construction, and operation of a similar plant in France, and they reused and adapted the plans and technical data from this project in Rumania.[2] According to the contract, HRI turned over to the Rumanians "all of the process designs, plans and specifications" of the plant, "thus enabling the Rumanians to duplicate it elsewhere."³ Additionally, Western engineers were dispatched to make things work on the spot.

The president of Hydrocarbon Research Incorporated, Percival C. Keith, was perfectly aware that, in the context of the Cold War, Rumania and France were separated by the Iron Curtain, and he knew of his obligation to follow US export control regulations. He consulted early on with the Department of Commerce, which was in charge of controlling the exports of dual-use technologies. The DoC informed Keith "that HRI could engage in the transaction only if HRI, its associates and any HRI personnel who might go abroad in connection with it, were to use no unpublished United States technical data" as defined by the regulations.⁴ But even though Keith attempted to tread carefully in the bureaucratic minefield of export control rules, he did not grasp the far-reaching implications of their sophisticated understanding of knowledge flows. Keith relied on his own commonsense understanding of knowledge, with the result that serious penalties were imposed on him and on his company.

The Consent Denial and Probation Order that summarized the affair in order to clarify the official reading of export controls stated that Keith's offense was that he "construed unpublished data in the engineering field to be limited to fundamental, secret know-how held by persons or company." For him "published data," that is, data that was exempt from export regulations, meant "any information or documents which a competent, experienced engineer thought he could furnish or prepare with his own mind using well-recognized engineering principles, texts, technical articles, and patents."⁵ For Keith the concept of exempted published data also covered the export and disclosure of all the general knowledge acquired by the HRI engineers when they built the plant in France. This plant served as the blueprint for the subsequent Ploesti project that Keith thought could be shared with Rumania without seeking an export license.⁶

Keith's defense revolved around the dichotomy of openness and secrecy. In his reading, export controls dovetailed with the concept of trade secrecy, focusing only on very few crown jewels of company knowledge, but leaving everything else unregulated. This wide sphere of public domain included for Keith not least the minds, skills, and experience of people. Apart from some closely-to-be-guarded secrets, the knowledge an individual had acquired in the course of professional practice and education had to be beyond the reach of government regulation.

Keith was mistaken. For the US government, the duality of "secret" and "open" was not as clear-cut as Keith obviously assumed. For US export regulators, there was an expansive "gray zone" in between, in which information was, without being classified, not openly accessible but under government control. In this zone, transmissions of knowledge were not tantamount to general publication. Even an export control license for a set of data did not free knowledge from its regulatory shackles; each license covered only specific, clearly delimited acts of communication. Moreover, the brains of engineers were as a matter of principle subject to export regulation wherever they carried knowledge that originated in the United States. Thus, people who traveled were conceived of as data exporters, and knowledge was ascribed a national identity just like an individual's citizenship.

In practice, these principles had far-reaching consequences as the HRI case made clear. The Commerce Department subscribed to the "mosaic theory," a basic concept rather than a full-fledged theory used in the realm of classification and intelligence analysis. It assumes that unrelated and innocuous pieces of information could in combination yield a bigger picture that could be useful to foreign intelligence services and that was so dangerous that it required classification. The Commerce Department argued in the HRI case that "whether the technical details contained herein could separately be found in published literature or worked out by simple engineering did not deprive them of their unpublished status when combined and incorporated in the overall plans and specifications." The lines between "open" and "controlled" were thus decidedly blurred.[7]

For the government, US technicians traveling to Rumania were not free to share their knowledge even if they did not divulge clearly defined trade secrets. Without a Department of Commerce license even "the application by such persons of their United States–origin know-how and experience constituted an unauthorized use and export of unpublished data." Nor did the theory of export controls target only US citizens. It was assumed that American knowledge kept a stable national identity that remained unchanged no matter where it had traveled and whoever handled it. As the Commerce Department stipulated, the use by HRI and its affiliates of "the unpublished US technical data in the French plant, the subsequent use of the *redrawn* French plant documents by United States and *foreign technicians* employed . . . to design the Rumanian plant, the *commingling* hereby of the unpublished US data with other published (or unpublished) technical data of whatever origin, did not serve to deprive the original data for the French plant of their status as unpublished technical data nor their United States origin."[8] In short, a German engineer, who had never set foot on US soil, working in Ploesti and using a blueprint drawn in New York but received from France, was still under the umbrella of US export controls, specifically of regulations that targeted reexports.[9] Their reach transcended the US border and applied extraterritorially to any geographical area to which US knowledge traveled. Knowledge had a "US character,"[10] or an imprint of "Americanness," which it did not lose by moving abroad or when it was appropriated by a foreign national's mind. Even products made using US data still carried this imprint.[11]

As esoteric as these considerations may have seemed, they set a legal precedent and had considerable economic impact on HRI. Even though the Department of Commerce conceded that the violation of the law was not willful but due to negligence, its sanctions were "inordinately severe."[12] HRI was barred for five years from any trade with the Sino-Soviet bloc and Cuba. Keith personally lost his export control privileges entirely for six months, that is, he was excluded from any foreign trade. The company and its president were placed under probation for three years, at risk of losing all their US export privileges if they made the mistake again. To enforce these penalties, Keith and his firm were placed under special surveillance for two years, forcing them to submit all relevant documents and even information on oral communication for any transaction subject to export controls to the Department of Commerce. The DoC had the right to stop any further transaction if it had merely the suspicion of an irregularity.[13] In addition, HRI faced considerable reputational damage. Some negative press, but more importantly being blacklisted in the *Federal Register* and in the export control report of the Commerce Department to the president and the Congress, left the impression that Hydrocarbon had violated the rules of trading with the enemy beyond the Iron Curtain.[14]

In 1962, the technical data regulations were not at all new, and HRI was not the first company that had to deal with the complexities of data export control regulations. But apparently, the Department of Commerce deliberately used the Rumania deal to make an example of Hydrocarbon and to establish a legally strict, far-reaching reading of its export control regulations for technical data. The DoC's investigations spanned more than one and a half years,[15] and the closed-door hearings with HRI involved no fewer than thirty people, inflicting high costs on the company. In the eyes of J. Forrester Davison, who was one of HRI's lawyers in the case, the DoC tried to overwhelm the company by amassing a huge amount of evidence in order to reach an out-of-court consent order.[16] Such an order would provide the DoC with the "opportunity to give a *unilateral* interpretation of its own regulations," without the intervention of the courts.[17] But more importantly, it had the effect of a warning signal.[18] Since US controls over goods and data cast such a wide net, they were in practice enforceable only if the business community cooperated and complied. Legal deterrence was supposed to impose a kind of self-censorship.[19]

Stimulated by the Hydrocarbon case, technical data export controls became widely discussed, especially in scholarly law journals.[20] The contemporary literature clearly understood the incredibly far-reaching implications of the legal precedent especially for academic freedom and the freedom of travel. And yet the public and scholarly interest in data controls was short-lived and had effectively petered out by 1967.[21] Data export controls, however, did not vanish in the late 1960s. They were continuously (and mostly quietly) administered behind the closed doors of the US federal bureaucracy.

More importantly, the literature we have on data export controls in the 1960s—and for that matter, the entire scholarly output up to the present day—has not understood that the data controls that defined the Hydrocarbon case were the outcome of a policy-making process and control practices that began in the late 1930s. The first two chapters of this book are the first historical account that we know of, of the origins of export controls over "technical data." It will show how World War II changed the way bureaucrats in the federal government thought about the circulation of scientific-technological knowledge and why they decided that knowledge flows needed oversight. The subsequent chapters explore the expansion and implementation of instruments to control the global circulation of sensitive knowledge throughout the Cold War, and up to the present day.

The developing concepts and practices of export controls were shaped by the ideology of "national security," an intellectual innovation of the 1940s. It fostered the perception that certain kinds of knowledge posed a danger to the very existence of the United States if shared too freely. The national security thinkers in the federal government developed a highly sophisticated idea of knowledge and a complex understanding of technology transfers, based on the experience of the mobilization of science and technology for World War II. A successful technology and knowledge transfer was conceived as the interplay of the movements of information, people, and things.

Export controls are usually described as regulating the mobility of goods. By contrast, this book will show that the idea of information and knowledge control was at the very core of the US export control regime. Scientific-technological information had to be regulated because it was the foundation of all military and economic artifacts. Conversely, trade with these artifacts had to be supervised because they embodied the knowledge used to produce them. Reverse engineering could make the embedded information accessible again. And finally, people were probably the most effective transmitters of information: indeed, not only their brains, but also their hands carried tacit knowledge. Hence they also had to be placed under surveillance.

The export control community within the US government understood all this in the 1940s. It was an enormous challenge to translate these concepts into policy within the framework of a liberal democracy. Moreover, the policy-making process was constantly propelled and catalyzed by the pull and push of the larger forces of total war, the Cold War, rapidly changing US foreign policy, and technological change. The overall trend and trajectory was an incremental expansion of export controls not only over goods, but also over information and people during the 1940s and 1950s. It is only against this backdrop that it is possible to make sense of the Hydrocarbon Research case of 1962 and of all the developments that came afterward.

Every single day beginning in the 1940s, US export controls have intervened in the global sharing of scientific-technological knowledge. They subjected universities, research institutions, companies, and foreign national individuals to national security and foreign policy considerations, circumscribing their latitude in their international communication of knowledge, know-how, and technical information. Export regulations demarcate the borderline where the reach of "free-market" ideology and liberal democratic freedoms end and the realm of national security begins. Such regulations hem in First Amendment rights and academic freedom, infringe on the freedom of travel, and inhibit the flow of foreign direct investment. Their main goal is to target enemies and adversaries and to deny them access to scientific and technological knowledge and know-how. However, they also affect allies as well as US citizens and institutions. Export controls were instrumental to waging economic warfare in World War II. After 1945 their main target was the Eastern bloc, and export controls played a crucial role in constructing and cementing the Iron Curtain and in isolating Mao's China. They played a prominent role in managing and policing détente in the 1970s and then in biting deeply into Soviet ambitions in the Reagan era (while simultaneously being relaxed

INTRODUCTION   7

to encourage China's modernization, inaugurated by Deng Xiaoping in the late 1970s). They deeply affected not only East-West trade, but also inter-West relations, and time and again caused harsh conflicts and deep rifts in the Western alliance, most prominently during the so-called Second Cold War in the 1980s.[22] Moreover, in the relations between the West and economic and strategic competitors in the global arena, export controls were and are being used to balance national and international trade interests with concerns of the military and national security. This is nowhere more visible today than in the US trade war with China, which is above all a war over access to advanced knowledge in which export controls play a crucial role to mediate what can cross borders and what cannot.[23] This book charts these developments by focusing on sites that have had a major impact on the evolution of the American export control system over knowledge flows from the 1940s to the present day.

The Place of Export Controls in the Regulatory Landscape: A Primer

Export controls are just one of an increasing, and increasingly invasive, regulatory system devised by the architects of the US national security state to restrict the flow of information, people, and commodities across the national border to friend and to foe alike. They occupy a space alongside more familiar and widely studied restrictions on the proliferation of weapons of mass destruction, although they were increasingly integrated into that realm after the Cold War to deal with asymmetric threats from "rogue states" and terrorist nonstate actors. They coexist with realms of secrecy and classification that build high walls around select kinds of information and knowledge deemed crucial to the security of the nation that have been intensively studied by historians of science and technology. They provide criteria for visa and immigration policies that restrict the traffic of brainpower into and out of the country. Above all, they help police a vast gray zone lying between total restriction and unimpeded circulation, regulating transnational traffic in the name of national security.

The export control system has two main pillars. The first is the regulation of dual-use technologies (i.e., technologies with a civilian as well as a military application), which is administered by the Department of Commerce. Its first statutory basis was the Export Control Act of 1949. It has gone through numerous renditions and variations all the way up to today's

Export Control Reform Act, which was signed into law in 2018. The Department of Commerce implements this law through the Export Administration Regulations (EAR), whose core element is a list of controlled dual-use items, the Commodity Control List, or CCL. This pillar and the messy world of dual-use technologies is the main focus of our book.

From time to time, however, we will also touch on the second pillar, which comprises control regulations and lists for military items. The statutory basis today is the Arms Export Control Act, implemented by the International Traffic in Arms Regulations (ITAR), including a detailed control list (the United States Munitions List, USML). ITAR was and still is administered by the Department of State in close cooperation with the Department of Defense, which also plays an important role in the realm of EAR. There are other pillars we touch on only sporadically, even though they grew more important in the course of the Cold War: export controls for weapons of mass destruction and their delivery systems, and the implementation of the sprawling US use of economic sanctions (supervised by the Department of the Treasury).[24]

The US government complemented these national institutions by international agreements, like that establishing the Coordinating Committee for Multilateral Export Controls (Cocom) in 1949–50. Multilateral export controls were and still are the key tool for the implementation of nonproliferation policy. The Nuclear Non-proliferation Treaty, signed in 1968 by Britain, the United States, and the Soviet Union, complemented national and international efforts to prevent the spread of atomic weapons without stifling the sharing of knowledge for peaceful purposes. The Zangger Committee and the Nuclear Suppliers Group are important, albeit loosely organized multilateral bodies that help to enforce nuclear export controls centered around the International Atomic Energy Agency.[25] The Australia Group, founded in 1985, attempts to coordinate national export control policies to frustrate the spread of chemical and biological weapons.[26] Impeding the spread of missile and drone technology is the goal of the Missile Technology Control Regime, which has existed since 1987.[27]

Export controls should be understood as a powerful instrument in a large, diverse toolbox built by the federal government to regulate the flow of information, people, and commodities across its borders. All the tools in the box (see table 1) complement each other and have considerable overlap in bureaucratic practice.

At the heart of all export controls are lists enumerating the technologies and goods whose circulation must be regulated. To add or to elimi-

TABLE 1. **Toolbox of National Security Regulations**

Export Controls
EAR
ITAR
Controls over technologies related to weapons of mass destruction (for example, nuclear export controls)
Economic sanctions (Office of Foreign Assets Control, OFAC, part of the Treasury Department)
Multilateral export controls (Cocom, Wassenaar, Nuclear Suppliers Group, MTCR, etc.)

Government Secrecy
Classification, based on Executive Orders and statutory laws (Espionage Act, Atomic Energy Act, Invention Secrecy Act)

Foreign Direct Investment Screenings
CFIUS

Travel Restrictions
Visa regulations

Criminal Law and Law Enforcement
Enforcement of export controls and classification
Economic Espionage Act of 1996
Customs

---

nate items—and, by extension to determine whether or not to issue a license—is a complex decision-making process that is tied to the dominant strategic concepts that shape the perspective on the state of international relations and define the overall goals of export controls.

To assess how dangerous a technology, if exported, could be, is incredibly challenging.[28] Setting up an export control list demands a high level of technical expertise as well as information about the global distribution of technology, the flows of global trade, and the composition of foreign military machineries.[29] Accordingly, export control lists are highly technical and very detailed and can be properly understood only by experts in the respective technical field. The regulation of high-performance computers (HPCs), for example, relied on the assessment of computing power. The liberal export policies adopted for them in the 1970s defined the "dangerous" threshold for HPCs that could be exported to Communist countries using a Processing Data Rate (PDR) of thirty-two million bits per second, considerably above the state of the art in the Eastern bloc. HPCs beyond this threshold were "reviewed on a case-by-case basis and strictly limited to demonstrably peaceful applications."[30] There were several other PDR tiers below this threshold, marking a gradual fading out of controls from specific ("validated") licenses with some restrictions, to computer sales without license requirements ("general license").

In technological fields with a high innovation rate, like the computer industry, these export limits were revised on a regular basis to reflect the shift of the cutting edge of development, the "diffusion" of technology to other countries, and the changes in global technology markets. Hence, control lists are not static, though their revision often proceeds more slowly than the changes in the international technological landscape. The export control community has constantly to make complex judgments. Can a specific technology realistically be controlled, or is it so widely available from other than US or Western sources that it would be futile to even try to do so (export controllers call this the "foreign availability" problem)? What is the "end use" of the technology, and can it be "diverted" from a civilian application in, say, a high-energy physics laboratory to model nuclear weapons explosions for the military? What is the state of the art in the recipient country? Can it reverse engineer the technology in question, incorporate it into its technological systems, and thus catch up to the technological level of the United States? If the receiving country is technologically less advanced, even US technology far removed from the cutting edge could advance its technological capacity considerably. In such cases even "obsolete" technology can appear to be in need of tight control.

Export control lists discriminate between the countries of the world on the basis of their relationship to the United States. There would be no export controls without competitive rivals in a world system. The main goal of export control lists is to deny these rivals—be they allies or enemies—substantial technological advantage as best as one can without harming national exports. Not all targets of export controls are equally threatening, however. Even at the height of the Cold War the United States implemented controls to some Iron Curtain countries less stringently than it did to others, not least in order to incentivize changes in political behavior and to sow discord in the Communist world. In the 1960s, for example, exports to the Soviet Union and China were stricter than those to Poland or to Rumania.[31] In short, export controls not only are often technically highly specific but also always reflect the political relations within the international system from the hegemonic point of view of the United States. Depending on the recipient country, export controls are invoked "for all goods [and technical data] shipped to certain countries, or for certain types of goods [and technical data] shipped to certain countries, or for certain types of goods [and technical data] shipped to any country."[32]

What Export Controls Are—and What They Are Not

There are tenacious public and scholarly misconceptions about what export controls are and what they do. It seems therefore just as necessary to point out what they are *not* as it is to explain what their scope and functions are. Often export controls are perceived and treated as an instrument of trade policy. That is not wrong, as export controls are indeed intertwined with US trade policy. The perennial debates about the goals, advantages, and drawbacks of controls are closely linked to concerns about US economic interests abroad, the international competitiveness of US companies vis-à-vis their foreign rivals, the US trade deficit and the balance of payments. But export controls usually work within a different political framework than do, for example, most tariffs. Export controls are an instrument of national security and foreign policy; their trade dimension is subsidiary to this main objective. We will show how, why, and when the lines between national security and trade policies get blurred in theory and practice, but we cannot stress enough that the raison d'être of export controls is to protect national security.

Treating export controls only as a form of trade policy can also easily distract one from one of their main aims, which is also the focus of this book: they regulate the sharing of scientific-technological knowledge and know-how related to high technology at the cutting edge of research and development. Export controls are usually understood as affecting the movement of goods. That is certainly not wrong. Indeed, since the 1940s, export controls have been focused on the entire gamut of high-technology artifacts from jet engines and machine tools to lasers and computers. But the export control regulations reach beyond the physical objects themselves, which can be shipped in crates and containers. In fact, we argue that the control of *intangible knowledge and know-how* was the most important function of export controls. They target objects as carriers of "embodied" knowledge that can, for example, be extracted by means of reverse engineering or actively transferred in face-to-face interactions between an American donor and a foreign receiver. Moreover, these targets are assessed not in isolation, but in relation to the technological system in which they are, or will be incorporated. The overriding question is: "would this export help the foreign country to close a gap it cannot fill with its own knowledge?"

From here it is only a short step to inhibiting the circulation of "technical data." Ever since the 1950s, export controls have increasingly affected the transmission of information—indeed, the terms "technology" and "information" were and still are used interchangeably in the export regulations. They impact all forms of communication, from the shipment of technical manuals to a conversation between an American engineer and his or her foreign colleague. In fact, the verb "to export" means both to send something across the physical border of the United States, and to transmit controlled knowledge from an American person or entity to a foreign person or entity.

Up to the present day, export controls are a policy area in which the US federal executive wields enormous powers over global exchanges since its reach does not stop at the geographical borders of the United States. Export controls defend the American border wherever scientific-technological knowledge is shared. Hence, export controls even monitor communication with foreigners *within* the United States, treating the foreign national as if he or she were an exclave of her home country. The EAR and the ITAR regulate not only the shipment of physical goods but also the transmission of scientific-technological information in all possible forms, from blueprints and manuals to the oral conversations of scientists at a corporate or university laboratory. Hence, export controls even impinge on interpersonal contacts, going so far as to impact US visa legislation.

An emphasis on the impact of export controls on the movement of intangibles makes it easy to confuse them with intellectual property rights (IPR). These are very different regimes. IPR protects the property rights of private actors like companies against the infringement by other private actors in national and international markets. By contrast, export controls protect scientific-technological knowledge in the name of national security. They protect the state. Hence, IPR and export controls operate on different political planes, have different aims and justifications, and, most importantly, are based on completely different laws and regulations.

Intellectual property rights and export controls overlap, however, since private intellectual property is a key regulatory object of export controls. A patent or a trade secret, even though the property of a private company, is subject to export regulations, if its owner wants to share it with a foreign national or foreign government that the American government defines as an actor of national security concern. In fact, the US government uses IP as an instrument of national security policy by forcefully recruiting domestic companies and research institutions for its denial policy against foreign actors. The US government also uses control over privately held IP as leverage to influence the behavior of foreign nationals and governments. In short, in the world of export controls, companies are an extension of state power, and their intellectual property is an instrument of national security and foreign policy.

Export controls should also not be confused with government secrecy.[33] Even though ITAR also covers the export of classified information, the main target of export controls is the regulation of *unclassified* information. They cut through the dichotomy of open versus secret information and carve out a vast "gray zone" (a designation already used during the Cold War) between the realm of government classification and the public sphere. Export controls play out in the ambiguous world of what is today called "sensitive but unclassified" (SBU) information or "controlled unclassified information" (CUI).[34] This information is neither freely and openly communicable nor strictly shielded by secrecy regimes but indeed "controlled" by means other than formally putting a "top secret" stamp on it. Export controls regulate the communication of SBU or CUI through its licensing system: it can be shared after a case-by-case risk assessment. This epistemological and legal gray zone has massively grown since the terrorist attacks on 9/11 and pervades the US government's approach to information control far beyond export controls and in wide swathes of its bureaucratic practice.[35] In fact, "SBU" was not invented in the age of the "war on terror" but was, as we will show, conceived and implemented as early as in the 1940s and was intensely discussed during the 1980s. Moreover, a look at the history of export controls shows that the reach of the US government's control over scientific-technological information goes much further than the already vast system of classification. Thus export controls have complex implications for First Amendment rights, for the values of academic freedom, and for American democracy.[36]

Classification and export controls are, however, complementary. Export controls begin where secrecy ends and carry the logic of information control well into the sphere of what is often mistakenly perceived as open and free of government influence. Moreover, the classification of scientific-technological knowledge and its regulation by means of export controls both have one main goal in common. They defend the US technological lead over both its adversaries, and its rival allies, in the competitive international system.

What Does This Book Add to the Existing Literature on Export Controls?

It is striking that, notwithstanding the range and scope of export controls, historians and political scientists as well as economists seem to have underestimated or completely overlooked the importance, significance, and impact of export controls.[37] There are many reasons for this. One of them is that export controls are closely associated with notions of state regulation and so appear to be irrelevant to widely shared notions and narratives of globalization. Discussions about globalization, even if they emphasize the highly heterogeneous outcomes of global integration, are often dominated by ideas of an increasingly free flow of ideas, goods, people, and money. They stress how global interchanges have made national borders more and more porous and hollowed out the power of nation-states to pursue national policies and protect their national interests and security.[38]

Export controls, by contrast, are about the *inhibition* and *denial* of transnational transactions. They are a border regime par excellence. Historically, the main objective of the multilayered US export control system was to cut off the Soviet Union and its allies from the benefits of international trade and from access to Western technology and knowledge. In the first postwar decade the export control system pursued this course with missionary zeal, using its economic and technological clout as weapons in a campaign of "economic warfare." This aggressive policy, in combination with the building of new institutions in the West, divided the globe into two economic, political, and technological spheres. The United States shaped them using two different sets of rules. Whereas "free trade" characterized the US approach to organizing its economic relations with its allies, export controls defined the United States' relations to the East (and to "the rest" of the globe).[39]

Whereas globalization narratives often underline the waning strength of state sovereignty, US export controls are an unabashed tool to assert, defend, and enhance state power.[40] As we will show in this book, time and again, there has been a debate on how successful the United States was and is in this regard. But the fact remains that it has used export controls since 1945 to translate knowledge into international power by tying decisions over sharing it to national security and foreign policy interests. Indeed, US technology policy has a decidedly "realist" impetus that often interprets sharing knowledge as a zero-sum game and stands in direct opposition to the principles of a liberal international order.

This seems to be another reason for the lack of attention paid by scholars to export controls. The fields of economic history as well as economics in general still struggle to firmly integrate national security concerns within their intellectual frames. Even though this book can draw on a large body of sophisticated literature, it still holds true for existing research what Susan Strange, Michael Mastanduno, and others have complained about since the 1970s: that economics, international relations, and security studies are basically separate scholarly universes that barely overlap. This is highly problematic because national security considerations have, as we will show, a deep impact on global economic activities and international technological relations.[41]

That granted, export controls should be familiar territory to diplomatic historians, with their strong interest in power relations within the international system. But their scholarship has not paid much attention to the regulation of exports. One of the reasons is certainly that the field still has not really discovered science and technology as not only a legitimate but a central object of analysis.[42] Our book will show time and again how questions of the sharing of scientific-technological knowledge were not just peripheral to the "high politics" of international diplomacy but indeed a concern of immense importance to the most senior representatives of national governments. This book strongly argues that diplomatic historians and the political science field of international relations cannot ignore export controls any longer.

Historians of science and technology and their colleagues in the broader science and technology studies world are no strangers to thinking about the relationship between scientific-technological knowledge and national power. There is a rich literature on the complex relations between national security policies and knowledge production and protection. The bulk of this literature, however, is preoccupied with national, especially US, developments and neglects the inter- and transnational dimension. Nor has it grasped that export controls reach far beyond trade as such and quite explicitly deal not only with technical data, but also with knowledge and even with intangible know-how.

There is another probable reason for the lack of scholarly attention to export controls: researching them is challenging. The relevant literature is scattered across a wide array of disciplines. To learn about export controls, historians have to turn to the work of political scientists who draw on historical analysis. The best starting point for a further exploration of export controls is a handful of studies.[43] But even within political science, the available studies are dispersed over several subdisciplines like political economy, international relations, security studies, and nonproliferation studies. Export controls are often discussed as part and parcel of economic sanctions. But they are also dealt with in studies on research and development, academic freedom, and government control over information sharing and secrecy. In short, scholarly research on export controls demands a decidedly interdisciplinary approach that combines historical analysis not only with the sophisticated insights of political science, but also with economics, and national and international law.

Existing research on export controls is also quite spotty. It has developed in distinct waves followed by times of relative neglect. While there is a lot of literature available from the 1980s and early 1990s (reacting to the Reagan administration's push for much stricter national and international export controls), there has been a relative dearth of research in the last twenty years or so.

It must also be admitted that the regulations and their implementation are very technical, vexingly nuanced, and often contradictory. These complexities are enhanced by intense political clashes between the US government agencies on the one hand and domestic and foreign companies, governments, and scientific communities on the other. Different approaches to security interests and alliance politics, differing perceptions of international threats, and competing economic interests fueled neverending debates about the goals, the course, and the proper assessment of the effects of export controls. But even though they have been the bone of contention for decades now and have been repeatedly attacked by the business community as ineffective, unnecessary, and damaging to US competitiveness, international relations, and academic freedom, the United States has not abandoned export controls because they serve key national political, economic, and strategic interests. In short, export control policy and practice are technically challenging and politically highly controversial, and this often makes it difficult to identify the regime's contours. It is often even difficult to identify what the main points of export control debates are. Be that as it may, our book will show that delving into the world of export controls is not only possible but truly rewarding. Export control debates shed light on the political intersection between national security, foreign policy, economics, and the workings of the American innovation system.

A more mundane reason for the intellectual neglect of export controls is that they barely graze the lifeworld of scholars in the humanities and the social sciences. Ironically this is because such spheres of academic enquiry are, formally at least, among those that have been specifically excluded by the government from their reach, first by mapping the basic/applied science distinction onto the distinction between knowledge that can circulate freely and knowledge that must be controlled, and then, in the 1980s, by carving out a legal space that, along with the First

Amendment, exempted fundamental research from government control (see chapter 6). That said, no academic or corporate researcher in science and engineering today can ignore the importance of export controls, and both devote considerable time and effort in ensuring compliance with them. Major firms can have over one hundred legal and administrative staff dedicated to dealing with the government's export control system. Small firms live in dread of violating export control regulations, on pain of being denied further government contracts. Universities have now set up their own export control offices, which train all science and engineering research faculty and students in export control policy. A professional Association of University Export Control Officers, set up in 2008, now has over 270 members, and it lobbies to protect the freedom of university research from unwarranted government intervention.[44] An understanding of the role of export controls in restricting the circulation of knowledge across borders, as discussed in this book, is essential to an understanding of the political economy of the production and circulation of knowledge in the United States, and indeed in the world.

Export controls are also invisible because they are embedded in the complex bureaucratic structures of the federal government, where they are not exactly easily accessible. The US system is an arcane, highly complex, conceptually sophisticated bureaucratic regime that appears even to experts to be a veritable labyrinth, making research incredibly time-consuming and often dauntingly dry. Export controls are devised and implemented by a technically proficient core of scientists and engineers in various departments and agencies of the executive branch, in consultation with the intelligence community, with inputs from industrial leaders in key technological sectors, and, periodically, in discussion with senior university officials. They are debated by Congress, proffered for public comment in the *Federal Register*, defined by law, and applied with reference to detailed lists that specify what is controlled, to what destination, and for what purpose. Failure to comply can lead to stiff penalties that can be ruinous. The Export Control Reform Act of 2018 allows for criminal penalties up to $1 million per violation of the export regulations and twenty years of imprisonment. Violations can also lead to the withdrawal of export privileges for the firm concerned or to the withdrawal of funding by a federal agency to a university research group. Many corporations and, increasingly, university administrators are constantly discussing and negotiating the application of export controls to their transnational activities to ensure that they are compliant. In short, the system is to a large extent predicated on cooperation and self-policing of companies and R&D institutions. Export controls are not invisible because they play a minor and insignificant role in the management of the national security state. They are invisible because they are one of its fundamental, taken-for-granted raisons d'être. Corporations that depend on export markets for their profits and university administrations that hold international cooperation as a core value go out of their way to comply with them on pain of grievous sanction. We don't see them at work because most people impacted by them make sure that they comply with their injunctions.

How can we study export controls when most export control decisions—thousands every year, hundreds of thousands since the 1940s—are made behind closed doors? Rarely do the details of mundane export control cases reach the public, though sometimes highly controversial decisions are discussed in the pages of daily newspapers or of the *Congressional Record*. That said, the prevalence of bureaucratic secrecy does not mean that there is a lack of sources. On the contrary, one could argue that there is too much material available and it can easily paralyze any scholar. Congressional oversight over export control policy has produced hundreds of hearings. There is an incredible number of reports on export control issues authored by such diverse institutions as the Department of Defense, the National Academy of Sciences and other institutions of higher education, think tanks, and business associations. It is easy to get lost in this ocean of material, and it can be a struggle to find the red thread and the nuggets needed for a coherent analysis. To give it a positive spin: there is more than enough material available waiting to be mined. And we are not even talking about the archival record, which we consulted in a rather unsystematic fashion but that also needs to be explored by future scholars who are persuaded of their importance. Our study offers a gateway to the world of export control sources by showing their richness, diversity, and depth above all when they deal with the circulation of knowledge.

In this book we argue that US export controls have, more than any other national or international knowledge control regime, profoundly shaped the global sharing of science and technology. Even though many states have institutionalized export controls, since the 1940s the United States has been the central player in the global regulation of scientific-technological knowledge. Without the United States' determination to combat Communism, defined as an existential threat to its core values of freedom and democracy, the global movement of goods, information, and people would be much less regulated today. When in the 1940s the

United States became the global hegemon with its preponderance of military and economic power, it reshaped the international system according to its concepts of national and international security, "free" trade, and global financial order. But equally important was the construction of a highly complex system of interlocking export control regimes that span the entire globe and that was modeled on the national US regulations that had been formulated during World War II and then in the years immediately after V-J day. The United States translated its system into the international realm by means of bi- and multilateral agreements that were incorporated into national law by the participating countries, fostering a dense web of regulations, albeit one with marked differences in control practices from state to state.

Although complaints by US companies and research institutions have animated a constant debate over export control reform, they have not led to a dismantling of the regime. The longevity and remarkable stability of the export control system bear witness to the enormous weight that the appeal to national security has in the US political system. For the world's most expansive global power, with the most powerful military and the biggest national economy, tensions between national security and economic interests are unavoidable. With both the global economy and the development of the military largely driven by innovation, technology is squeezed between conflicting administrative spheres that follow different rules and goals, but that are nevertheless densely intertwined. The tensions between the advocacy of free markets, and the concern to protect national and economic security, are not going to fade any time soon in a world of capitalism and military conflict driven by ceaseless technological innovation. Export controls on the production and circulation of knowledge are here to stay for the time being. This book provides a historical introduction to this rambling, arcane knowledge-regulating structure that is deeply ingrained into the political, military, industrial, and academic complex that constitutes the United States as a dominant world power in the twenty-first century.

National Security: The US Way of Seeing the World (and Knowledge)

Export controls are historically deeply rooted in the development of modern warfare. The more that war came to depend on the mobilization of national economies, the more the international system forged economic interdependencies between states, and the more science and technology became key to military planning, production, and battlefield action, the more the regulation or even interruption of trade and of knowledge flows appeared to be an important or even necessary extension of warfare. Thus, complex relations between military, economic, and technological changes on the national level and within the framework of the international system shaped the understanding of the functions, aims, and practices of export controls. Moreover, export controls always reflected economic and political theory as well as military strategic thinking.[45] We have to keep all this in mind lest we carelessly jumble together the export control regime after 1945 with its predecessors like the naval blockades of the eighteenth and nineteenth centuries.[46] They are separated by the trajectory of warfare from relatively limited "cabinet wars" to "total war" and the age of nuclear war; by the effects of global integration with its oscillation of expansion and contraction since the last third of the nineteenth century; and by profound technological changes that bound together economic and military development ever more closely.[47]

The American export control system has its origins in World War II and the early Cold War. Before 1945 the United States did not make systematic use of export controls in peacetime. The US government interfered in the foreign trade of US companies only in times of war and otherwise haphazardly imposed restrictions in the form of economic sanctions in times of international crisis.[48] For example, during World War I, the Trading with the Enemy Act of 1917 and other legislation interdicted trade with foes unless licensed by the US government.[49] And in the years leading up to World War II, export controls played a role in maneuvering US foreign policy through an increasingly crisis-prone international system. In 1937 Congress authorized the president in an amendment to the Neutrality Act to control the export of goods, especially arms, to "belligerent states, or to a state where civil strife exists" if restrictions were deemed in the security interests of the United States.[50] But when in 1938 the State Department proclaimed a "moral embargo" for airplane technology against states that perpetrated air raids against civilians, thus targeting Japan, Germany, and the Soviet Union, this was little more than an appeal to the US airplane industry to cooperate voluntarily.[51] In July 1940 the Act to Expedite the Strengthening of the National Defense imposed new mandatory controls on arms exports. Extended and amended several times, this act became the frame for the all-encompassing control of US foreign trade during World War II.[52]

When the war ended, the US government did not abolish export controls and returned to the status quo ante of peacetime noninterventionism. The Cold War justified the passing of the Export Control Act of 1949, the first American peacetime legislation that subjected outbound trade to governmental regulation. In 1949 it was widely hoped that these measures would be temporary—they were slated to expire in mid-1951. The outbreak of the Korean War was used to justify the maintenance of controls. And even though the export control legislation after Korea always had sunset provisions, in fact they became permanent by periodic extensions. The Export Control Act granted the president far-reaching authority to regulate trade, stating that he "may prohibit or curtail the exportation ... of *any* articles, materials, or supplies, including technical data."[53] It was a major turning point in US economic history because it defined trade not as right of every citizen but as a privilege imparted by the president. Thus, in theory, *every* export needed to be licensed by the government.[54]

This new approach to trade reflected the hegemonic position that the United States had assumed in the postwar international system as well as the redirection of the American ideological compass toward national security. Indeed, "National Security," from then on often spelled with capital letters, came to be the dominant concept that shaped America's perception of the world and saturated all actions that the US government took in the international realm. National security was born from the experience of total and global war with its far-reaching mobilization of national manpower, the economy, science, and technology. It was further catalyzed by the rising tensions of the early Cold War that suggested the need for continued military preparedness and even for a new kind of state.[55] The "free security" provided by having vast expanses of ocean east and west, and benign neighbors north and south, had evaporated with the development of long-range bombers and the first Soviet A-bomb test in August 1949. National Security Council document NSC/68 of April 7, 1950, defined the Soviet leadership as driven by a "new fanatic faith, antithetical to our own, [that] seeks to impose its absolute authority over the rest of the world." National security concerns seeped into every nook and cranny of government and society and into policy fields that as recently as the 1930s had seemed at the fringes of security thinking at best. It was especially significant that the new way of American thinking established a close link between security, economy, and ceaseless technological innovation. The national economy, which was fueled by the national scientific-technological knowledge base, was seen as a core asset that needed to be carefully protected, not least by means of export controls.[56]

In 1949, along with the passing of the Export Control Act, the United States prodded its NATO partners to establish Cocom, the Coordinating Committee for Multilateral Export Controls, in 1949–50. Cocom was a response to the increasing amount of critical technology being transferred by the United States to its NATO allies, and by their increasing capability to produce sophisticated weapons themselves.[57] It aimed to ensure that the restrictions imposed on American business activity in the Communist world were also imposed on allied governments and their firms. European countries on the front lines of the Cold War were less constrained by national security concerns to trade with the Eastern bloc. Cocom was established to harmonize these conflicting approaches: in effect it sought to sustain American hegemony by imposing Washington's conception of national security on its allies. Inevitably the United States and its partners squabbled constantly over the aims and scope of export controls. During the Cold War, the United States generally advocated more radical and more intrusive regulations than its partners, who preferred limited, carefully defined areas of control. They took an approach that was geared much more toward trade and business interests than toward national security concerns.[58] US policy was always much more ambitious. In the early Cold War the US waged "economic warfare," reducing its already quite limited trade relations with the Eastern bloc to a trickle. The Cocom partners were willing to follow this total embargo policy only in the face of the direct military confrontation in Korea. Before and after they advocated a "strategic embargo" instead, which limited export controls to militarily relevant goods and technologies.[59] Many Western European countries like West Germany (FRG) and Great Britain were keen to promote their historically well-established ties to Eastern markets in order to address the urgent needs of economic recovery after World War II. They also argued that East-West trade was a means to exert political influence on the Iron Curtain countries on the basis of shared economic interests. This "linkage" strategy would be most vigorously pushed by the West German "Ostpolitik" of the 1960s and its program of "change through rapprochement," but it was a strong current in the thinking of many Western politicians.[60] Whereas the United States time and again emphasized the national security risks of trading with the enemy, the allies stressed economic and political opportunities and benefits. In practice, however, each Cocom member weighed the factors differently, contributing to the emergence of distinct national "styles" of export control.⁶¹ The US style was characterized by export control lists that were longer and more complex than those of all other countries. Its bureaucratic apparatus was bigger, the role of the Defense Department was more pronounced (in other states, like Germany, the defense ministries were not involved), and the enforcement, by customs and the intelligence services, was stronger (even though with significant oscillations over time). The United States was also the only state in the Western alliance that implemented far-reaching export control regulations for technical data and intangible knowledge and know-how.

The very different weight given to national security by the United States and its Western partners has had one very important and highly contested effect. The federal government imposes strict reexport rules on all its trading partners, including its closest allies. Every shipment from a foreign country of American technologies and goods whose direct export to the client country needs to be controlled needs a reexport license. That also goes for products made by a foreign company if they incorporate American know-how, technical information, or spare parts. This principle of "contamination" assumes that a national identity can be attached to technologies and goods: reexport controls are triggered if a foreign shipment embodies a certain percentage of "Americaness."⁶² The extraterritorial application of American law in the name of national security has been repeatedly criticized by European trading partners as infringing on their national sovereignty. Tensions ran particularly high during controversies over the export of oil pipeline parts and associated technology to the Soviet Union in the early 1980s.⁶³

European allies were not the only social actors to object to the scale and scope of the federal government's application of export controls in the name of national security. It was also constantly contested by key sectors of American business—for example, firms producing semiconductors, satellites, machine tools, and advanced materials—in the name of free-market ideology and by their determination to secure a major share of global export markets. Indeed, since the 1970s, the high-tech business community has constantly criticized the considerable economic repercussions export controls have had on their revenue streams and international competitiveness.

The arguments against export controls have virtually not changed in the last four decades or so. They are criticized as cumbersome, overly complex regulations that no other country forces on business, thus imposing on US companies a serious competitive disadvantage. Indeed in 1969, with the onset of détente, the balance between national security and "free" trade shifted briefly in favor of increased trade with the Soviet Union. The Export *Control* Act of 1949 was renamed the Export *Administration* Act of 1969. This was no mere semantic shift. It was intended to replace the near-embargo aspects of the 1949 act by liberalizing export controls on American industries that were under increasing competitive pressure from Western Europe and Japan. It was updated and further liberalized in 1979, when a new version of the act was passed in Congress, which remained the pillar of export control legislation for the next twenty-five years.

The Soviet invasion of Afghanistan in 1979 and the imposition of martial law in Poland in 1981 dealt a deathblow to détente and to the argument that liberal trade relations encouraged liberal political reforms. A "Second Cold War" against the "evil empire" was waged by a new administration, which took a wide range of measures to clamp down on what they saw as threats to national security. The government's reach challenged traditional American values so seriously that a Subcommittee of the Congressional Committee on the Judiciary held a number of hearings provocatively entitled "1984: Civil Liberties and the National Security State." Trade relations with China were liberalized to forge a Sino-American alliance against the Soviet Union, plans were made to erect a vastly expensive technological shield against a possible Soviet attack (SDI, or the Strategic Defense Initiative), and steps were taken to tighten export controls on militarily critical technologies, knowledge, and know-how. This included using export controls to deny Soviet scientists and engineers visas to attend international conferences held on American soil and to restrict what they could do and where they could go as invited guests to American research university campuses.[64]

In the late 1980s and early 1990s the American concept of national security was radically revised. The proximate cause was the Japanese conquest of the semiconductor manufacturing industry at the expense of the United States, along with widespread global production of increasingly powerful computers. With the collapse of the Soviet Union the United States emerged as the single most important military power on the globe. Yet its military-industrial base was heavily dependent on foreign suppliers in key high-tech sectors. A spirited debate ensued over the limits of basing national security on military superiority. National security had to be extended to include "economic security": a single national techno-industrial base had to replace parallel military and civilian production systems. With the liberation of civilian industries developing dual-use commodities from the burden of export controls, they could secure a greater share of world markets, and pump their profits back into R&D that would provide the military with the most advanced technology at a fraction of the then-current cost.

American business weighed in. The perception of decaying industrial competitiveness and the dramatization of an "American decline" became the rallying point for critics.[65] Export controls were attacked as slowing down innovation and thus curtailing the United States' ability to compete with Japan or West Germany as well as undermining the technological superiority of the US military. One of several reports commissioned by the National Academies that was published in 1987 claimed that export controls led to a direct export loss to US business of $7.3 billion and claimed that this was the equivalent of more than 188,000 American jobs.[66] Regulating the economy would inflict damage to the very economic basis that national security had to be built upon. By contrast, an unfettered technological sector would set free innovative energies in the civilian sector that would allow the United States to return to economic and military preponderance.

The Clinton administration embraced this approach wholeheartedly: national security was not possible without economic security and so demanded liberalizing export controls that discriminated against US firms in the world market. The Chinese market beckoned, and business boomed in a whole range of advanced sectors (like high-performance computers, space industries and advanced machine tools). The administration's argument was summed up using a nice example by William Reinsch, of the Department of Commerce, in congressional testimony in 1998. "As the lines between military and civilian technology become increasingly blurred," Reinsch said, "a second-class commercial satellite industry means a second-class military satellite industry as well. The same companies make both products," he went on. "And they depend on exports for their health and for the revenues that allow them to develop the next generation of products."[67] The number of dual-use licenses applied for annually to the Department of Commerce decreased precipitously from nearly one hundred thousand in 1989 to about 10 percent of that number in 1996–97, in line with trade liberalization.[68] Cocom was dismantled in 1994, and replaced with the Wassenaar Arrangement, which expanded the number of participating states in the multilateral export regime, but severely reduced the power of any state to stop others trading with partners of their choice.[69] Despite an arguably feeble institutional framework and seemingly endless internal frictions and controversies, Cocom had played a crucial role in the regulation of international trade that the Wassenaar Agreement could never match.

In a spirited backlash, "control hawks" in Congress accused Clinton of putting corporate profit ahead of national security, allowing China to modernize both its economy and its military sectors in return for political favors. These criticisms fed into a broader political assault on the Clinton administration that eventually led to the president's impeachment in 1998. Their impact on export control policy and practice was more muted, though highly visible in the domain of space cooperation with China, and indeed with the rest of the world.

The expansion of the concept of national security to embrace economic security is now taken for granted. The Trump administration's *National Security Strategy* for 2017 cited the president himself as saying that "economic security is national security." In this document, which advocates an overall strategy to put "America First," economic security now has two complementary dimensions. It continues to stress the need for deregulation to stimulate exports. But it combines the use of export controls to *deny* sensitive technology to rivals with a raft of new measures to erect a protective wall around *sharing* emerging critical technologies and related knowledge with China, in particular. Beijing is accused of combining unfair trading practices, with IP theft and cyberattacks on American industry. These forms of "economic aggression" will be countered, we are told, "using all appropriate means, from dialog to enforcement tools."[70] Export control loopholes not covered by EAR and ITAR have been closed by imposing national and economic security screenings of foreign direct investment through the interdepartmental Committee on Foreign Investment in the United States (CFIUS).[71] In 2018, this connection became even more pronounced by tying together the Export Control Reform Act (ECRA) and the Foreign Investment Risk Review Modernization Act (FIRRMA), as we will show in our epilogue to this volume.

These protectionist measures have had a dramatic effect on Sino-American cooperation in both corporate and academic settings. The expansion of national security to embrace economic security has given the federal government immense powers to intervene in the "free market" by tightening up exports of sensitive technologies and by denying Chinese authorities and entrepreneurs (and students) access to advanced knowledge in American firms (and universities) at the cutting edge of the "Fourth Industrial Revolution."

Defending America's Technological Lead

"Lead time" is a central concept of export control and national security thinking. Since the 1940s, the idea of technological superiority has been the core of US military and national security policy. After the victory over Germany and Japan in World War II had demonstrated the enormous potential not only of mobilizing US industry but also of a massive state-driven national innovation system, the maintenance and widening of an American technological lead became a central preoccupation of the burgeoning national security state—a shorthand for the complete restructuring of the federal government in the context of the global projection of US power.[72] The postwar "cold" confrontation with the quantitatively vastly superior military forces of the Soviet Union cemented the conviction that technology had to be mobilized as a "force multiplier" to "offset" the enemy's capabilities.[73] Deterrence as a foundational building block of the Cold War order depended on keeping a technological gap so wide as to make the Soviets sufficiently fearful of American military power as to never dare launch an attack on their rival superpower. The US goal was to achieve strategic asymmetry through technological predominance. This was, of course, the historical logic of arms races, but it was not fully adopted in US policy making until the 1940s. It irreversibly changed the American way of doing science, waging war, and organizing government institutions. Beginning with the very early years of the Cold War, maintaining a comparative technological advantage was seen as the most effective way the United States had to survive in a struggle against the numerically superior military forces of the Soviet Union and its allies. It substituted firepower for manpower, capital for labor, quality for quantity, relying heavily on science and technology to overwhelm its enemies and to secure a major share of world markets.[74]

Ceaseless competition for scientific and technological preeminence, or "leadership," in the development of more effective, more sophisticated, and deadlier weapons systems was and remains one of the driving forces of the US export control system. It is the often forgotten and ignored flip side of the vast expenditures that the federal government and especially the Department of Defense have poured into research and development since the 1940s. The technological dominance that the United States had reached at the end of World War II—signified by accomplishments like the atomic bomb, radar, and penicillin—was to be extended indefinitely into the future by combining relentless innovation with strategies to keep the enemy behind.[75] The United States institutionalized export controls in order to regulate the dissemination of scientific-technological knowledge across the globe. Even though the technological breakthroughs during World War II had been the result of substantial international cooperation with US allies, the new postwar US hegemony, the Pax Americana that reshaped the international system from the ground up, was based on the idea of an American-based technological lead.[76] Export controls were institutionalized to defend this lead. They were an openly protectionist instrument of a burgeoning techno-nationalism that positioned the American state against the outside world.[77] Their aim was to protect the gap between the United States (and, by extension, its Western allies) and the Soviet Union—not to say the rest of the world—on American terms. France's quest for an independent nuclear deterrent, for instance, and West Germany's interest in buying natural gas from the Soviet Union were treated as intolerable challenges to the United States' vision of the international order and its authority to shape it as Washington saw fit.

Defending the gap through export controls meant, ideally, the complete denial of access to cutting-edge technology. However, the national security community always realized that there were limits to stopping the spread of technological knowledge and capabilities. Thus it was clear from the get-go that export controls—like secrecy—were less water tight than Fort Knox and more akin to a leaky bucket. They attempted to "control the uncontrollable" even though critics argued again and again that this was futile.[78] However, while it was almost inevitable that technological rivals would gradually learn all that they needed to know about a new technological breakthrough, using every legal (e.g., buying on regular markets) and illegal (not least economic espionage) means available, one could at least slow them down and make them invest heavily in their indigenous capabilities, even deliberately leaving them to follow less-than-optimum research paths.[79] Tempering the pace of technology transfers through export controls would in turn buy additional time for the innovation of new, once again superior technologies. Keeping economic and political rivals behind while innovating more rapidly and effectively than anyone else would allow the leader to stay ahead.

The national security community usually measured American lead time in years that it would take its adversaries to close the gap in the event that the US lowered its export control guard.[80] This was true not only for the well-known nuclear domain. The entire international high-tech sector was assessed this way, and myriads of reports and studies calculated for how many years America could maintain its technological advantage. It's not entirely an oversimplification to say that, after 1945, US national security was defined as a direct function of this logic of scientific-technological lead time.

The concept of a technological gap between the United States and its rivals caused intense debates in the 1960s when the Western Europeans were famously chastised for falling so far behind the United States in the development and management of technology that they risked being reduced to an American technological and political colony.[81] In a similar vein, the perceived loss of the competitiveness of the US high-technology sector in comparison to Western Europe and Japan in the 1970s and 1980s stirred deep-seated fears of American decline and the end of US global leadership (see chapter 7). Indeed, export controls were understood by both the American national security community and its Western counterparts as an instrument to defend and to maintain US hegemony.

Export controls not only defended the United States' technological lead; their efficacy also depended on it. Regulating and denying technology make sense only if you have something that others do not have, but want. This also implies that the gap and its regulatory management offered considerable political leverage to the technological leader. The leader can use access to, and denial of, knowledge as an incentive or as a sanction, for enemies and allies alike. In some cases technology gaps can lead to international dependencies that can be exploited in favor of the hegemon. In such constellations, export control licenses can be turned into bargaining chips (in the sense of a "linkage strategy": *do ut des*) or used to exert considerable political pressure on a receiver entity. As early as the 1940s the US government understood that the asymmetric relationship between the technological haves and have-nots could be "weaponized" to its advantage.[82]

Bureaucratic Structures

Export control policies are formulated, implemented, and administered by a complex bureaucratic apparatus. Authorities and competencies are not concentrated in a single department or agency but are widely dispersed over several departments of the federal government. Even though its bureaucratic structure has been criticized as inefficient and confusing, it has demonstrated a remarkable historical stability and resilience against change. The basic administrative structures have been tinkered with in countless reform initiatives. All the same, in the early twenty-first century they are strikingly similar to how they were when they were established in the 1940s.

The main actors in the densely layered export control community are the Department of Commerce, the State Department, and the Department of Defense. They form a highly dynamic triangle, characterized by close cooperation, lively communication, and collective policy and decision making, but also acrimonious turf wars and mission-driven clashes over the right balance between economic interests, foreign policy, and national security considerations.[83] Indeed, conceptual and political tensions and contradictions are a prominent feature of the US export control world, leading to considerable confusion among "customers" in industry, research institutions, and analysts alike.

The direction of controls was time and again influenced by changing coalitions between the members of the big three within the context of overall "grand strategy," often pitting two of them against the third. One should not assume that the Department of Defense always played the role of the hardliner while the Department of Commerce was an unflinching advocate of trade liberalization. During the early Cold War, for example, the Commerce Department was a staunch proponent of economic warfare while in the 2000s, the Defense Department stood for a relaxation of export controls. Their positions depended very much on the changing personnel who stood at the helm of the respective departments. Moreover, the president, who is directly involved in the everyday administration of export controls only at rare moments of irreconcilable interdepartmental conflict, can shape the dynamic inside the executive branch by his "grand strategy," which can have major implications for export control policy and practice: we think of George H. W. Bush's emphasis on the importance of nonproliferation and Bill Clinton's leitmotif "it's the economy, stupid!" Indeed, in the last few decades the White House has played a prominent, even dominant role in shaping regulatory policy and pressing for its implementation.

Congress is another big player.[84] Its engagement surged whenever there was the need to renew, reform, or amend the statutory framework of export controls. Each time the legislation was about to expire, Congress reassessed policy and scrutinized practice in lengthy hearings, thus becoming one of the main producers of historical source material. Not surprisingly,

Congress also became one of the main arenas in which control "hawks" and trade liberalization "doves" clashed, the balance of power between them shaping the overall direction of control policy. In the 1960s, for example, Congress played a key role in the liberalization of US trade, in the 1970s it was the scene of much acrimony over how to deal with détente, and in the late 1990s it steered US export controls into an open confrontation with China.[85]

The dynamics of the separation of powers, with confrontations and alliances between hawks and doves in Congress and the executive branch, have a strong impact on export controls. But even though Congress has considerable clout in the formulation of policy, the everyday work of assembling control lists and making license decisions lies in the hands of a highly specialized group in the federal bureaucracy. Since changes in the administration have only limited effects on the composition of this group and since its members have accumulated expertise in a highly complex field, the export control bureaucrats in the Commerce, State, and Defense Departments have a high degree of autonomy and are more or less impervious to external reform efforts, including the lobbying efforts of industry. They are assisted by multiple other sections of the state apparatus. The Treasury Department for the implementation of economic sanctions, which leverage the power of export control regulations; the Department of Energy and the Nuclear Regulatory Commission for licensing nuclear technology, equipment, and materials;[86] the Department of Justice and the FBI for the enforcement of export controls; the intelligence community for assessing the foreign availability of new technologies developed in the United States, and for monitoring compliance with licensing stipulations. In short, the executive dominates the export control world. The US national security state apparatus is vast, rambling, and remarkably powerful and autonomous.[87]

The Outline of This Book

This book is the first historical study that systematically looks at export control regulations from the perspective of *the sharing or denial of knowledge and intangible know-how*. To date, we know little about how the production and circulation of knowledge in and by universities and corporations was affected by export controls, how these actors perceived the regulations, how they implemented the rules, how their cooperation with the government export control agencies looked, or how they lobbied against these inconvenient restrictions. Thus, a crucial dimension of the national innovation system as well as of business-government relations is insufficiently understood and, for historians, virtually invisible. This lack of knowledge and awareness is especially troublesome for research on the histories of specific high-technology sectors that were the main, but not the only, sources of American global power, and that were (and still are) tightly regulated.[88] This book will show that export controls have, in fact, had an enormous political relevance for American debates about national security, foreign policy, and trade policy since 1945. Indeed from the 1940s to the present day, how to control the transnational movement of information, as well as of people and things that are "bearers" of advanced knowledge and know-how has been absolutely central to the thinking and actions of the guardians of the American national security state.

Taking the big picture, we will argue that there has been an overall shift in the regulatory regime from the control of trade in the first two or three decades after World War II toward the regulation of knowledge flows from the 1970s up to the present day.[89] The application of export controls to technical-scientific information in all its shapes and forms has been a key element of the export control system since its establishment. But it has become more intense since the 1970s with further vigorous pushes in the 1980s, the late 1990s, after the terror attacks of September 11, 2001, and in the context of the intensifying competition between the United States and China in the last two decades. This expansion of control over knowledge and know-how—mirroring the rise of communication and information technology as the engines of innovation and economic growth—is most visible from the increasingly systematic inclusion of universities and research institutions into a system that in the 1950s and 1960s mainly targeted business activities. Today, the activities of foreign scientists, especially from countries seen as adversaries like China, are policed by US "deemed export" regulations that treat every communication of formally unclassified but controlled technical information as if a physical export had occurred.[90]

The book is organized into four main sections whose foci reflect this trajectory. Part 1 describes the historical roots of the export control regulations in the early twentieth century, and their embodiment in legal instruments with a focus on the Export Control Act/EAR in the first decade after World War II, with particular reference to the measures taken to control the flow of technical data. Part 2 deals with the 1970s and the 1980s. It describes the emphasis placed on integrating controls on the circulation of knowledge and know-how into the export control regime. The stimulus for this came from a report prepared for the Department of Defense in 1976 by a select panel of industrialists chaired by Fred Bucy, the then-CEO of Texas Instruments. Bucy's fears were confirmed by the dramatic revelations in 1981 of the organized attempt by the Soviet administration to acquire advanced knowledge and technology from Western and US corporate and academic research sites. This brought research universities into the crosshairs of the federal government's regulatory system. A report commissioned by the National Academies in 1982 (the Corson Report) exonerated research universities from any responsibility for leaking sensitive knowledge to the Communist bloc. It also laid the ground for a core policy statement, adopted by the Reagan administration in September 1985, that built a protective wall around "fundamental research" in science and engineering that, under reasonable conditions, would be exempt from export controls. This Fundamental Research Exclusion remains in force today, though it is seriously imperiled by charges that China is exploiting its provisions to climb the economic ladder at America's expense.

Part 3 focuses on the expansion of the concept of national security to embrace economic security in the 1980s and 1990s. It analyzes in detail the complex debate—taken for granted today—around the implications of Japan's success in wiping out the majority of the American semiconductor industry. The recognition that laissez-faire economic policies were allowing highly competitive foreign firms to dominate manufacturing in sectors that were crucial to the US economic competitiveness and military strength caused a fundamental reassessment of the role of the American state in the economy. Military power without economic strength was a limited asset in a global, interconnected economy: national security was hollow if it was not constructed on a techno-industrial base that ensured economic dominance in global high-tech markets. The Clinton administration absorbed the lesson and sought to overhaul the military procurement system with a view to producing a single, dual-use techno-industrial base that relied on conquering export markets in information technologies. Trade with the People's Republic of China, which had gathered momentum during the Reagan years, blossomed. Chapters in part 3 briefly summarize the emergence of China as a most-favored-nation trading partner for the United States, the setback caused by the repression of prodemocracy protestors in Tiananmen Square in June 1989, and the subsequent enthusiastic expansion of Sino-American relations espoused by the president himself. His impeachment in 1998 occurred in a toxic political climate that included charges made in the so-called Cox Report that the Clinton administration had put business interests ahead of national security. One particularly onerous accusation—that his administration had recklessly allowed American satellite manufacturers to share missile technology with Chinese scientists and engineers—is analyzed in some detail.

Part 4, the epilogue, brings us up to 2020. It quickly surveys the measures taken by the Trump administration to use all the available instruments of the American regulatory state—augmented by new measures to close every loophole in the system—to limit China's access to new knowledge being generated in corporate and academic research environments. These measures include imposing broad export control restrictions on Chinese high-tech companies, including punishing firms in other Western countries for trading with them; restricting the possibility of knowledge transfer back to the mainland by Chinese foreign direct investment in American corporations; limiting visas for Chinese graduate students who want to pursue their studies in American universities; and penalizing US-based researchers who have accepted funding from China's Thousand Talents Program to develop research programs there.

Historically export controls and related regulations have been deployed alongside building alliances and constructing multilateral regimes that played along with Washington's determination to secure its global leadership by controlling the circulation of sensitive information, people, and things. The Trump administration's strategic agenda of "America First" went along with the rise of a protectionist techno-nationalism that sought to deny Chinese entities access to twenty critical and emerging technologies on which future economic and national security depended.[91] It also used its immense economic power, and the dependence of other national economies on it, to pressure them into acceptance of its confrontational policies with Beijing. As always, when national security is involved, corporate and academic actors have sought ways to adjust their relationships with China in line with the overall thrust of national policy, while protecting their core missions as best they can. While we are too close to events to evaluate the long-term effects of these policies, it is likely that, having been institutionalized, they will leave an indelible mark on the political economy of knowledge circulation in the global arena for many years to come.

PART ONE

CHAPTER TWO

The Invention of Export Controls over Unclassified Technological Data and Know-How (1917–45)

Putting Legislation in Place in World War I

As we emphasized in chapter 1, the perceived threat of the emerging Cold War catalyzed the institutionalization of permanent peacetime regulations over the movements of information, things, and people within the framework of the burgeoning national security state. Previously the US government exerted controls only in times of war or in emergencies that threatened peace. Indeed the path toward the controls discussed in this book was paved by the two world wars. World War I produced some of the conceptual and legislative cornerstones that the US bureaucracy would begin to build on when World War II cast its ever-growing shadow over American foreign and domestic politics. The onset of the Cold War did not represent a sharp break with previous regulatory practices. In the 1940s the American government pursued a line of thinking that was put in place toward the end of World War I, which was dramatically extended to cover the circulation of knowledge during World War II, and that—and herein lay its novelty—after a brief stutter in 1945 was institutionalized in export controls over people, information, and things *in peacetime* in the name of national security as the Iron Curtain descended in Europe. This and the following chapter will tell the complex story of how the Cold War export controls system came into being.

The Espionage Act of 1917 was one of the cornerstones of the regime put in place in the early 1940s. The scope of this law was broader than its title suggested. Its first section gave the law its name. It prohibited the transmission of a broad range of information pertaining to national defense to the enemy. Even though the act and its amendment, the Sedition Act of 1918, were called on to stifle public dissent, the text of Title I targeted only classified information and indeed remains the heart of the US classification system up to the present day.[1] Technological information was clearly included but not singled out and limited to descriptions of the military infrastructure of total war like airplanes, ships, and communication lines. The act enumerated the carriers of information: "any sketch, photograph, photographic negative, blue print, plan, map, model, instrument, appliance, document, writing or note of anything connected with the national defense."[2]

The fears of uncontrolled communication of dangerous information were closely related to concerns regarding the mobility of people.[3] Accordingly, travel documentation like visas and passports became national security control tools par excellence in the course of the twentieth century. Before World War I, US citizens did not need a passport to leave the country. Historically, the United States used passports only in wartime and abolished them when the guns fell silent.[4] As a matter of fact, around 1900 not only Americans but also Europeans could travel internationally without any "documentary surveillance" of nation-states if they were not immigrants. With the onset of World War I, all Great Powers reintroduced passports in order to secure their borders, keep tabs on "enemy aliens," and enforce the draft. These steps were, however, regarded as temporary wartime, emergency measures. Reacting to these developments, the United States followed suit with an Executive Order in December 1915 that made the passport mandatory for every person who left the country.[5] Title IX of the Espionage Act stipulated passport regulations, laying out rules for the application process and listing the penalties for making false statements in the application.[6] The Espionage Act was also the basis of the control by US Customs over foreign travel. A staff of five hundred officers was fielded for this task alone.[7]

Finally, the Espionage Act established a statutory basis for the implementation of export controls. In practice, in 1917 the United States was already using the trade controls that the Allies had developed to wage economic warfare against the Axis powers. Title VII empowered the president to impose export controls over "any article" he would specify in a proclamation if he found "that the public safety shall so require." Additionally, Title VI regulated the export of arms and munitions, and Titles II, III, and V addressed the control of shipping from and to the United States.[8]

Thus, the Espionage Act placed export controls in the context of a much broader complex of national security concerns with border security as their common denominator. The dangers the act envisioned were the uncontrolled movements of *people, information, and things*. This triad of targets of national security control also shaped the second statute that US export controls were anchored in, namely, the Trading with the Enemy Act of 1917. Obviously, its main purpose was to curtail trade relations with enemy states and impose export controls on goods (section 3a). Yet in section 3, the act also prohibits (b) "to transport . . . into or from the United States . . . any subject or citizen of an enemy or an ally of an enemy nation" as well as (c) "to send, or take out of, or bring into . . . the United States any letter or other writing or tangible form of communication." Paragraph (d) was about censorship, providing the statutory basis for the US Censorship Board.[9]

The impact of these and other wartime regulations on the dissemination of scientific and technological knowledge is difficult to assess on the basis of the available scholarly literature. This is no small lacuna. World War I was widely understood by contemporaries as a watershed moment because the belligerent governments on both sides began more than ever before to harness research and development for the production of increasingly sophisticated and deadly weapons. Scholarship has time and again reinforced this line of argument. But what were the implications for controlling and safeguarding scientific-technological knowledge?[10]

The military subjected research data to military security classification. A typical example is the Naval Consulting Board, one of the most prominent US institutions for the mobilization of science and technology for war. Founded in summer 1915 to address the challenges of the German submarine, in 1917 it became the Board of Inventions of the US Council of National Defense, the World War I equivalent of the National Security Council. Chaired by inventor-hero Thomas Edison, this short-lived institution searched for ways to improve the transfer of scientific-technological knowledge to the military. From the very beginning everybody involved agreed that secrecy was supposed to be "a governing factor." To what extent the broader US innovation system—including, for example, the National Research Council and NASA's predecessor, the National Advisory Committee for Aeronautics—was affected by classification is, however, unclear.[11]

The United States also cast a net of control over patents. Publication of specifications was traditionally integral to the issuance of a patent. The

Trading with the Enemy Act saw a potential leak in this practice and stipulated that the president may classify inventions for the duration of the war if publication could be "detrimental to the public safety and defense, or may assist the enemy."[12] This stipulation was legislatively amplified, for the duration of the war, by the first US Invention Secrecy Act of 1917, which would become the basis of similar laws in World War II and the Cold War.[13]

Moreover, the Trading with the Enemy Act allowed for the expropriation of all US patents held by companies and citizens from enemy countries and sold them, via the Alien Property Custodian, to US companies. This forced technology transfer targeted above all the German chemical industry and brought it under national control. The Germans had become the predominant nation in this high-technology field since the second half of the nineteenth century and demonstrated its strategic impact by using the Haber-Bosch process in the production of explosives and the use poison gas on the battlefield. The United States as well as the other Great Powers used their victory in the war as a welcome pretext to acquire a vast amount of scientific-technological information.[14] Clearly, knowledge had become a precious resource that World War I was fought with and fought for.

It is unclear to what extent export controls were systematically used to regulate knowledge flows at this time. In any case, the scope of the export control bureaucracy was impressive: The War Trade Board had no fewer than 1,526 employees and processed 425,000 export and import applications in six months. The board's list of items requiring an export license included eight hundred items in September 1917.[15] But impressive as all of this is—did the export control regime include any considerations of technology transfers? Or was the trade with raw materials and foodstuffs, in practice, the predominant object of Allied economic warfare? We face here a yawning research gap.[16]

The same can be said about the regulation of unclassified information. Censorship, voluntary self-censorship of the US press, and controls by the postmaster general over the exchange of mail certainly had an impact on the dissemination of scientific-technological data. In an appeal published by its propaganda office, the Committee on Public Information, the US government asked the national press to voluntarily refrain from publishing "military information of tangible benefit to the enemy." The guidelines that the committee offered treated information on military technology as a secondary matter, focusing mainly on information on troop locations and movements. But it also asked the press and people in the know to keep confidential aircraft, ship, and artillery types and "information of all Government devices and experiments in war material."[17]

In addition, the US censors screened technical periodicals like *Motor Age* and the *Air Service Journal* "for leakage of militarily significant information." An issue of *Scientific American* was held up because it contained an article giving the capabilities of a machine for straightening and checking the bore of rifle, shotgun, and machine-gun barrels. But overall, in the opinion of scholar Harold Relyea, the "vast majority" of US scientists and engineers "were probably more aware of wartime censorship as citizens" than because of their professional work.[18] The same can be said for travel restrictions. Nothing suggests that scientists were prohibited from leaving or entering the country because of the special knowledge they carried.

It seems safe to argue that compared to World War II and especially the Cold War, the impact of World War I controls and censorship on scientific-technological knowledge was relatively limited, commensurate with the smaller scale of the mobilization of the scientific, technological, and business communities between 1914 and 1918. But the legislation, experience, and intellectual concepts the United States developed during World War I provided the matrix for the control system as it unfolded from the late 1930s to the 1950s.

From "Moral Embargo" to Systematic Control

Most of the regulations described above fell dormant when World War I ended. After the ceasefire with the Axis powers, the US reluctantly continued to participate in Allied trade regulations against the Soviet Union. But in 1920, the United States stopped the controls and returned to peacetime exchange practices.[19] In the late 1930s, as the escalating international crises in Europe and in the Far East set the scene for World War II, export controls reentered the picture. The United States engaged in so-called moral embargos to sanction aggressive behavior. It supported, however meekly in practice, the League of Nations' oil embargo against Italy after Mussolini sent troops to neutral Ethiopia in 1935. Three years later, in June 1938, the State Department called on airplane producers to refrain from doing business with governments—that is, Japan—that flew air attacks against civilians. The airplane industry heeded the call and its exports to Japan quickly dropped by 90 percent. This moral embargo on aircraft was extended to the Soviet Union in December 1939 after it attacked Finland. Also, in November 1939 the Roosevelt administration began to limit trade with Germany.[20]

Increasingly, the export control policy reflected the tensions between an "isolationist" minded Congress and the Roosevelt administration more willing to engage in international relations. Congress hoped that export controls could help to keep the United States out of any military involvement. The two Neutrality Acts of 1935 and 1937 restricted the export "to belligerent countries" of armaments, munitions, and dual-use goods like airplanes and chemicals that could be used for the production of poison gas. Roosevelt came to criticize the measures because in his view they actually benefited the aggressor states by denying help to the victims. Just before the international crisis in Europe came to a head in 1939, it became obvious that the Neutrality Act would cut off Great Britain and France from the US armament market and weaken them at the very moment when they would have to fight the heavily armed German military.[21]

Accordingly, the Neutrality Act was revised in 1939 to allow for arms exports to Great Britain and France on a cash-and-carry basis, ushering in the Lend-Lease Act of March 1941. Whereas the United States thus facilitated exports to the Western Alliance, it geared up its own armament production and tightened its export controls. On May 16, 1940, six days after German troops had begun their campaign against the Netherlands, Belgium, and France, Roosevelt asked Congress in a joint session for military appropriations of $896 million, stressing that the new dangers to the United States and "vital American zones" were the result of the technological developments, especially in regard to airplanes. "The Atlantic and Pacific oceans," the president said, "were reasonably adequate defensive barriers when fleets under sail could move at an average speed of five miles an hour. Even in those days by a sudden foray it was possible for an opponent actually to burn our national Capitol. Later, the oceans still gave strength to our defense when fleets and convoys propelled by steam could sail the oceans at fifteen or twenty miles an hour." But, the president emphasized "the new element—air navigation—steps up the speed of possible attack to two hundred, to three hundred miles an hour." In response Roosevelt wanted to gear up the entire US weapons production, calling especially for a massive expansion of the air force. He envisioned an industry that could churn out fifty thousand planes a year.[22]

One of the key measures authorizing this military buildup was the Act to Expedite the Strengthening of the National Defense of July 2, 1940.

Its section 6, soon called the Export Control Act, established the place of export controls within the framework of the defense program and would, amended several times, become the legal basis of the control regime during World War II and the immediate postwar years until 1949. Section 6 had a clear focus on military exports and targeted explicitly technology in the form of machines, stating, "Whenever the President determines that it is necessary in the interest of national defense to prohibit or curtail the exportation of any military equipment or munitions, or component parts thereof, or machinery, tools, or material, or supplies necessary for the manufacture, servicing, or operation thereof, he may by proclamation prohibit such exportation."[23]

The aims of US export policy were threefold, all of them closely related to the overall goal to win the war. As a tool of economic warfare, it denied technologies, merchandize, and raw materials to enemy states to exacerbate bottlenecks in their war production. Second, export controls played a crucial role in the national bureaucracy of resource management for US military production, the needs of the fighting troops and of the population on the home front. Here export controls were supposed to avoid short supply problems above all. And finally, the resource management and exports of goods were channeled to help the war efforts of the US allies. In short, export controls were "established to ensure that our critical supplies flowed to the right places at the right time and did not go to the wrong places."[24]

On the same day that the July law was enacted, a new control list was published, supplementing the official US list of Arms, Ammunition and Implements of War of May 1937 that enumerated guns, cannons, battleships, military airplanes, and chemicals especially for the production of explosives and poison gas.[25] The new list added forty-three categories to its barely two-page-long predecessor, twenty-six of which were raw materials—from aluminum and asbestos to hides and industrial diamonds to silk and toluol. Eleven chemicals were added; and two further subsets aimed at five technologies: aircrafts parts and equipment; armor plates; "optically clear" plastics; bulletproof and nonshatterable glass; and "optical elements" for fire control and aircraft instruments. Machine tools for melting, casting, pressing, cutting, grinding, and welding metals were in a separate category.[26] The list grew constantly until December 1941.[27] Then the export controls were at one fell swoop extended to *all* exports leaving the United States to *all* destinations, except Canada. The export control practice still focused on defined categories, which, however, went far beyond the rather modest beginnings on the eve of the war. By 1943 the Foreign Economic Administration controlled some twenty-five hundred commodities and commodity groups, involving the commodities of approximately sixteen thousand US companies and thousands of individuals and firms in more than 140 different foreign countries. Several thousand applications for export licenses came in every day, and the annual volume ranged between 1.5 and 2 million applications.[28]

The regime incorporated control over the export of technical data into this program from the outset. Echoing Roosevelt's stern warning of the dangers of air warfare and the need to build a huge air fleet, at first information controls focused on aircraft and aviation fuel technology. In September 1940, the controls of equipment for the petrochemical production of aviation fuel and for the production of tetraethyl lead (which is used to boost the octane of fuel) explicitly included "any plans, specifications, or other documents containing descriptive or technical information of any kind (other than that appearing in any form available to the general public) useful in the design, construction, or operation of any such equipment or in connection with any such processes." This extended also to documents "setting forth the design or construction of aircraft or aircraft engines."[29] A presidential proclamation from December 1940 mentioned "equipment and plans for the production of aviation lubricating oil."[30]

These regulations harkened back to the "moral embargo" against Japan that had targeted its ability to execute bombing campaigns by restricting exports of planes and their engines and equipment.[31] In December 1939 the State Department, after consulting with the War and Navy Departments, had widened the embargo's scope to include "technical information for the production of high quality aviation gasoline" and added, by way of explanation, "This decision has been reached with a view to conserving in this country certain technical information of strategic importance."[32] This is a remarkable statement. It is to our knowledge the first public document that explicitly lays out the idea of export controls as a measure to stop the flow of unclassified knowledge, understood as a national resource, in order to secure a technological advantage over an adversary. During the Cold War this was usually referred to as technological "lead time."

The US authorities quickly realized that, to secure this lead, it was not enough to keep control over aviation technology. Even before the country officially entered the war, it became obvious that this war would, to an even greater extent than that of 1914–18, be fought with technology built on the largest scale by the entire national production system. It would be a "total war," encompassing the full spectrum of technologies available—and these would have to be produced by the national innovation system. The "Great Arsenal of Democracy" that Roosevelt famously invoked in his radio speech of December 1940 was a technological arsenal: "American industrial genius, unmatched throughout all the world in the solution of production problems, has been called upon to bring its resources and talents into action. Manufacturers of watches, of farm implements, of Linotypes and cash registers and automobiles, and sewing machines and lawn mowers and locomotives, are now making fuses and bomb packing crates and telescope mounts and shells and pistols and tanks. But all of our present efforts are not enough. We must have more ships, more guns, more planes—more of everything. And this can be accomplished only if we discard the notion of 'business as usual.'"[33]

To secure this technological arsenal, Presidential Proclamation No. 2465 of March 1941 prohibited the export except under license of "any model, design, photograph, photographic negative, document or other article or material, containing a plan, specification, or descriptive or technical information of any kind (other than that appearing generally in a form available to the public) which can be used or adapted for use in connection with any process, synthesis or operation in the production, manufacture or reconstruction of any of the articles or materials the exportation of which is prohibited or curtailed in accordance with provisions of section 6 . . . , or of any basic or intermediary constituent of any such articles or materials."[34] This was certainly not "business as usual." This proclamation covered "technical information of any kind" related to controlled merchandize. In December 1941 all exports of goods were subject to export regulations. Thus when America entered the war, export controls pertained in principle to *all* unpublished technical data in the United States—or, in practice, to data related to twenty-five hundred commodities and commodity groups.

Both in their scope and in the concepts they applied, these new regulations went markedly beyond the aircraft data controls of September 1940. The expansion of controls in 1941 embraced information conveyed on a wide variety of platforms, included the upgrading of artifacts, and broke down the end product into its constituents. Obviously, in their attempt to fill all potential loopholes to avoid unwanted technology transfers, the controllers had thought more broadly about what constituted and what was needed for the successful communication of knowledge. Even though the concept of "information" was closely tied to goods on the control list, the new data controls recognized "that 'technical data' was as essential to national defense as commodities themselves."[35]

The new status that information had acquired is manifest from the specific instructions on the legal definitions and the licensing processes regarding technical data, published for the first time by the administrator of export controls in April 1941. This marked the beginning of the systematic enforcement of export controls over scientific-technological information.[36] As the categories were translated into bureaucratic practice, they were further refined. Three different kinds of licenses were introduced. The "Special License" covered the "export of a single item of Technical Data from one specific sender to one specific addressee." The "General License" covered "the export of specific Technical Data as such, and not a license to a person or firm" in certain destinations friendly to the United States.[37] And a "Blanket License" was necessary for "sending routine 'technical data' continuously to [US companies'] subsidiaries or affiliates in large volume." The important qualifier here was that the Blanket License was not applicable to "materials which will include any innovations or essential changes in current processes."[38] In 1942 a fourth category was introduced, the "Unlimited License," to facilitate the exchange of printed mass publications like catalogues or advertising materials—while periodicals and magazines fell under the Special or the General License. By stressing that "the material exported under Unlimited License must be of no military importance," this addition emphasized once more the intimate interrelations between security, the national economy, and the production and dissemination of scientific-technological knowledge.

In February 1942 the War Department's Policy regarding Dissemination and Publication concerning Contracts, Production, Site Locations, Etc. reinforced innovation as one of the categories for assessing sensitive data. In practice, a wide swath of information was singled out as being excluded from a Blanket License: information of "such novelty and completeness as to be the subject of an actual or of a proposed patent application," patent applications themselves, including amendments and related abstracts or other papers; "research, laboratory-progress, testing, or experimental reports"; "data pertaining to materials" intended for military use, except "widely-known commercial designs"; "data pertaining to materials to be used in connection with any new development, projects, or installations." Thus, patent applications and unpublished research data was treated in the same way as arms and munitions, and information bearing a military classification. Moreover, patents, munitions, and classified information were also excluded from the General License, that is, they were restricted even in communications with the United States' closest allies in Britain and the Commonwealth.[39]

Export Controls, Classification, and Censorship

The scope of export controls on knowledge during World War II was thus immense. They dovetailed with the sprawl of military classification during the war years. Most famous (not least because of its leaks!) is probably the system of secrecy that was set up on the sites of the Manhattan Project. But secrecy became a common phenomenon in huge areas of US research and development as it quickly grew, fueled by large expenditures from the federal government war chest. Government money also meant government control, and secrecy was one of its most powerful tools that could be combined with other forms of information management. Cold War secrecy had its roots here and was emphatically not just about securing "the Bomb."

Patents were an area of knowledge where secrecy and export controls overlapped—once again, as they had already loosely met during World War I in the Trading with the Enemy Act. This law was still on the books and came to be reused in World War II. Patent applications were closely guarded not only by export controls (which extended, as we saw, even to patentable knowledge!), but also, once again, by invention secrecy. The Invention Secrecy Act of 1917 was revived in July 1940, limited to two years. In 1942 the act was renewed, explicitly for the duration of the war.[40] The rationale for keeping patents secret was based on the assumption that the enemy was reading and exploiting all unclassified patents. The Army and Navy Patent Advisory Board, founded in 1940 to screen patent applications to support the national defense effort, explained in June 1945: "While the war continued in Europe and especially as Germany and Japan both had laboratory and manufacturing facilities for exploiting any disclosure which might have military value, it was clearly proper, in case of doubt, to impose a secrecy order." In a similar vein, Lieutenant Colonel Francis H. Vanderwerker of the Office of the Judge Advocate General emphasized in 1941 the dangers of patent publications, stating that "one erroneously issued patent may do more to injure national defense than the improvident holding of a hundred applications in secrecy."[41]

In World War I, owing to the short engagement of the US, patent secrecy did not have a strong impact.[42] However, between 1940 and 1945 no fewer than twelve thousand patents were subject to a secrecy orders.[43] Since the US Patent Office issued about 184,500 patents between 1941 and 1944, roughly 6.5 percent of all US patents were held secret.[44] In fact, patent secrecy cast an even wider net since the Invention Secrecy Act was only one of six different bureaucratic procedures that the Patent Office used to limit the circulation of patent texts.[45]

Patent secrecy affected a wide range of top-notch technologies with high importance to the US war effort. Especially closely guarded were nuclear and radar technologies as well as cryptology, though a much wider spectrum of technologies was covered, from "petroleum and synthetic rubber technology to torpedoes and nylon (when it came to be used in parachutes)."[46] In the case of atomic patents, the security screenings even extended to patents that were issued before the war to retroactively classify them if they contained sensitive information.[47]

It was not only patent secrecy and export controls over patents that complemented each other. There was also considerable overlap with wartime censorship. As an extensive report on World War II controls put it in 1950, "Export control of Technical Data is quite evidently a Censorship function."[48] Indeed, both targeted the dissemination of information beyond the realm of classification and dealt not least with the regulation of printed papers. Accordingly, the Postal Censorship Regulations of April 1942 referred directly to the data export regulations and listed categories of technical information prohibited from "all communications to foreign countries." Also, any piece of mail to an enemy national required a license.[49] Export controls were thus a subset of censorship, specializing in control over scientific-technological information. It is of little surprise that they were also organizationally closely intertwined.

The new Office of the Administrator of Export Control in Washington had a specialized unit for technical data from the outset, and when the export control bureaucracy branched out to other cities in the United States, data units spread as well. New York was the first export control office with such a unit outside Washington, DC. Early on, this expansion was closely tied to the Office of Censorship, which had been established on December 19, 1941, in reaction to Pearl Harbor. After the attack, the United States immediately shut down the export of *all* newspapers and periodicals, creating a huge backlog for the new office even before it started its work in March 1941. When the Censorship Office opened branch offices in Chicago and Los Angeles, it included technical data control personnel that reported to the Export Control Administration. When the censorship office in New York followed suit, the majority of the data export controllers already working there moved over and joined the censors. At the height of the war, twenty-three staffers in New York, whose work was mainly to screen technical publications, worked on data controls. "At all times, there was at least one official with engineering experience and considerable acquaintance with the various technical associations and societies that could aid in the identification and classification of data when required." However, in contrast to the later Cold War practices, there was no systematic communication with industry advisory committees. The controllers communicated usually with individual experts to assess license applications.[50]

Spanning from the shipment of goods to the sending of letters, the concept of what constituted an "export," albeit never defined, was wide. It included also exports "by cable, telegram, or radiogram"—and the movement of people, even though their control was not systematically enforced. Following the historical pattern, the US passport had a comeback when the prospect of war became ever more threatening. For the first time since the end of World War I, the freedom of travel was curtailed, now by the Neutrality Act of 1937, forbidding American citizens from traveling "on ships of belligerent states" and on any ship if its destination was in an area declared to be a combat zone by the president. Furthermore, security checks for passport applications were reintroduced. The Passport Office screened applicants "in cooperation with the various intelligence officers of other Government agencies to determine whether the public safety would permit the granting" of a passport.[51] In November 1941 the president proclaimed that every US citizen "or person who owes allegiance" to the United States should not leave the country without a passport.[52] At the same time, the data export control bureaucracy attempted "to screen technical personnel moving to operations" but seems to have given up on this quickly—probably because of the huge number of people traveling for the military. "There was at no time any coordination of action and of objectives in this respect between technical data controls and the Passport Division of the State Department."[53] As we will see, the idea to combine these two regulatory spheres was on the agenda again in the early Cold War.

World War II had not even reached its fiercest and bloodiest phase, when American bureaucrats began to ponder the direction of US technology and information policy in peacetime. Against the backdrop of the enormous expansion of the US innovation system in the war years, it was clear to policy makers that technology had become a crucial asset that would define the position of the United States in the global order. Recent technological progress provided the foundation for conquering civilian world markets after the war and would also smoothen the period of reconversion from war to peace economy. "Industrial technology" was seen as "our potent trading weapon." It was also to be expected that the United States, the strongest economy in the world, would play a crucial role in the reconstruction of many foreign countries. They would not only seek US capital, raw materials, machinery, and consumer goods but would also be interested in acquiring "technical developments, i.e. those of a special and continuing nature as opposed to technical information usually supplied to all customers" (or "know-how") as well as "patents and secret processes."[54]

For US policy makers, there was an apparent tension between winning technology markets and sharing technology with war-torn countries: "While the development of an extensive foreign trade is a most desirable post-war objective and essential to secure maximum employment it is felt that completely unsupervised private trading at this time might result in dissipation of certain national assets without adequate quid pro quo." Export controls over patents, trade secrets, and know-how were supposed to reconcile and balance trade interests and technological advantage. While allowing for exchange, they would insert "restrictions necessary for the national welfare" and for national security.[55]

Freeing Information

From mid-1944 to the end of 1945, the general thrust of US information policy favored quick and far-reaching liberalization. At its height in summer 1945, this trend even led to policies of radical openness—only to be followed by a violent backlash. Three intertwined strands in the debate about technical data control show that these developments were intended to establish a predominantly liberal and postwar order in which only narrow national security considerations would curtail some information exchanges. These strands were the question of trade with the Soviet Union; the question how to disseminate technologies captured from the enemies in Germany and in Japan; and the question of how to regulate nuclear knowledge.

In the last year of the war, the United States nurtured hopes that the Soviet Union would become a vast peacetime market for American technology exports. For example, when the State Department's Munitions Control Section met on December 30, 1944, to discuss the "subject of the transfer of technological information to foreign countries for post-war use," the conversation revolved exclusively on trading relations with the Soviet Union.[56] The recent past showed that such hopes were not unrealistic. In the 1920s and 1930s, a huge number of US companies, from Ford to RCA, had done business with the Soviets on the basis of technical aid contracts. The Soviet Union had sought the influx of Western technology to support its ambitious industrialization program. The contracts it consummated with its foreign partners entailed usually cash payment for elaborate packages of technology transfers, including blueprints, specifications for production processes, and, most importantly, the movement of people with technological know-how. The Soviets sent large numbers of engineers for sometimes very extended stays—in the case of Ford, for years—to the United States. In exchange, American experts traveled to the Soviet Union to help implement knowledge transfer on the ground by showing local technical personnel how to do things and monitoring production processes.[57] These intense exchanges had petered out in the 1930s, though the close technological relationship was renewed in World War II. The United States, as part of the Lend-Lease program, shipped enormous amounts of weapons, merchandize, and production facilities to its ally to help it fend off the brutal onslaught of the German army. Under the umbrella of Lend-Lease, a lively technology exchange unfolded, bringing thousands of Soviet officials to US plants and research and development facilities. These visitors were not welcomed without mistrust. The FBI sounded the alarm time and again, pointing out that it had strong evidence that the Soviets engaged in widespread industrial espionage. These concerns were, however, overridden by the US government's interest in keeping the anti-Hitler coalition strong and on track.[58]

Lend-Lease and the interwar contracts appeared to be a good starting point for closer trade relations when the end of the war was in sight. Anticipating a wide interest of US companies in trade relations with the Soviet Union, the Department of Commerce started work on a brochure entitled "How to Do Business with the USSR" in March 1945.[59] There were intense debates for that entire year about the renewal of commercial relations.

One central strand of these debates was the question of to what extent technological information should be shared and how it could be safeguarded. Matters of intellectual property (IP) protection took center stage. In contrast, data export controls, even after the fighting had stopped in Europe and in the Pacific, were exclusively seen as a measure against the war enemies until peace treaties had been signed. Exports to all other destinations were quickly decontrolled.[60]

Intellectual property was also the focus of the December 1944 Munitions Control Section's meeting on the prospects of trade with the Soviet Union mentioned earlier. The controllers deemed that the most serious hurdle to regular trade relations with the Soviet Union was that it gave "practically no protection to holders of American patents." The Soviets had in many instances shown their disrespect for IP protection by "infringement of patent rights, the purchase of articles in this country for the sole purpose of copying and reproducing them in Russia, and the use of technological information and of trade secrets, in contravention of the terms of the agreements which had been entered into." Cash payments, which covered the price of the commodity as well as the value of the technological information transferred, were therefore currently the only basis for US companies to sell technology to the Soviet Union. Remarkably, these discussions did not dwell on the security implication of technical information. They touched only briefly on the "release of such information from the standpoint of military security" and only with reference to the ongoing war.[61] That did not mean that security concerns vanished entirely.[62] They were, however, narrowed down again to control over classified information, and when US officials outside the FBI spoke during 1945 of "protecting US technology conveyed to the USSR," it did not have the nefarious Cold War ring it would acquire only a year later. It simply meant IP protection.[63]

At first, the doubts about Soviet respect for IP did not affect the general direction of US policy, spearheaded by the Department of Commerce: to open up toward the Soviet Union in order to sell goods and exchange technical information. In this liberal climate, the Soviets began to seek new technical aid contracts with US companies.[64] At least thirty-nine US firms had consummated or negotiated such contracts with the Soviets in 1944–45, among them some household names of US capitalism. Standard Oil, for example, was hammering out an agreement with Soviet officials on the exchange of information on synthetic rubber. International General Electric began in summer 1945 to negotiate over the export of technical information.[65]

At the same time as the United States opened up to technology trade with the Soviet Union, it was involved in one of the biggest campaigns of technological information transfer in modern history. Technical intelligence teams were embedded in the troops that fought their way to Germany after their landing in Normandy. They systematically explored every German and German-run factory and research facility along their path. With acquisition lists in their pockets, they got hold of every shred of information they could about the technology the Germans had developed and used, running the whole gamut of military and civilian technology from nuclear research and rockets to the accomplishments of the German chemical industry, like synthetic rubber and the production of fuel from coal, to mundane technologies like machines used for the production of pencils or packaged food. This was no less than an attempt to acquire Germany's entire corpus of scientific-technological knowledge.[66]

In an attempt to allow for a dense flow of knowledge, the technical intelligence teams in Germany collected not only information (like blueprints, laboratory records, and every patent specification in the archives of the German Patent Office), but also "things" (like machines that would be reverse engineered) and people (technicians and scientists who ended up working in US factories and research institutions like the rocket program in Huntsville, Alabama).[67] Military and civilian interests went hand in hand. The military provided the support and put representatives of American industry and science into uniform for a mission that was, from a German perspective, industrial espionage on the largest scale. The knowledge thus acquired was fed into weapons research to fight the war more effectively and to enhance the US stance in peacetime. In particular the acquired knowledge was supposed to strengthen the competitiveness of US industry, enabling it "to maintain its place in world trade after the war," as Vannevar Bush, the mastermind of US science mobilization for war, put it.[68] For him, strengthening the military as well as US industrial power seemed all the more pressing since Great Britain, France, and the Soviet Union ran similar programs of "intellectual reparations."

In order for US industry to make use of the technological knowledge imported by this program, the knowledge needed to be communicated to potential users. The US government had a radical idea: it would move all available information, after a military security screening, into the public domain, bluntly and openly suspending all German intellectual property rights. This would, as government officials hoped, make possible a quick utilization of the knowledge and forestall its "rapid obsolescence" due to technological-scientific progress.[69] In its most sweeping rendition, this public domain strategy extended beyond US borders, in line with the State Department's enthusiasm for a new open world order that called for "full dissemination among all the United Nations."[70]

Having been discussed as early as January 1945, this policy was implemented through two Executive Orders signed by President Harry Truman in June and August 1945.[71] Even though they did not support the State Department's UN angle, they were testimony to the spirit of openness that now dominated the US discourse on technical data. First, Executive Order 9568 set in train the declassification of information pertaining to R&D that "has been, or may hereafter be developed by, or for, or with funds of any department or agency of the Government" for "maximum benefit to the public." The order was amended in August by Executive Order 9604, which sought in addition the "prompt, public, free and general dissemination of enemy scientific and industrial information."[72] Even though national security screenings were to cushion this push for declassification, there was no doubt that open sharing was the chief goal. The policy was quickly implemented by the Army and Navy Departments under the supervision of the Committee on Release of Scientific Information (CORSI). Material freed of its security markings was made publicly available by the newly established Publication Board within the Department of Commerce (in 1946 renamed the Office of Technical Services).[73]

Just how powerful the declassification trend was, was emphasized by the fact that the very first item CORSI discussed was one of the crown jewels of military technology: radar. The development in cooperation with the British of 150 different radar systems was one of the great triumphs of the OSRD, outsized only by the Manhattan Project.[74] Similar to the nuclear bomb project, radar was regulated by strict secrecy. In 1943 all public dissemination of radar information had been stopped, and all patent applications in this field were subject to invention secrecy screenings, adding up to a "total blackout."[75] In September 1945, CORSI started, in agreement with the military and with the British government, the process of "prompt and wide-spread declassification" of radar information to "assist" US industry "in the reconversion."[76]

Executive Orders 9568 and 9604 did not extend to the nuclear field. But even here the declassification trend made inroads during the second half of 1945. Although the military head of the Manhattan Project, General Leslie Groves, was a fervent hardliner when it came to information control, he made concessions in the name of the greater public interest in nuclear technology after Hiroshima and Nagasaki. As early as September 1945, Groves allowed for the publication of the first official report on the science behind the construction of the atomic bomb, *Atomic Energy for Military Purposes*, published by Princeton University Press.[77] Usually called the

Smyth Report after its author, the physicist Henry DeWolf Smyth, who had worked for the Manhattan Project, the report was meant to celebrate the accomplishment of the US government's largest science project, explaining to the American public just what the government did with two billion tax dollars. The book, an instant bestseller, suggested a great degree of candor and transparency. In fact, however, it seamlessly continued the strict security regime that had characterized the Manhattan District. In all its details, the report went through a fastidious process of censorship. It gave nothing away that an expert audience did not know already. For Groves, the Smyth Report was essentially a safety valve: through the controlled disclosure of some information it was supposed to prevent uncontrolled leaks.[78]

A few months later, in November 1945, Groves made a greater concession to industry and the scientific community, who both had called for more openness to make use of the knowledge amassed during the war. Groves established a Committee on Declassification, representing the cream of American science: Robert J. Oppenheimer, Ernest O. Lawrence, Arthur H. Compton, Harold C. Urey, Frank Spedding, Robert F. Bacher, and Richard C. Tolman as chairman. They began to develop criteria to separate knowledge that could be used without danger and to the benefit of US welfare from militarily relevant information as such.[79]

The committee was part and parcel of a much wider debate in the Truman administration as well as in Congress about how to deal with nuclear technology after its awe-inspiring military power had been displayed. Obviously, the technology had to be controlled, but in the second half of 1945 this did not mean a continuation of the impervious secrecy of the war. Rather, the challenge was to unleash the atom's scientific and economic potential without curtailing the nation's security. In late 1945, it seemed that this balance would be decided in favor of openness and knowledge sharing—in spite of the opposition of hardliners like Groves. The McMahon Bill of December 20, 1945, the result of months of intense deliberations in Congress and with a good chance to be enacted, opted for a liberal information policy to foster scientific progress. Indeed, it advocated "the free dissemination of basic scientific information and for maximum liberality in dissemination in related technical information." National security was not obsolete but secondary—the bill's short preamble, "Findings and Declaration," does not even mention the term. Information deemed militarily sensitive was, as during the war, to be shielded by the Espionage Act of 1917. Moreover, the government would have full control over all fissionable material and its production and use.[80]

The liberalization trend also had an international dimension. Parallel to the debates in Congress on a nuclear bill, there were negotiations about the international control of atomic technology. In December 1945, the foreign ministers of the United States, the UK, and the Soviet Union agreed to establish a United Nations Atomic Energy Commission, which would "control atomic energy to the extent necessary to ensure its use only for peaceful purposes." It would work toward nuclear disarmament and would, for this purpose, establish safeguards to detect evasions. The commission would also facilitate "between all nations the exchange of basic scientific information for peaceful ends."[81]

Invention secrecy was also largely dismantled. Congress had limited its patent secrecy legislation to wartime, and even though there were no peace treaties with Germany and Japan as yet, the commissioner of patents issued a rescission order in November 1945 that immediately declassified 6,575 patents. At the end of the year, only 799 classified patents remained—probably the lowest number in all US postwar history.[82]

Finally, in December 1945 the controls over exports of technical data to friendly countries were officially liberalized. The Office of Censorship had closed shop in August, and data controls had not been strictly enforced anymore anyway.[83] The State Department's official position summed up the general picture, stating "that the exchange of technical information in general in a peace-time economy should be as free as consistent with military security." Ideally, privately held technical data should circulate with "complete freedom."[84]

Thus, over the course of 1945, many signs pointed to a profound liberalization of the US stance toward sharing scientific-technological information and a return to a peacetime status that imposed only a few restrictions in the name of a narrow concept of military security. Even the very flagship of military science, nuclear technology, and radar technology were opened up.

Cold War Backlash

But there were storm clouds gathering on the horizon. For some exponents of the security establishment that had grown in power and size during the war, the McMahon Bill reached too far. Moreover, to influential players like FBI director J. Edgar Hoover it seemed utterly dangerous to trust the Soviet Union. As relations with the war ally in Moscow became slowly more complicated, first and foremost because of clashes of interests in regard to reordering Germany and Europe, the course of openness was more and more questioned—becoming untenable in early 1946.

This climate change made itself felt in the nascent trade relations that had unfolded since late 1944. Within less than a year, enthusiasm for trade with the Soviet Union tapered off. US officials increasingly worried about the flows of technological knowledge. The lack of IP protection was compounded by the lack of reciprocity. Knowledge flowed in one direction only, giving the Soviet Union an advantage that appeared economically unacceptable and, increasingly, dangerous to US security. As US officials came to realize in the course of 1945, the Soviet Union, while restrictive within its own borders, was taking advantage of the openness of US society: "The Soviet Union has relatively free access to the technology of the United States; Soviet representatives are able to purchase copies of all patents issued by the Patent Office; Soviet engineers are able to travel freely around the country and visit American plants and obtain ideas regarding construction and plant layout." By contrast, US engineers were "greatly restricted in their movements in the Soviet Union." They did not have access to Soviet plants and could not obtain patent specifications.[85] It also worried the State Department that US companies had in the recent past been "extremely lax in protecting their own interests . . . and that for some reason they seem to disclose valuable information to representatives of the Soviet Union in this country without taking proper safeguards."[86]

Owing to the lack of IP protection and the openness of the US system, US companies did not have much leverage in dealing with the Soviet Union. The only bargaining chip they seemed to have left was "know-how."[87] Indeed, the term "know-how," while not widely used in the 1930s, had by 1945 become shorthand for government officials to denote an idea of knowledge that went beyond technical information written down on paper and was central to the successful transfer of technology. Never clearly defined, it was nevertheless used by government agencies working on matters of science and technology in a manner that implied common understanding. Indeed the concept was widely used by the US government before it started its steep career in postwar business and legal thinking and before scholars paid attention to it. Moreover, know-how became an important category for the national security community in its attempt to curtail and control the flow of technology.[88]

The term *know-how* combined ideas of trade secrecy, "unpatentable knowledge," and what scholars would later call "tacit knowledge." A strong emphasis was placed on the importance of direct communication between people for a successful transfer of technological knowledge. As a State Department memorandum put it in April 1946: "It is a commonplace among technicians that the working of important inventions usually requires a considerable body of information which does not appear in published form and which can be acquired only through plant visits and intimate consultation with knowledgeable engineers concerned. Even where secret processes are not involved, information falling under the general heading of unpatentable and unpublished industrial experience must be acquired in order to reduce technology to practice."[89] Arguing along the same lines, an earlier memorandum stated that these ideas actively shaped US policy in Germany: "This principle was also acknowledged in our 'reverse looting' program in Germany, where it was deemed necessary to allow consultation between German and visiting US technicians."[90] Know-how was, however, not only carried in "someone's mind."[91] It was also incorporated in physical objects: "Prototypes and technical data may be regarded as a substitute for know-how; instead of procuring the services of experienced engineers, one gets working models or operational information" from which Soviet technicians could extract the embodied knowledge through reverse engineering.[92]

The 1946 statement also suggested that "know-how" dissolved the open-secret binary because the difficulties of sharing know-how could hamper technology transfers even without the impact of formal classification or confidentiality. Once disclosed and shared, however, know-how lost its exclusivity and slipped beyond control: It was impossible to "unscramble the eggs. 'Know-how' cannot be withdrawn."[93]

Thus, the debate about US-Soviet "technological relations"[94] had a complex structure. It blended reflections on the role of intellectual property protection in economic international relations, on the dangers posed by the openness of democratic political systems, and on the nature of technological information and "know-how." All three strands were tied to the concept of "reciprocity": an equal, two-way flow of information and people transferring technological knowledge. The United States was distressed that the Soviet Union denied equity in their technological relations, leading to a constant relative loss of US knowledge.

In the course of 1945, the State Department thought to address this imbalance by including the interchange of technology in a "general treaty of friendship, commerce and navigation" with the Soviets. Its aims were to establish satisfactory IP protection and the exchange of printed patent specifications and to establish for US businessmen "the right to enter the Soviet Union on a comparatively free basis in order to promote the exchange of information."[95]

By fall 1945, the State Department had grown increasingly disillusioned by the lack of reciprocity. In a letter to the secretary of commerce in October, Henry A. Wallace wrote: "There is no desire on the part of this Department [i.e., State] to object to American firms transferring on a reasonable basis technological information abroad. However pending clarification of the policy of the Soviet Union regarding access to its technology, this Department is unable to suggest that the transfer of information to the Soviet Union is within the national interest."[96] Six weeks later the State Department still did not "interfere at the present time with the export of unclassified technology to the USSR" but stated, with growing urgency, "Should attempts to obtain reciprocity fail, consideration will be given to the desirability and feasibility of restricting or discouraging the transfer of technology and know-how to the Soviet Union." Moreover, the US government had begun to ponder "the security aspects" of its "policy toward the release of unclassified technology."[97]

Thus by late 1945, the intense push toward more openness had begun to seem ambivalent, and there were more voices that were skeptical about the far-reaching liberalization of wartime information and export controls. The chief of the Patent Section in the Office of the Judge Advocate General (War Department), for one, wanted to retain, in the interest of military security, at least some of the controls over the export of unclassified technical information and also over patent applications.[98] That went together with the desire in military circles to continue the control over "eight or ten strategic materials which are in short supply and which are not produced in this country and which military authorities wish to stockpile."[99]

In February 1946 the first atomic espionage case, the Gouzenko affair, gave the public a glimpse into Soviet infiltration of the Manhattan Project. It became one of the catalyzers of a profound change in how the US government discussed the control and dissemination of scientific-technological information—and not just in the nuclear field. Contrary to what historians usually write about this spy case, the first press reports lacked the hysteria that would become a trademark of McCarthyism.[100] But the fallout from the affair coalesced with a growing pessimism in the Western world about its relations with the Soviet Union. Just how downbeat the international climate was became clear in the four weeks from early February to early March 1946.[101]

Early in February Stalin delivered a startling speech about the inevitability of war in a capitalist world economy. A mere week later, on February 16, the readers of the *New York Times* learned from the headline on page 1 about Gouzenko's revelations of atomic espionage. Against the backdrop of growing tensions with the West over Soviet interests in Iran and Manchuria, US pundits interpreted this speech as dangerous saber rattling, if not a thinly veiled threat of war.[102] Shortly before the speech, the State Department had assigned George F. Kennan, chargé d'affairs in the US embassy in Moscow, to write a report that outlined the goals, strategies, and ideological foundations of Soviet foreign policy. The result was the so-called Long Telegram, which depicted the Soviet Union as a neurotic, expansionist, and subversive adversary plotting to weaken the West everywhere in the world. Even though Kennan spoke against attacking the Soviet Union militarily, he advocated a policy pursued "with [the] same thoroughness and care as [the] solution of major strategic problem in war, and if necessary with no smaller outlay in planning effort." The Long Telegram, at the time widely read in Washington, was the first coherent formulation of the idea of "containment" that was to shape US Cold War policy for decades.[103] No less influential was Winston Churchill's "Iron Curtain" speech in Fulton, Missouri, two weeks after Kennan's telegram. Churchill coined a powerful metaphor that encapsulated the new state of international affairs only seven months after the end of the war in the Pacific.[104] The Cold War was taking shape, soon calling for the construction of a national security state that would completely redefine the transnational circulation of ideas, people, and things across American borders. Already by the end of the 1940s there would be little left of the short-lived liberalization trends of 1944–45. As the Soviet-American relations deteriorated the discussions over IP protection, the sharing of German technologies and nuclear secrecy would take a decidedly restrictive turn. This paved the way to the conceptualization and implementation of the American postwar peacetime export control system over scientific-technological knowledge, as we will show in the next chapter.

CHAPTER THREE

The Cold War National Security State and the Export Control Regime

World War II saw an unprecedented mobilization of the national innovation system that changed it for decades to come. In only four years, 1941 to 1945, the state had become the main driver of R&D, providing the lion's share of funding, defining the direction of research in the interest of winning an industrial war militarily, and building the organizational framework to boost knowledge production. This completely changed the balance within the US innovation system. Whereas in 1940, two-thirds of all American research expenditures had been covered by industry, the federal government paid—even excluding the $2 billion spent on the Manhattan Project—for 83 percent of all wartime R&D. Industry's share fell to 13 percent.[1] At the same time, war spending massively expanded the overall expenditures for science and technology.

World War II showcased what was possible if the federal government took an active role in the organization of science and technology. When an end to the fighting in Europe and in the Pacific came in sight, it seemed only logical to continue the success story in peacetime. In November 1944 President Franklin D. Roosevelt asked OSRD head Vannevar Bush to undertake a thorough study of the role that science should play once the war was over:

> The Office of Scientific Research and Development . . . represents a unique experiment of team-work and cooperation in coordinating scientific research and applying existing scientific knowledge to the solution of technical problems paramount to war. . . .
>
> There is . . . no reason why the lessons to be found in this experiment cannot be profitably employed in times of peace. The information, the techniques, and the research experience developed by the Office of Scientific Research and

Development and thousands of scientists in the universities and in private industry, should be used in the days of peace ahead.²

Bush's report, the famous "Science—the Endless Frontier," published between V-E and V-J Day, presented his vision of a federally funded system of basic research that would accomplish no less than the institutionalization of scientific-technological progress. National security as well as national welfare depended on science: "New products, new industries, and more jobs require continuous additions to knowledge about the laws of nature, and the application of that knowledge to practical purposes. Similarly, our defense against aggression demands new knowledge so that we can develop new and improved weapons. This essential, new knowledge can be obtained only through basic scientific research. . . . Without scientific progress no amount of achievement in other directions can insure our health, prosperity, and security as a nation in the modern world."³

While Roosevelt and Bush both placed the emphasis on the peaceful uses of science, national security became the guiding principle of the national innovation system. Science became embedded in the new institutional structure of the national security state as it was built in the immediate postwar years.⁴ Decidedly turning against isolationist ideas, the strategic thinkers in the Truman administration insisted that the United States could be safe only if it reordered the world according to its ideals of liberal democracy and free-market capitalism; built a system of global alliances under US leadership, containing and isolating the Soviet Union; and organized the strongest standing military machinery in the world, which would be permanently prepared for war on the largest scale.

With the deterioration in US-Soviet political relationships, the circulation of scientific-technological knowledge was reassessed—and not just in the nuclear realm but in general and including unclassified information. Five days after Kennan sent his Long Telegram to Washington, representatives of the State and Commerce Departments and the military met with representatives of the firm RCA to reconsider its negotiations with the Soviets about a technical assistance agreement since the "political and economic conditions had evolved considerably since the original submission of the RCA contract." Nervously RCA asked, "what the present feeling of the Government is regarding Soviet collaboration in research." In the midst of an unfolding political crisis, government officials struggled to find a clear-cut answer "regarding the advisability of such close collaboration," but apparently "a new approach to the subject is now necessary." The baseline, however, was that in their "personal opinion" the government officials "did not like" the idea of technology transfers to the Soviet Union. Nevertheless, they allowed RCA to do business that excluded classified information because the contract would then "be viewed as a purely commercial transaction upon which the Espionage Act would have no bearing." The following discussion, however, showed that things were much more complicated and that there was no sharp demarcation between classified and unclassified research. The research process might start off in the unclassified field but some of the side products may turn out to have "military significance" and thus be in need of classification. As a solution to this problem, the government representatives suggested that any research shared by RCA should be subject to military oversight. RCA also promised to closely control Soviet engineers coming to its facilities, giving them access to production methods but not to the laboratories.[5]

The tone of the debate about scientific-technological information very quickly became much more alarmist because American officials began to see technical aid contracts, and for that matter atomic espionage, as part and parcel of a Soviet campaign of "'total exploitation' of all American technical information. This is gathered here indiscriminately from all available sources and sifted and analyzed in the Soviet Union." Suddenly it seemed dangerous that the Soviets bought technical journals in the United States whereas it was difficult to acquire such literature from the USSR. It seemed even more threatening that the Soviets purchased patent specifications from the US Patent Office "on a wholesale scale."[6] US officials estimated that the Soviet Union had "recently" bought several hundred thousand patent copies with large orders pending. These activities were understood as connected to the Soviet program of accelerated industrial development that also had grave military implications.[7] Drawing on the central lessons of having fought and won a total war, one State Department official stated that "wars of today are fought on the basis of technology and there is very little technical information which does not have a more or less direct relationship to the prosecution of war."[8] Moreover, the US analysts were convinced that the Soviets had fought World War II "on the basis of technology obtained from America in the years prior to the war."[9] Thus, it seemed wise to be careful about what knowledge to share.

Tightening Restrictions over Exchanges with the Soviet Union

This stance hardened in the following weeks. By early spring 1946 a conviction had taken hold that would become the basis of US Cold War technology policy. "The only future conflict which may be considered possible," a State Department memorandum stated in April, "will place the United States on one side and Russia on the other. Therefore, whatever is done to protect American 'know-how' or technical knowledge must be done with this in view. On the one hand, this country must place itself in a position to benefit by the technological advances made by as many other countries as possible and, on the other, do what it can to prevent knowledge which might affect the national security from becoming known to the Union of Soviet Socialist Republics."[10]

Thus, scientific-technological knowledge was seen as a central resource of national security and military power that had to be guarded closely. The US relationship with the Soviet Union was understood as a race for knowledge accumulation. The United States had to learn more than its prospective enemy and to deny it knowledge that would allow it to catch up. This was the idea of scientific-technological lead time that was so central for the US postwar regime of knowledge regulation and control.

This way of conceiving scientific-technological relations with the Soviet Union closely codeveloped with the new policy framework of "containment." It gained currency in the same few weeks early in 1946. At the same time, the reassessment of the benefits and dangers of scientific-technological knowledge reflected the lessons learned in World War II. In 1945 it seemed beyond doubt that the United States' ability to produce and apply knowledge had secured victory.

Military power was now defined by the availability of the most modern and effective weapons technology. This would offset the advantage the Soviet Union had due to the much larger manpower it could muster. What would become "the doctrine of comparative technological advantage" assigned science and technology a central role in national security thinking. Only the steady accumulation of new knowledge could ensure that the United States would always be one step ahead of its enemies.[11] Accordingly, the Steelman Report of 1947—a top government report, published after the announcement of the Truman Doctrine and the Marshall Plan, and also as a follow-up and as countermodel to the Bush Report—spoke of a close "relation of science to military preparedness" and stated apodictically: "A nation which is backward in fundamental knowledge would be severely handicapped in any future war." In expectation of a war that would begin at any moment, Steelman saw knowledge as well as the scientists and engineers who produced it as resources that had to be "stockpiled" like oil or steel or weaponry like planes and shells. Moreover, innovation and military power were closely related to economic prowess. Only economic strength could provide the industrial base the US military, and indeed US power in the international realm, relied on: "If we are to remain a bulwark of democracy in the world, we must continually strengthen and expand our domestic economy and our foreign trade. A principal means to this end is through the constant advancement of scientific knowledge and the consequent steady improvement of our technology." Indeed, the very status of United States as "a world power" was built on science and technology.[12]

Fostering science and technology became an integral part of the architecture of the national security state, and the Department of Defense became the core of the national innovation system. Even though in the immediate postwar years federal R&D expenditures experienced a reduction, in 1948 the Department of Defense still spent about $430 million on thirteen thousand individual projects. As the Cold War picked up steam, the expenditures climbed steeply. When the Korean War broke out, the Department of Defense's R&D budget had reached $900 million; at the end of the Korean War, it had spent $1.8 billion. Moreover, in competition with the Soviet bomb program, the expenditures for nuclear research tripled between 1949 and 1952 and had reached $2 billion.[13]

By 1953, the military had become the predominant force in US science and technology. While 54 percent of all R&D expenditures in the United States were paid for by the government, the Pentagon commanded 90 percent of this federal money.[14] Even though the relative share of military spending decreased over time, the Department of Defense remained one of the biggest players for decades to come.[15]

If knowledge was a crucial source of American power, how could it be safeguarded against an enemy who was both spying on the secret US nuclear program and reading openly available patent specifications and publications? In 1946 government officials underscored that US companies could not be trusted to protect information since in the past they had shown the inclination to "dispose of it for substantial advance dollar payments without much regard for long-run social consequences."[16] The cooperation of firms like RCA notwithstanding, the state had to take responsibility for national security and for "the economic welfare of this country as a whole."[17] Policies addressing knowledge circulation were urgently needed since members of both Houses of Congress had publicly expressed concerns about sharing technological information with the Soviet Union. The recent declassification of radar technology appeared now as a particularly dangerous mistake.[18] Several congressmen quickly condemned technology trade with the Soviet Union in general, publicly attacking the State Department for its complacency.[19]

The State Department USSR Committee, meeting in March 1946, tried to find a measured answer to the Soviet challenges without embarking on "commercial warfare with a country with whom we are at peace." Official policy was to increase world trade. The State Department called for "a systematic campaign" to reduce and "slow down the flow of US technology," assuming that this would pressure the Soviets into more reciprocity in the exchange of information. The department shied away from openly confrontational steps, advocating direct negotiations with the Soviets while "very informally" discouraging US companies from doing business with the USSR. The main focus of the countermeasures was to cut back the access of Soviet engineers to US facilities, seen as the "most important single source of information" for them. It was proposed to use the Foreign Agents Registration Act, as well as government contracts with companies, to regulate foreign engineers' access to US facilities. More importantly, the State Department suggested using visas to restrict travel into America and to avoid technology losses.[20]

These debates were only the beginning. In spring 1946, concerns about the sharing of nonnuclear and nuclear technological information were aired everywhere: in Congress, in the newspapers, in the Truman administration. And they led to a far-reaching rollback of the policy of openness that had been dominant as late as December 1945. Less than a week after the meeting of the USSR Committee in the State Department, the Senate's special committee working on the McMahon Bill of December 1945 began to revise the bill's section 9, entitled "Dissemination of Information," which called "for the free dissemination of basic scientific information and for maximum liberality in dissemination in related technical information." National security concerns, of course, did limit this liberality, but they were clearly not at the center of the McMahon Bill. As mentioned earlier, its preamble, "Declaration and Findings," did not even mention national security.[21]

In April, this section received a new title that showed the general thrust of the revisions: "Control of Information."[22] Every reference to free communication was dropped, and in the committee sessions of April 9 and 10 a new term was introduced: "restricted data." It was defined as "all data concerning the manufacture or utilization of atomic weapons, the production of fissionable material, or the use of fissionable material in the production of power." This new category included literally every shred of information related to nuclear technology and research, no matter if basic or applied, if produced by a federal or a private entity. "Restricted data" was to be safeguarded by the strictest possible secrecy regime. This information was deemed classified at the very moment it was "born." In order to be shared it had to be declassified by the Atomic Energy Commission.[23] This reversed the very logic of the established classification system that treated all information as open until specifically declared secret.[24] Accordingly, a new preamble emphasized the bill's "paramount objective of assuring the common defense and security."[25] Enacted in August 1946, the Atomic Energy Act was the most radical information control law in US history.

Also in April 1946, three weeks after the "invention" of the concept of "restricted data," Senators Kenneth S. Wherry (R-NE) and James O. Eastland (a conservative Democrat from Mississippi) sponsored a bill to amend the Espionage Act that was supposed to prevent the dissemination of radar and electronic technology, thus reversing the declassification trend in this field. The bill, obviously modeled on the upcoming Atomic Energy Act, demanded that it should be unlawful "for any person, firm, partnership, association, or corporation knowingly to sell, lend, lease, exchange, give, transfer or otherwise make available to any foreign country . . . any radar or electronic equipment" or any information related to it. Breaking the law would be punished by up to ten years in prison and/or a fine up to $10 000.[26] The text was deliberately expansive in order to raise the much broader implications of the export of technological know-how to the Soviet Union. During a hearing about this question in April and May 1946, Wherry and Eastland along with their colleague Ernest W. McFarland (D-AZ) raised the possibility of introducing a control board to screen exports before they left the country for possible detrimental effects to the United States.[27] This idea was obviously applicable to technologies other than radar and electronics.

Just days before the senators began their hearings, the military made proposals for "legislation to protect industrial 'know-how' and scientific knowledge." The War and Navy Departments recommended making permanent the wartime Invention Secrecy Act and also the technical data export controls as administered during World War II.[28]

This surge of initiatives to control the circulation of scientific-technological knowledge in 1946 targeted, once again, all three dimensions of technology transfers: information, people, and things. Export controls of technical data were accompanied by the regulation of the export of artifacts like electronic equipment and the control of the movement of engineers through visas. Secrecy was complemented by controls over unclassified information. Classification was only one option for knowledge regulation. And finally, the nuclear field was only one area of knowledge falling under the purview of regulation—a fact that is usually overlooked in the historiography of nuclear secrecy. US regulators understood nuclear technology as part and parcel of a national scientific-technological system and arsenal. Its power and destructiveness demanded the most restrictive, even radical safeguards, but it was only one element in a wider control regime. The officials in the State, War, and Commerce Departments, members of Congress, and even the press talked about technology in this expansive sense. In contrast to many historians of science and technology, they were not followers of the rather narrow nuclear paradigm that has dominated much thinking about state control of knowledge circulation.

The "Gray Zone"

In the immediate postwar years, distinctions that had for decades guided US policy makers dramatically dissolved. The broad scope of technologies seen as in need of protection corresponded with the expansive reach of the concept of national security that began to restructure US politics and the state itself. As the term "national security" superseded the previous more common language of "national defense," the borderline between military and civilian technologies became blurred. This dissolution of a clear-cut distinction, today usually referred to as "dual use," did not happen by accident. It was the result of extensive debates within the community of bureaucrats who thought about the safeguarding of American technology. Their deliberations not only reflected the changing role of the United States in a new unfolding global order. They also translated the experience of "total war" into peacetime. As State Department officials reasoned in May 1945, technology "used in total war, virtually encompasses the total US technology."[29] In practice, as we saw, this meant that the distinction between classified and unclassified became fuzzy and, from a security standpoint, unreliable. And finally, it became doubtful if there should be a distinction between knowledge produced under government contract and that generated by private industry since both were equally dangerous if the enemy acquired them.

While national security thinkers advanced this dissolution of borders between war and peace, military and civilian technologies, and classified and unclassified information, they did not unequivocally embrace the new ambivalence. On the contrary, as clear lines turned into "gray zones" of unknown dimensions, a sense of unease pervaded their discussions. It was clear to everyone involved that national security and information control came at a price. Safeguarding technological and scientific knowledge with governmental instruments could have serious repercussions for American aspirations to openness and to democracy. The precepts of liberal capitalism, freedom of speech, and the freedom of science were at stake, both as a matter of principle and as a matter of security. Discussing the military's proposals to extend export controls of technical data and invention secrecy, Robert G. Hooker from the State Department succinctly explained in April 1946:

> The danger of a possible enemy building up its war potential by helping itself to the greatest intangible asset that makes for the superiority of our scientific knowledge, is a peculiarly difficult one to deal with. Our really important asset lies not so much in the superiority we may have at any given time as in the ability to maintain our superiority. In a totalitarian society it may be possible to direct great energy to absolute security controls. It is relatively easy to build a Chinese wall around a totalitarian country and still maintain a free interchange of scientific and industrial knowledge within the enclosed area. But in a private enterprise economy there is a real danger that measures intended to prevent the unauthorized outflow from the country of certain knowledge may have the effect of checking its flow at the source, thus impoverishing our own supply.[30]

Many experts within and outside the government concurred with this analysis. Scientists, and not only they, pointed out—and would do so again and again during the Cold War era—that control of information would slow down the progress of science and technology, resulting in inferior weapon systems, thereby weakening the United States' stance against the Soviet Union. The United States did not have "a monopoly of knowledge"[31] and was therefore dependent on the exchange of ideas within the international scientific community. Safeguarding information in the present could mean missing the benefits of future exchanges. Obviously, openness in a democracy was inevitably ambivalent. It was both an important precondition for containing this enemy by threatening it with high tech weapons, and an opportunity that could be exploited by the totalitarian and reclusive Soviets, who denied reciprocity of knowledge exchange. Even more disturbingly, overly restrictive controls threatened to turn the United States itself into a "garrison state" or worse, a totalitarian society.

To address these complex problems, in November 1946 the State Department established an interdepartmental working group, the Unclassified Technological Information Committee (UTI).[32] The UTI was the first of a string of committees that discussed ways to control the circulation of unclassified scientific-technological information.[33] These committees formalized and channeled the discussions on technology regulations that had picked up steam during 1946 and continued them until the late 1950s. Completely overlooked in historical research, they played a key role in the postwar development of technical data export controls.

These committees were charged with offering advice to the US government for the development of national security controls on exports of unclassified "technological information," "consonant with this country's domestic and foreign policy, particularly in the commercial field." They systematically scrutinized the adverse effects of the circulation of unclassified information through publications, patent disclosures, the mobility of people, and technical aid contracts. They also discussed ways to improve the flow of information to the United States, striving for greater reciprocity with Iron Curtain countries. It was clear from the very beginning that it was not possible to regulate all forms of information without endangering American values and interests. It was the goal of the committees' work to find the right balance between freedom and security.[34]

The rather cautious approach toward data export controls, reflecting concerns over their similarity with censorship, stood in contrast to the increasingly aggressive stance that overall US export control policy took against the Soviet bloc. In the immediate aftermath of the war, export controls—which were still seen as a wartime measure—were not allowed to expire so as to avoid short supply bottlenecks of raw materials and agricultural products. But very soon, facing a string of Cold War crises in 1947 and 1948, the United States, embarked on a course of "economic warfare" against the Soviet bloc.[35] Export controls became the main instruments of this policy—and a key element of how the United States fought the Cold War.

This move toward economic warfare was part of the unfolding of a very broad concept of national security that increasingly dominated US foreign and domestic policy and propelled a powerful trend toward a profound institutional restructuring of the federal government.[36] The result of this process was the National Security Act of 1947, which became the foundation of a new state structure built around a large and powerful military apparatus, embedded in the first US peacetime intelligence institutions, and having the National Security Council as the locus of policy making.[37] The idea of national security was not just a military doctrine. Reflecting the experience of total war, it envisioned the whole of US society and the state as being both in perpetual danger and a resource to fend off all threats that might come from without and within. For the new breed of national security thinkers, war and its sibling, political subversion, was not a distant possibility but a near and likely event that the United States had to be ready for. Calling for "permanent preparedness," this thinking deliberately blurred the line between war and peace.[38] Even though often limited by antistatist ideology,[39] national security now seeped into all nooks and crannies of US society, completely redefining its relation to the federal government.

This did not happen overnight, but still surprisingly swiftly. By the late 1940s, the defining institutions of the national security state were in place. By the late 1950s, with the Korean War as a crucial moment of solidification and further expansion in the early years of the decade, the US state had been transformed into a structure to project military, political, and economic power on a global scale and to fight internal political enemies. The new form taken by the US federal government would have been hardly imaginable as recently as 1940.

We already alluded to the impact that the buildup of the national security state had on science. Indeed, the dominant role of the Pentagon in the sustained financial support of science and in defining its mission can only be understood as an expression of the policy of "permanent preparedness." It also changed the way economics and trade were understood and organized. The United States pressed for the establishment of new international institutions like the Bretton Woods system and the World Bank not only to further its economic interests as the largest national economy with global ambitions and an insatiable appetite for foreign markets. The flow of goods and money were also deeply embedded in the ideology of national security. Since a strong military depended on a stable and highly productive industrial base, economic prowess was a precondition of security. The blending of security and economics was also the driving force behind the Marshall Plan. Reconstruction of the European markets would strengthen the US economy, reinvigorate the economies of the American allies, and thus enable them to rebuild their militaries. Collective and US national security stood on an economic foundation.[40]

Trade policy became a function of national security as well. Export controls were the chief instrument that welded both policy fields together. Whereas the Marshall Plan aimed at increasing the trade volume with the allies, export controls denied the revenues of trade to the enemies. Building on the experience of economic warfare in two world wars, the United States wanted to weaken simultaneously the Soviet economy and its military by isolating them from the world markets. US export control policy carved out two spheres with different sets of economic rules. The Western sphere adhered to free-market ideology. The Iron Curtain countries were shut out from this sphere and "contained" in their own Communist sphere—a structure reinforced by Soviet attempts to turn the Eastern bloc into a self-sufficient zone of planned economies.[41] Thus, containment became an important part of US free-market ideology. It defined the political as well the geographic border where the freedom of the market ended and the rules of national security took over.

To define and fortify this border, the United States pushed its allies to subscribe to a system of multilateral export controls. In 1949, the NATO allies (except Iceland) founded the Coordinating Committee for Multilateral Export Controls (Cocom). This organization had an informal structure. It was not based on a treaty according to international law but on a "gentlemen's agreement" that derived its power from US hegemony, political-moral suasion, and peer pressure. And yet, Cocom played a crucial role in the control of international trade until its dissolution in 1994.

Cocom served as the international extension of US national export control policy, as enacted in the very same year, 1949, in the Export Control Act. This act replaced the export control legislation of 1940 that was supposed to expire at the end of the war, but had been extended several times after 1945. The new act was the first comprehensive US export control legislation in peacetime. It was however, never meant to be permanent and was set to lapse after two years. In the end, this provisional arrangement was repeatedly extended and so stayed on the books for decades.[42]

The Export Control Act endowed the president with extremely far-reaching powers. To secure the "welfare of the national economy," to protect the US economy from the negative impacts (like inflationary pressure) of the excessive outflow of scarce materials, to further foreign policy, and in the interest of national security, the president could "curtail or prohibit the exportation . . . of *any* articles, materials, or supplies, including technical data."[43] The definitions and standards the act provided were broad and vague, giving the executive great leeway in their application. In fact, according to the Export Control Act, foreign trade was not a right, but a privilege that was granted (or taken away) by the government.[44] Which merchandize and technical data had to be controlled was a decision made by the executive alone on the basis of its understanding of "foreign policy" and "national security." The national security argument was applicable only to "materials" with "potential military significance," a criterion that was open to different interpretations and interdepartmental negotiations, and whose meaning was established by the president in the last resort. The enormous breath of the act included the communication of technical data.

Voluntary Data Export Controls

These broad powers notwithstanding, the US government still hoped not to impose in practice the tight regulation of data circulation. Instead, it appealed to the patriotism of US businessmen and scientists, calling for their voluntary cooperation in the control of potentially dangerous technical data. In November 1949, nine months after the Export Control Act was enacted, the Commerce Department presented to the public its first voluntary plan to control the export of nonsecret technical data, the dissemination of which abroad might be harmful to national security. This program asked every US citizen who contemplated sending "abroad information on new technical procedures, industrial know-how, chemical processes, or even technical consultants themselves" to seek a national security assessment from the Department of Commerce's Office of International Trade, which would cooperate with other federal agencies and the military. The office would then give a recommendation to the company or the scientists if it deemed it desirable to share the information in question. The company could decide whether or not it would heed this advice.[45] In the first quarter of 1950, the department assessed 303 inquiries; in thirty-three cases it advised against an export.[46]

To narrow down its scope, the Commerce Department focused on "advanced technological developments . . . and 'know-how,' including prototypes and special installations."[47] Also covered was data related to arms and ammunition as enumerated in Executive Proclamation 2776 and regulated by the State Department.[48] This proclamation, issued by President Truman in March 1948, listed objects and technologies that the United States considered to be military items, providing a framework for export controls. Even though striving for a clear focus on the military, it also affected the civilian sector in at least one large technological field. The proclamation defined all and every aircraft and related "components, parts and accessories" as weapons subject to export controls.[49] Thus, from then on the sale of commercial planes, even small models like crop dusters, depended on a license from the State Department and the Pentagon.[50] At the same time information related to crop dusters now became a potential national security risk. The voluntary program also singled out patent specifications, recommending patent holders to contact the commissioner of patents in the Department of Commerce before sending technical data abroad.[51]

Unclassified "technical data for use in educational or scientific research," as well as information "generally available to the public" like catalogues and pamphlets, was explicitly excluded. The Commerce Department was optimistic that the US business community would cooperate. It even claimed that the voluntary program was not least a reaction to "numerous inquiries" of "firms and individuals for guidance on the export of technical data of possible security importance."[52]

Even after the onset of the Korean War, as the West began to aggressively wage an economic war against the Eastern bloc, the US government stuck to voluntary self-control of the business and scientific communities as the preferred way of regulating unclassified technological data. To encourage voluntary compliance on an expanded scale the Department of Commerce embarked on a public relations offensive. In January 1951 it touted its program in a brochure, circulated in forty-five thousand copies.[53] With war in Korea underway, the government felt the control of unclassified information had become even more urgent. Explicitly subscribing to the mosaic theory, because "all major powers depend on published data for a great share of their strategic intelligence," it pointed out that the present state of emergency "had directed attention to the security implications of imprudent release of technical information." To win voluntary assent to an unpopular program, the Commerce Department emphasized again that the government was "fully aware of the dilemma presented by any limitation . . . on the flow of information among private citizens" not least because the "free exchange of information contributes to rapid progress in science and industry."[54]

In one important point, however, the 1951 version of the program further developed the ideas of 1949. Its precautions went beyond the communication of information to recipients abroad. It now included information released *within* the borders of United States—in order to forestall its *export*. Indeed, the US government deemed every publication in the United States to be a "publication to the world," to be an act of sharing information with foreign intelligence services that were working hard to strengthen the enemy's military capabilities by taking advantage of American technological knowledge.[55]

This variation on the intricate problem of what constituted a data export, and where and when regulation was supposed to set in, alluded to the constantly shifting landscape of export controls beneath the surface of a voluntary self-control program. Early on, there was palpable uncertainty about the extent to which the private sector bearers of knowledge could be trusted in their efforts at self-policing. As early as December 1949, just one month after the start of the program, the Department of Commerce announced that in special cases with serious security implications it would make its "recommendations," that is, its adverse decisions, binding.[56]

Controlling Traveling Knowledgeable Bodies

After years of discussion about information control, a distinct chilling effect became noticeable that clearly went beyond the classified sphere. In a high-profile analysis of the new postwar role of science for US foreign relations written for the Department of State, the physicist Lloyd V. Berkner stated in spring 1950 "that to an alarming extent arbitrary restrictions are placed upon activities in unclassified fields of scientific enterprise, through such means as restrictive visit clearance requirements, unwarranted controls over the exchange of unclassified materials, etc."[57] Indeed, the crucial role of information flows for international science was the "most important" and central concern of the Berkner Report. Berkner subscribed emphatically to the ideals of scientific internationalism. Quoting from a 1943 book by Raymond B. Fosdick, the former president of the Rockefeller Foundation, the report proclaimed: "Although wars and economic rivalries may for longer or shorter periods isolate nations and split them up into separate units, the process is never complete because the intellectual life of the world, as far as science and learning are concerned, is definitely internationalized, and whether we wish it or not an indelible pattern of unity has been woven into the society of mankind." Berkner added: "The fundamental truth and impact of this statement is in no way diminished by the fact of the cold war." Accordingly, he advocated that the international flow of scientific information should be as free of state interference as possible.[58]

All the more worrisome for Berkner was that government restrictions seemed to mushroom everywhere. He appeared not to know about the nuts and bolts of the bureaucratic debates since he summarizes some trends only perfunctorily and selectively, but his report mentioned the work of the interdepartmental working groups that had been discussing the problem of how to deal with unclassified information ever since 1947, the Export Control Act of 1949, and the voluntary program, stating that the Department of Commerce had used its powers so far rather "sparingly."[59]

The report directed special attention to travel restrictions on foreign and US scientists as another form of control over the international flow of information.[60] Berkner clearly registered the recent trend to increasingly subject scientists who wanted to cross the US border to special scrutiny regarding their political leanings, but more importantly as regards their expertise and the knowledge they potentially could share or acquire. In January 1951 a report to the president and the National Security Council stated that, for controls over technology "to be meaningful," "they must also extend to the movement of persons. The inspection of the plants of American companies and overseas subsidiaries by suspected foreign agents must be prevented. Moreover, travel abroad by American scientific and technical personnel likely to impart information to Soviet bloc agents should be carefully limited."[61]

Visas and passports, governmental control tools par excellence to keep tabs on its citizens, became key instruments to police national borders so as to keep track of individuals carrying scientific and technical knowledge and skills with them. Passport and visa policies were complementary, and together they constituted a "documentary regime" of border security that directly answered to the challenges posed by Soviet espionage and open-source intelligence.[62] Early on, export control considerations began to shape the administration of travel restrictions. In 1948 the Interdepartmental Committee on Industrial Security, for example, a division of the State–Army–Navy–Air Force Coordinating Committee, discussed the export control of "unclassified technology," the regulation of "technological publications," and "visits of scientists" in the same context. The committee asked: "Should any passport or visa rules be amended in order to restrict the movements of alien and American scientists during the present 'peace-time'? . . . What standards should be used in determining those scientists who can move freely and those who cannot?"[63] US physicists and chemists were in fact the only professional group that was soon singled out for a particularly close security screening by the Bureau of Security and Consular Affairs within the State Department's Passport Office.[64] Scientists applying for a passport were "screened as to their loyalty to the United States, keeping in mind what the individual knows."[65] If doubts existed about the political beliefs of a scientist and if his or her knowledge seemed to be too sensitive to be shared abroad, the State Department could deny a passport or issue only limited passports allowing travel only to certain, clearly stated destinations. This happened time and again, not surprisingly mainly to scientists with leftist political leanings.[66]

In the same vein, visas were used to control and restrict the mobility of foreign scientists to and within the United States.[67] As Berkner reported, after World War II, the Immigration and Naturalization Service of the Department of Justice increasingly used its regulatory authority to prevent Communists from entering the United States. In the late 1940s this began to visibly affect US scientific exchanges with foreign countries. For example, after a French scientist did not receive a visa to attend a conference, the Executive Committee of the International Council of Scientific Unions submitted a resolution to UNESCO in September 1949 to counter the US obstacles to free scientific exchange.[68] These travel restrictions became even more pronounced with the enactment of the two McCarran Acts (after their main sponsor, Senator Pat McCarran, D-NE),[69] passed during the Korean War in a climate of strident anti-Communism, to replace the Immigration Act of 1918. The new legislation, officially called the Internal Security Act of 1950 (McCarran Act) and the Immigration and Nationality Act of 1952 (McCarran-Walter Act), used visa denials and delays to fight Communists, as well as against threats of sabotage and espionage.[70]

These stipulations complicated the lives of the international scientific community even more. The Federation of American Scientists reported in 1952 that "at least 50 per cent of all the foreign scientists who want to enter the United States" faced visa denials or delays of their visa applications from four months up to one year. The Federation of American Scientists had collected information on about sixty cases but estimated that altogether three times as many scientists had had visa problems.[71] Conferences were particularly affected by these difficulties, and from 1950 onward some scientific organizations began to plan to hold their meetings outside the United States to spare foreign colleagues the "embarrassment" of the visa procedure.[72]

Even though businessmen, journalists, artists, and—also of interest for our argument—technical assistance team members felt the effects of the stricter visa policy, it seems that no professional group was affected by visa restrictions as much as were scientists.[73] However, in contrast to what we know about US passport policy, we know very little about the extent to which US visa practices of the late 1940s and early 1950s established formalized mechanisms to address the dangers of unwanted knowledge transfer as such. But there is some evidence. For example, the guidelines of the Office of Defense Mobilization for the Control of Visits to Industrial Facilities in the Interest of Internal Security informed its readership that "through visa, immigration and naturalization, and related procedures, the Government endeavors to exclude from the United States foreign nationals whose background indicates that they might engage in espionage, sabotage, or otherwise present a threat to the national security."[74] Similarly, the FBI also referred to visa restrictions to all Iron Curtain country citizens as well as the denial of passports to US Communists as countermeasures against Soviet espionage. It considered all Soviet travelers, explicitly including scientists and students, as agents sent by the Soviet intelligence services with specific intelligence collection missions, including scientific and industrial espionage.[75]

In addition, as early as the late 1940s, the Department of Commerce had established a special clearance procedure for foreigners who wanted to visit unclassified research facilities. Every scientist had to submit a request detailing "just what it is he wishes to see." If a clearance was issued, it was "limited to stated areas," and a new clearance became necessary "if other subject matters prove[d] to be relevant to the visiting scientist's interest." This is the earliest reference we know to the application of export controls to communication within the United States to restrict the export of knowledge via the minds of foreign individuals—what today's control bureaucracy calls a "deemed export."[76]

The scientific community clearly felt the effects of these regulations rooted in the logic of counterintelligence. In a special issue of the widely read *Bulletin of the Atomic Scientists* on the repercussions of visa restrictions on the scientific community, Edward Shils described how in the enforcement of the McCarran Acts, "scientists in certain fields of work like nuclear physics, electronics, and other fields are especially suspect. Some of the recent victims of the American visa policy . . . have remarked how alarmed consular officers have become when they learned that the applicant was a physicist." The visa problems encountered by the University of Chicago when it organized an International Congress on Nuclear Physics in 1951 were for Shils "only one more illustration of this contradictory belief in the supreme importance of scientific knowledge and the terrible fear of scientists as unreliable, untrustworthy vessels of this crucial knowledge."[77] Shils called the visa and passport hurdles that the State Department put in the way of exchange a "Paper Curtain."[78] British and French scientists grasped the knowledge focus of the regulations much better. They referred to an "Uranium Curtain,"[79] hinting at a connection between visa policy and the information control regime of the Atomic Energy Act.[80]

The Berkner Report summed up this developing regulatory landscape of information and travel restrictions stating that they "in effect comprise a morass of practices, policies and procedures that are frequently conflicting, usually uncoordinated, and generally burdensome. This is vexatious to individual scientists and administrators interested in exchanging unclassified technical information and is dangerous to the progress of our own science."[81]

The Effects of the Korean War

The early 1950s were not the moment to limit the reach of export controls over scientific-technological information. To the contrary, the Korean War ushered in a further expansion of the regulations. At the very same time as the Department of Commerce touted its voluntary information export control program in a new brochure, it recommended in interagency negotiations "that mandatory controls over the export of high level technical data become operative as soon as possible." With the fighting in Korea in mind, the DoC stressed the "importance of technical data originating in this country that has high strategic value in so far as a defense program or war effort is concerned." The scope of such mandatory controls was controversial within the national security community. The CIA supported mandatory controls citing "the efforts of the Soviet industry [sic] espionage system to secure" technical data from US industry. Yet the agency was also rather circumspect, demanding that everything generally available to the public should not be controlled. More importantly, the CIA objected "to the effort to apply export controls to intangibles, and to technical know-how of personnel which could be controlled only by censorship, the Espionage Act [i.e., classification] and the control of passports and visas."[82]

This was a direct repudiation of the Department of Defense's position. The DoD wanted the mandatory controls to reach beyond written and printed paper related to the controlled goods enumerated on the Positive List (the key list of items controlled by the DoC) and advocated "that export controls are now or shortly will be exercised over all other types of technical data, including oral transmission or revelation." The DoD went even further and demanded controls of unclassified "technical data related to non-munition articles and materials" not just to the Sino-Soviet bloc but "to all destinations except Canada."[83]

The regulations that came into force in spring 1951 extended mandatory controls markedly but stayed well below the regulatory level the DoD envisioned. After March 1, 1951, *all* exports of data and goods to the Soviet bloc, regardless of inclusion in the Positive List, were in need of a license. Exports to all other destinations were still covered by the voluntary program.[84] In September of the same year, publications were exempted from licensing—unless they went to North Korea, in which case they were still tightly regulated.[85]

The mandatory controls stayed in place after the Korean War had ended.[86] Indeed, they became a permanent feature of the US export control system. This development was driven by increasing disenchantment with the idea that US industry and science could police themselves. While the US government had commended the cooperation of American companies,[87] a reassessment in mid-1954 was much more sobering: "The fact that some American firms do not see fit to voluntarily obtain Commerce Department clearance on their export of technical data opened the way for a potentially serious loophole in the United States defense program. BFC [Bureau of Foreign Commerce, the successor of the OIT within the Department of Commerce] experience . . . has demonstrated the need for expansion of mandatory controls."[88]

The supposed scope of such mandatory controls remained, however, quite controversial within the export control community that met in the Advisory Committee on Export Policy (ACEP).[89] There were several problems.[90] First, controls that were too extensive would probably have a negative effect on the activities of US companies abroad, and hamper the government's efforts to promote American foreign investment. Broad general licenses for published material and the continuation of voluntary controls on data circulating within the Western sphere were supposed to create a climate that did not deter business. Security concerns and economic interests had to be balanced using the control lists that were supposed to focus on a "'hard core'" of especially strategically sensitive technical data like "electronics, petroleum, chemicals." It was clear, however, "that the administrative difficulties in drawing the line between the types of technology covered and not covered under the registration were substantial."

Second, since the United States' Cocom allies pressed for a relaxation of controls after the end of the Korean War, and for the reduction of the number of controlled items, how could the United States protect technology that was exclusively produced in the United States?[91] The export controllers advocated unilateral regulations more extensive than those internationally agreed on; the implementation of reexport controls to keep tabs on the shipment of US data from other countries; and finally controls over the sharing of data with the allies, especially of information pertaining to military technology but also knowledge in the "unique possession by [sic] US engineers or technologists."

Third, the ACEP discussed whether data controls had to be tied to commodities—or if there should be controls over data without any direct connection to controlled physical objects. There was no doubt that data control was "a necessary adjunct to the security export control of commodities," but there was the challenge of what the Department of Commerce called "the frequently intangible nature of technical data."[92] In 1951 the DoD had argued against a separation of the controls over data and over commodities in order not to burden US business with two sets of regulations.[93] Also the Department of Commerce did not deny that "administratively ... the export control of technical data is more complex than that in the commodity field."[94] Nevertheless, the ACEP called for controls over data "not necessarily keyed to specific commodities," particularly if it had military relevance and was in US possession only. This arguably also included "know-how" as suggested by the reference to "US engineers and technologists."[95]

And finally, there remained the crucial questions of what constituted an export, that is, which forms of communication of technical data should be regulated. The ACEP opted for a very broad definition: "Any release for foreign use constitutes an exportation."[96] The Commerce Department's new extensive data export control regulation that went into force in January 1955 followed this very broad understanding of the scope of the field to be regulated. It defined an export as "any release of technical data for use outside the United States (except Canada) . . . includ[ing] the actual shipment out of the United States as well as *furnishing of data in the United States* to persons with the knowledge or intention that the persons to whom it is furnished will take such data out of the United States."[97] Examples of exports included "blue prints, specifications, technical aid contracts, manufacturing agreements, patent license agreements, instructional or training material, training of foreign personnel, personal delivery by US personnel sent abroad" and using and seeing technical data.[98]

This broad definition dovetailed with a similarly wide-ranging definition of controllable technical data as "*any* professional, scientific or technical information, including *any* model, design, photograph, photographic negative, document or other article or material, containing a plan, specification, or descriptive or technical information of *any* kind which can be used or adapted for use in connection with *any* process, synthesis, or operation in the production, manufacture, or reconstruction of articles or materials."[99] This, it should be kept in mind, targeted *unclassified* dual-use information. The export of data related to arms, munitions, and implements of war were at the same time implemented separately by the Department of State (even though there was in practice a considerable overlap); data on foreign affiliates and subsidiaries of US persons and firms was under the control of the Department of the Treasury.[100]

Classified data was covered by several separate sets of rules. The largest part of the US classification system was based on Executive Orders. The first one was issued by President Truman (EO 10290 of September 24, 1951) to be replaced on October 15, 1953, by Eisenhower's EO 10501.[101] Nuclear data was classified under the Atomic Energy Act of 1954 (which replaced the 1946 version). It was the basis of export controls over unclassified nuclear data.[102] And finally, as of February 1952, the new Invention Secrecy Act of 1951 allowed for the classification of patents with national security implications for the first time in peacetime.[103] In sum, the new data export control regulations were part of a large cluster of information control legislation and regulations introduced in the first half of the 1950s.

Even though the new data export control regulations extended the areas (like publications) in which no license was necessary for an export (or, in the controllers' parlance, a General License was issued), they tightened the government's grip on the communication of scientific-technological knowledge. US companies could still share most information with partners in allied countries—preferably after they voluntarily requested the Department of Commerce's advice—but there were limits even among friends. Following the ACEP's lead, certain technologies "uniquely or predominantly possessed by the United States" fell under regulation, as did data related to fifty-three commodity groups encompassing chemical products, petroleum technologies (like drills, pumps, refinement equipment, but also petrochemical data) and an array of machinery like welding and rolling mill equipment.[104]

Export Controls and Academia

Obviously, these new regulations had potentially huge implications on several, very different, but closely connected levels: for the alliance policy of the "West," for businessmen in their activities in the global marketplace, and for constitutional questions about First Amendment rights and the problem of freedom of travel. The ACEP had not talked about the impact of the new regulations on the work of scientists. Yet the committee quickly understood that the new rules would hit the research community hard when faced by their push back.

In 1955 a protest note against the data control regulations by the Engineering College Research Council of the American Society for Engineering Education spelled out the possible serious consequences for the everyday activities of American universities. The society argued that much of the teaching and research done in universities and colleges was based on unpublished material and thereby fell under the jurisdiction of the new controls. Indeed, almost every kind of scientific communication would require a government license. Every conversation with a colleague who would leave the country afterward was an export. Every foreign student, every foreign visitor at a conference, but also every American who went abroad to present his or her research would be an exporter. And every letter to a colleague and every contract with a foreign institution would come under suspicion of being an export, which had to be approved by the government. In other words, the "Export Controls on Technical Data now issued, would require if conscientiously adhered to, that the entire programs of colleges and universities for teaching and research in science and engineering should be conducted under conditions of 'trade secrecy.' . . . Greater damage to higher education in this country, and to the country's international relations, can hardly be imagined." The engineers therefore requested that the Department of Commerce exempt universities and colleges from the export controls on data.[105]

On the surface, the protest was successful, as the export regulations were amended in April 1955 and explicitly excluded "instruction in academic institutions and academic laboratories." Also, they stipulated an exemption for "the dissemination of scientific information not directly and significantly related to design, production, and utilization in industrial processes. Information thus exempted, includes correspondence and attendance at or participation in meetings."[106] This was an attempt to distinguish harmless, even desirable exchanges of scientific knowledge from dangerous "'leaks' which informed unfriendly nations about developments of strategic importance in the United States."[107]

Despite these changes, many of the basic problems the protest note addressed were not resolved. What exactly was, in practice, the difference between "educational" information and technical data? At what point was scientific information "significantly related to industrial processes"?[108] The definitional fuzziness that would time and again cause US academia a serious headache was unavoidable, given the conceptual complexities of "knowledge."

And there were some other lingering issues. In April 1956 the National Council of the Federation of American Scientists demanded the abolition of the General License for information exempted from the need for a special ("validated") license. The council explained, "At present, the regulations require that shipments of scientific data bear the notation 'GTDS, export license not required.' This notation is actually self-contradictory," the council insisted, "because GTDS is a general export license and required." Since in practice items like letters US scientists sent to their colleagues abroad needed such a designation on the envelope, they "often require clearance of letters through a central office, with consequent delays." In the council's view this was a curtailment of scientific freedom through the back door: "The regulations themselves do not prohibit the flow of such information; yet the labeling requirements amount to a restriction." The license markings on the mail also had a negative impact on the US image abroad: "Even though the stamping procedure has barely begun, there are already unfavorable reactions from European scientists who assume that censorship is in force." No doubt, data export controls had a chilling effect on international scientific communication, even though in January 1957 private persons and philanthropic and academic institutions had been exempted from referring to the General License on their mail.[109]

The mid-1950s were a watershed moment for the development of data export controls. Concepts that had been under development since World War I and, with added urgency, since the beginning of the Cold War had become a fairly well established part of the architecture of the national security state. The US government regulated formally unclassified information on a routine basis and in peacetime. This did not mean that the regulations were set in stone. On the contrary, they would be further developed over time. The basic mechanisms and concepts, however, have remained in force up to the present day.

Cocom and the Problem of Multilateral Data Export Controls

In the second half of the 1950s and in the early 1960s, export controls played a crucial role in addressing the political and technological challenges faced by the United States after the Korean War and Stalin's death in 1953. The allies pressed for a reorientation of East-West policy and trade in a more relaxed Cold War climate to take advantage of the thaw in the Soviet Union. And the rapprochement between the United States and the Soviets in the form of scientific-technological exchanges carried national security risks with it that had to be policed and limited. Even though hardly visible to the contemporary public nor to today's historians, the United States used data export controls to manage these changes in the international framework.

The alliance dynamic within Cocom constantly challenged the goals and structure of US export control policy. The history of the multilateral controls is complex and has been carefully analyzed for the early Cold War.[110] For our purposes it suffices to say that the European partners sought much more limited controls than the United States. Their trade relations with the Eastern bloc countries had deep historical roots, and doing business across the Iron Curtain was also necessary for rebuilding Western European economies after World War II. This did not mean that they entirely disagreed with the US assessment of the dangers of sharing technology and trading goods and raw materials. But their understanding of what constituted "strategic trade" in need of regulation was much narrower because they did not share the expansive American concept of national security. Their calculus of how to balance economic and security interests was therefore much more tilted toward an expansion of trade. The US government understood this but still pushed its allies time and again to adopt more restrictive regulations, longer lists, and more rigorous enforcement measures. From the US point of view the Europeans' "lax" control practices left serious gaps in Washington's security perimeter. Time and again, the US export control community lamented the outflow of technologies they tried to deny to the Cold War enemy from West Germany, the UK, or one of the other Cocom partners.

The debates about the scale and scope of multilateral export control policy extended also to the question of the regulation of unclassified technological information. During the 1950s, the United States attempted several times to recruit its allies to accept policies for the international regulation of knowledge flows. The existing literature has virtually overlooked this dimension of US export control policy, and even though we cannot offer a full and detailed analysis, we can give a sketch of the general problems and trends using the sources available to us.

When the Korean War rattled the Western alliance, the United States was able to convince its allies to significantly broaden multilateral export controls. The prospect of a new world war made the very idea of East-West trade seem questionable. President Truman pressed in December 1950 for measures to "prevent the flow to countries supporting Communist imperialist aggression of those materials, goods, funds and services which would serve materially to aid their ability to carry on such aggression." He asked for recommendations on how to achieve this goal most effectively. In response at least two extensive intelligence reports discussed the vulnerability of the Soviet bloc to economic controls, advocating tighter measures. Recommendations developed in these reports were officially adopted in the National Security Council memorandum NSC 104 in February 1951.[111]

The report attached to NSC 104 acknowledged that the Soviet economy was largely immune to Western economic warfare because it was mostly independent of trade with the US allies. Hence, a complete embargo did not make much sense and was not even in the "best interest of the free world" because it needed certain goods from Iron Curtain countries. All the same the report demanded that allies cooperate in propping up US export controls to a point "short of full-scale economic warfare, which can materially retard the building of the Soviet war potential." Taking advantage of the partial dependency of the Soviets on imports from the West, the main targets were technological and resource bottlenecks "in the fields of machinery and equipment, precision tools, anti-friction bearings, electronics, certain non-ferrous metals, rubber and certain grades of essential materials. Selective controls directed at the vulnerable spots can achieve most of the results that could be achieved through a complete embargo." To target these technological vulnerabilities NSC 104 called, among other measures, "for more effective action by Cocom countries to implement their agreement that 'the object of the embargo or quantitative controls should not be defeated by the export of technical assistance, design data, manufacturing technique, and specialized tools for making any controlled items.'" Additionally, the "US should develop further programs to prevent the export of advanced technological information to the Soviet bloc, including export achieved through the movement of persons, and should enlist the cooperation of other countries in this effort."[112]

Indeed, as the State Department's Office of Intelligence Research acknowledged, the most important potential sources of Soviet knowledge acquisition—the procurement of prototypes, the collection of open-source and unpublished technical information, know-how licensing contracts, and the training of personnel in foreign universities and factories—were all not "formally controlled in the United States or Western Europe except [for] prototypes of items" explicitly included in the control lists. No Western government had anything like the limited and rather informal regulations of the American voluntary control program.[113] Thus, Western Europe—and especially West Germany—appeared to be a leaky bucket. Both knowledge controls and a tighter enforcement of US reexport regulations were necessary.[114]

As we saw, in 1954 certain data exports to friendly countries did indeed fall under control, not least to forestall the reexport of technology in the sole possession of the United States. Whether or not the Cocom members supported data controls during the Korean War is unclear. But even if they did, their support was short-lived. An American study in June 1955 came to the conclusion that "multilateral controls of unclassified technical data to the bloc are negligible . . . , and other countries do not have official controls over the export . . . to friendly countries."[115] The result was a gap between US and allied control practices. These differences grew after the end of the Korean War as the Europeans began to successfully press for a relaxation of trade restrictions and a significant reduction of the Cocom list. The United States had only two options. It could insist on imposing unilateral controls, which would have implied the increasing use of reexport regulations so as to maintain efficient control over American technology. Or the US controls could be lowered to the Cocom level. In the end, the second course prevailed. The United States did not want to "aggravate relations with our allies" because of disagreements about trade.[116] The overarching interest of a stable Western alliance trumped export control policy, and the United States soon also shortened its commodity control list to a level similar to the Cocom list. In September 1955 it abandoned the mandatory controls over data exports to friendly countries of more than fifty items established only five months before.[117]

This signified neither a departure from unilateral data controls—the export of unpublished data to the Soviet bloc was kept under surveillance—nor an end to attempts to secure a multilateral solution in the near future. As the zeal of the export control community to protect technical-scientific information grew in the second half of the 1950s, the United States argued again forcefully for the introduction of effective Cocom measures to "guard against the export of vital technological know-how." Accordingly, in 1959 Cocom began another discussion about the reformulation of its "administrative principles," which served as guidelines for the member states in making licensing decisions. Probably in response to an American initiative, in January 1959 the organization adopted the text of its "Administrative Principle No. 5," which stated that "so far as practicable, the object of strategic controls should be maintained by restrictions on the export of technical data, technical assistance, and any other technology applicable to the design, production and use of embargoed items."

In a second step, the United States also wanted tighter control over the knowledge embodied in technological components. The US representative reminded his colleagues that "now that the embargo list had been reduced it was generally agreed that enforcement should be stricter." For him, this entailed paying closer attention to technology sharing.[118] Hence, the United States proposed a new "Administrative Principle No. 3." It targeted items that were not on the control list but contained "one or more embargoed components." The United States specifically had such components in mind that could be "utilized for the acquisition of unique technological know-how."[119]

Several of the Cocom members objected against this paragraph for reasons of definition and in the national interest. The British and the French delegates pointed at the unclear meaning of terms like "unique" (or "vital") technology. The UK found the terminology "too sweeping and too vague" and predicted that "uniformity of application" would be "impossible to achieve." The US definition would likely "apply to the whole field of embargo" and was hence overly broad. Other delegates doubted that there was actually much technology worth protecting. The member from Japan summarized that within Cocom there were "widely differing opinions" on the "technical know-how question" and "recalled that some Delegations had held that there was little important technology which should be kept from the Soviet Union. The Japanese authorities had serious doubts that the United States proposal would prove satisfactory in practice."[120]

The existence of differing national "opinions" was closely tied to the question of national sovereignty. The Cocom rules were implemented by the national governments of the member states. Germany and France in particular feared that the American version of "Principle No. 3" would curtail their national discretion in interpreting the Cocom regulations.[121] Behind this position the contrasting US and allied interpretations of the right balance of security and economic interests loomed again. Neither the Germans nor the French wanted the US interpretation of know-how as a security issue to hamper their national trade policies. The final text of "Principle No. 3" watered down the US focus on know-how by adding other criteria that gave greater leeway to national interpretation. Not only did it not apply to technical components in general. It also excluded the "principal element" of a system. And in identifying this element, "the exporting country should weigh the factors of quality, value and technological know-how involved and other special circumstances."[122]

Such vague criteria notwithstanding, considerations about the protection of know-how frequently featured in the Cocom discussion, especially in the justification of national export control exemptions. In May 1959, the British delegate argued, for example, in favor of the export of civil aircraft equipment to the Soviet bloc "because the know-how [it] contained was compromised when 3 Convair aircraft were sold to Poland in 1957."[123] If the enemy had the knowledge already, there was no harm in sharing it once more. A year later, the British delegate wanted an exception for the export of infrared detector cells to Poland. Gauging the risk of "revealing technological know-how" he assured his fellow delegates that "expert opinion is that it is not possible to reproduce these cells merely by examination of the sample." Moreover, the "cell is a commercial type which is freely available in the West and full information about its characteristics and performance has been published."[124]

It seems then that even though Cocom members cooperated by and large with the US agenda, they were rather skeptical about the American concept of data export control. It is difficult to assess to what extent they actually implemented the idea of information and know-how regulation on the national level. But it is likely that from the standpoint of US hardliners, their compliance with these concepts was as unsatisfactory as were their export control systems in general. It was a Cold War trope that the allies never sufficiently regulated their trade, and that business interests almost always trumped multilateral and US national security. Deploring the "gradual erosion" of the Cocom list, the House Select Committee on Export Controls summed up the criticism in 1962: The Cocom partners "seem at times to be more preoccupied with the short-term advantages of the current trade balances than with the long-range objectives of the Group and the obvious dangers of strengthening the economies as well as increasing the military and industrial potential of the USSR and Communist dominated nations.... Competing producers in the ... free world, especially in Cocom nations whose industrial technologies are furthest advanced, are often willing and do provide items unilaterally controlled by the United States to anyone who wants to buy."[125]

The changing landscape of multilateral export controls after the Korean War posed great challenges. While the allies successfully pressed for markedly shortened control lists and the liberalization of East-West trade, the United States reluctantly followed this trend in the interest of the cohesion of the alliance. But at the same time, the United States tightened its grip on the very heart of the global knowledge economy by expanding data controls and tried to prompt the allies to follow the same course. Thus, in the late 1950s, there was a visible shift of export control theory and practice toward the regulation of intangibles.

Exchanges with the Soviets: Policing Visiting Enemies

This trend was reinforced by changes in US-Soviet relations. The inauguration of scientific-technological exchanges between the two countries posed challenges that contributed to an increasingly stronger and formalized link between technical data export and the travel control policies. As the United States vigilantly opened its borders, it also tightened its bureaucratic grip on the movement of people and information.

In the years following Stalin's death in 1953, the political, scientific and cultural relations between the United States and the Soviet Union became closer at a surprising pace. Milestones of this thaw in cold war tensions were the Four Power Conference in Geneva in 1955, the Atoms for Peace conference in the same city a few weeks later, and National Security

Council directive NSC 5607 of June 1956, which laid out the framework of East-West exchanges for the next two decades or so. The Statement of Policy that accompanied NSC 5607 made clear that exchanges were seen as an instrument to liberalize the Soviet Union through contact with Western values, thus weakening the stature of Communism in the confrontation between two world orders. The first US-Soviet exchange agreement, signed on January 27, 1958, was the culmination of negotiations that had begun only a few months earlier in fall 1957 that were stimulated by the Sputnik shock in October that year.[126]

Travel and export regulations needed to be adapted to allow for more Soviets to visit the United States and for a meaningful conversation about scientific and technological matters for the benefit of both partners. At the same time, visitors from the Soviet Union posed a security risk. Accordingly, NSC 5607 directed "the Secretary of State and the Attorney General to continue to cooperate in developing appropriate internal safeguards with respect to the admission of Soviet and satellite nationals to the United States."[127] The existing export and travel regulations were probably the most important of these barriers. Too much openness could put American knowledge at risk to the strategic advantage of the visiting enemy. The challenge for the national security administrators was to find the right balance between international exchange and security concerns. The main actors in this process were the State Department and the Department of Commerce. The two departments, step by step, paved the way for a closer coordination of travel and export controls simultaneously with the negotiations for the first exchange agreement with the Soviet Union.

In November 1957, members of the Departments of Commerce and State met to discuss the problem of visits to US plants under the auspices of the upcoming East-West Contacts Program "for the purpose of removing any obstacles to the progress of the program that might result from our present Department of Commerce technical data controls."[128] The State Department had brought up the idea of waiving the data regulations for officially authorized exchange visitors. Loring E. Nacy, director of the Bureau of Foreign Commerce (BFC, in charge of DoC's export control) was opposed. He warned of the dangers of a wholesale abandonment of data controls, fearing that data that would not be licensed by BFC under the established policy would be released during Soviet plant visits.[129]

In talks in December 1957 the Departments of State and Commerce came to an agreement. The export control regulations for technical data were to be amended by a passage addressing explicitly the visits to "US plants, laboratories and facilities for the purpose of general inspections, sales discussions, and private or government sponsored exchange visits." They were exempt of regulation, provided that visitors were "not furnished with detailed explanations, engineering drawings or models of such a specific character as to constitute the basis for copying of the product, design, etc., on the part of the visitor." This meant that only general information and basic science could be shared without a license. Correlatively, communicating applied, sophisticated technical data was still subject to licensing. This formula seemed to be a viable compromise for the State Department, which was "reluctant to take full responsibility for the elimination of technical data controls" for visits. However, it put the onus for the implementation of data controls on industry. Host firms had to define in practice where the line between basic and applied, innocuous and dangerous information was situated.[130] Export controls of technological exchanges were thus in fact dependent on self-censorship by US industry and on trade secrecy.

One main goal of this strategy was to alert the private sector of its obligation to take technical data control regulations seriously, especially since the "general unawareness throughout US industry" still continued to "plague" the Department of Commerce as late as 1960.[131] In the second half of the 1950s, the US government seems to have geared up its efforts to enforce the data controls and to make them thus better known to US industry. In May 1956 the Department of Commerce publicly announced that it had denied a license to Dresser Industries Incorporated for the export of technical information "related to the design, production, assembly and operation of rock drill bits" to the Soviet Union. This case made it onto the front page of the *New York Times*, which explained in quite some detail the objectives of data controls.[132] In the same year, the Department of Commerce denied a data license for know-how needed by the German Democratic Republic for the production of maleic anhydride, a petrochemical product used inter alia for the production of insecticides. The export of engineering data for a tire plant in the USSR was also prohibited.[133]

Another important tool of export control enforcement was "warning letters" sent to US companies when the Department of Commerce had received information (e.g., from US embassies) about a firm's activities that potentially violated the export control regulations. After issuing fifteen hundred such warning letters between 1948 and 1955, the Department of Commerce expanded their use, sending seven hundred of them in

1956–57.[134] In March 1956, for example, DuPont received a warning letter because it was expecting the visit of a United Nations Study Group interested in the progress of US color television that included representatives from Soviet bloc countries. The letter explained the data export control regulations and admonished the company to carefully limit the information shared with the delegation "to that which is generally available to visitors in a guided tour of your plant."[135]

The same enforcement tool was used in the case of academic and industrial conferences and symposia, trade fairs, and exhibits. Even though there were some doubts as to whether significant technology transfers occurred at such events—and the Department of Commerce wanted to avoid giving the impression of censorship—the department drafted a standard letter to be sent to conference organizers to inform them that oral and written presentations were within the reach of US export controls.[136]

Such measures were directed at Soviet bloc visitors who had already entered the country. In 1957, Loring Nacy also advocated a review of exchange proposals and visitors' itineraries in advance of the actual trip. The machinery was already available: the Standing Committee on Exchanges of the Intelligence Advisory Committee (IAC) vetted pretravel applications to enter the United States. The Department of Commerce wanted to become a member in order to give export control regulations their due weight and to voice "the Department's views on possible technology loss."[137]

The IAC Standing Committee on Exchanges was established in February 1956, a few months after the Four Power Summit and the Atoms for Peace Conference in Geneva and even before the promulgation of NSC 5607. The chairman, William Bundy, and the executive secretary, Guy Coriden, came from the CIA. Other members were from the State Department; the Army, Navy, Air Force, Joint Chiefs of Staff; and the Atomic Energy Commission.[138] Its main objective was to advise the Department of State "on all intelligence aspects" of exchange programs, to "assess the probable net advantages from an intelligence standpoint, considering both intelligence and technological gains or losses."[139] The committee had therefore both defensive and offensive functions. It not only tried to limit what visitors from Iron Curtain Countries could learn about US technology. Visits to those countries were also seen as missions to gather intelligence for the United States. For this purpose, travelers to the East were briefed and, after their return, debriefed about their trips. "Maximizing the intelligence yield from East-West delegation exchanges" was the avowed goal. Risks and advantages of the exchange program were weighed against each other.[140]

In July 1959, the Department of Commerce finally became part of the Exchange Committee after it had reminded CIA director Allen Dulles and the secretary of state, Christian Herter, of the necessity to monitor data export regulations in East-West exchanges. The Department of Commerce claimed not sole, but "principal responsibility of securing the necessary cooperation of US business in the implementation of any approved exchanges in the technical/industrial fields, including the itineraries of the visiting groups in the US."[141]

The Exchange Committee systematically screened exchange program proposals from the Soviet Union, Eastern Europe, and the United States, including US industry, for intelligence value and risk. Between March and September 1956 alone, it screened thirty-five exchange proposals, six of which the committee advised the State Department not to implement.[142] Whereas in the beginning the main concern was industrial exchanges, science entered the picture with the exchange agreement of January 1958.[143]

An example from 1962, related in the CIA's *Studies in Intelligence*, shows what effects these screening mechanisms of the State Department and the intelligence community had in practice. In that year, the Soviets proposed several different exchanges in the field of computer science and technology. The director of the USSR Academy of Sciences Computing Center lobbied the Computing Center at New York University to accommodate two Soviet students for a two-month visit. A Soviet student asked to attend the Western Joint Computer Conference in Los Angeles. The University of Illinois was asked to welcome a student in computer technology. The Soviet embassy in the United States, circumventing the Department of State, contacted IBM to ask if a Soviet educational exchange delegation could visit the IBM headquarters at Rochester. And an exchange visitor, sponsored by the American Council of Learned Societies, and with alleged ties to Soviet intelligence services, asked for permission to take a trip that would let him learn about the use of computer technology in the field of economic planning. "Although the Department of State of necessity handled each of these proposals separately vis-à-vis the Soviets, inside the government they were treated as a concerted Soviet effort to get needed information on all aspects of US research in automation and computer technology." After assessments by the intelligence community, certainly including the Exchange Committee, the State Department refused three of these requests, taking no action for the student who wanted to go to the University of Illinois and limiting the itinerary of the Council of Learned Societies guest to universities doing unclassified research only.[144] Similar exchange visits to IBM facilities in the following year were deemed acceptable only if IBM made sure that it protected its classified contractual work for the government and followed technical data export regulations.[145] The appeal to basic science provided the needed political instrument to make the judgment call.[146] The aim of these actions, one analyst wrote, was to "isolate our visitors from applied research and development and restrict their exploration to basic science" because "almost any scientific or industrial field can be related to war and weaponry."[147]

The Invisibility of Data Export Controls as a Daily Bureaucratic Routine

In the late 1950s and early 1960s, export controls over the communication of scientific-technological data had become an integral building block of the architecture of the national security state. The Hydrocarbon case of 1962 that we described in chapter 1 was the result of the successful establishment of controls over unclassified technological data and know-how that had begun in 1917, but made its greatest strides in the late 1940s and during the 1950s. In the early 1960s, however, this long-term trend had not come to end. Congress criticized the Commerce Department's data controls as inadequate and pushed in May 1962 for a further expansion of controls over technical data and prototypes. The Select House Committee on Export Control argued (quoting a DoC official) that data controls were "'even more essential to our economic defense effort than commodity export control.' The furnishing of plans, specifications, and production details of strategic items to the Soviet bloc in many instances have given as much as or more advantage to those countries as the shipment of commodities themselves."[148] This line of argument would, as we will see in later chapters, shape the export control debate and practice for decades to come.

The congressional critique notwithstanding, by the first half of the 1960s, data export controls had—as the Hydrocarbon case clearly demonstrated—an undeniably powerful impact on the flow of knowledge crossing the US border, whether in the form of printed paper, embodied in objects, or carried in the minds and hands of scientists, engineers, and business people. Export controls had grown into an extensive bureaucratic system that combined regulations for the movement of ideas, things, and people with far-reaching implications for the US political principles of freedom of speech, scientific freedom, free-market economy, and the freedom of travel. National security consideration regarding the communication of scientific-technological knowledge infringed on all of them, complementing other bureaucratic regimes like classification that also affected things, information, and people, the last via security clearance background checks, as well as the Atomic Energy Act of 1954 and the Federal Employee Loyalty program.

The scale of export controls on business and science varied along with geopolitical tensions. At first sight, it seems that the system markedly shrunk in the first two postwar decades. In the second half of 1949, measured in dollars, 30 percent of all US exports were subject to controls.[149] On the eve of the Korean War the number had dropped to 18 percent, only to rise again during the war to 43 percent.[150] As we have seen, after the war a strong trend toward decontrol set it. And yet, controls stayed on a surprisingly high level. In 1966, an estimated 12 percent of all US exports needed a "validated license."[151] These are global numbers for the entire national economy, and they do not say much about the impact of export controls on individual business sectors. But since one of the main objectives was to regulate the flow of technology the figures give a rough idea of the deep influence the control regime had on large swaths of the high-technology sector in particular.

These numbers are also remarkable because, during the early Cold War, US trade with the Eastern bloc shrunk to almost nothing. In 1953, the United States exported merchandize worth only $2 million to the Soviet bloc. After the Korean War the numbers remained minuscule.[152] The Soviet Union share of US exports was 0.2 percent in 1960, and 1.1 percent of all Cocom member states combined.[153] This means that the actual target of the export control system was the trade with friendly countries, driven by the fear of reexports. In the 1960s, more than 95 percent of the export licensing activities dealt with allies and neutrals![154]

It is difficult to quantify the importance of data export controls. Certainly, the regulation of goods took the lion's share of the Department of Commerce's workload. That said, in 1955 twelve people worked in the Bureau of Foreign Commerce on the administration of data controls. When in September certain mandatory controls to friendly countries were rescinded, this number shrunk to only two full-time and several part-time positions, together the equivalent of four full-time jobs. The entire machinery of the Department of Commerce's export controls sector employed 157 people in 1956.[155] It seems that ten years later, data controls made up a big chunk of the work of the Office of Export Control (as it was called in the 1960s). It had four licensing divisions: Production Materials and Consumer Products, Capital Goods, Scientific and Electronic Equipment, and Technical Data and Services. Even without having global numbers, it seems that the data division had considerable work to do. Of the 6,850 license applications the Office of Export Control received for exports to the Soviet bloc in 1966, 1,550, or 22.5 percent, pertained to data exports—that meant about four data cases every single day.[156] We assume that data controls played a similarly prominent role in the office's main business of regulating trade with the "free world."

All these figures, however, do not capture the significance and impact of export controls on the cross-border flows of scientific and technological knowledge. They do not cover the main dimension of export control enforcement: the self-regulation of the business and scientific communities. That is not to imply that export controls were a mere paper tiger. They had real teeth. Violations of export control regulations could be penalized with a fine of up to $10,000 and one year in prison.[157] But more important in practice and as a deterrent was the revocation of a US company's export privileges for a certain period of time or even permanently. Then of course there was also the threat of serious reputational damage to a firm that was blamed and shamed as a violator in the *Federal Register* and that was officially blacklisted.[158]

These harsh sanctions notwithstanding, the US government relied heavily on the cooperation and compliance of businessmen and scientists because of the simple fact that the all-encompassing control of the global movement of information, merchandize, and people was impossible. Like in the case of every other law, from traffic rules to the prohibition of theft, law enforcement worked only if citizens translated more or less voluntarily and automatically the given rules into their everyday activities. From time to time, the US government liked to send a reminder that the rules still existed and that it had the power to impose them on business and academia. For this purpose it periodically made the rules widely visible. The Hydrocarbon case had exactly this function, serving as a deterrent against future unregulated knowledge sharing.

That said, export controls were arguably the most effective when they were virtually invisible, when, for example, a businessman did not even apply for a data export license because he understood that a denial was likely or that the export was obviously not in the "national interest," or when a scientist, in an act of self-censorship, decided not to share his or her knowledge with a foreign colleague. Indeed, the export control regime slowly seeped into the nooks and crannies of the everyday handling of information in companies and universities. From this point of view, the export control regime's invisibility to historians actually is evidence of its successful implementation through routinization.

The detailed account in this chapter of controls over the global movement of technical data, of knowledge, and of know-how has two main purposes. Firstly, it has broken new ground by highlighting the significance attached to unclassified knowledge by the new US national security state, and by its conviction that the successful projection of American power abroad required that steps be taken to control its circulation within and beyond national borders. It is well known that the Cold War was a competition for technological superiority, and that knowledge flows in the nuclear field were restricted by draconian classification measures. This "nuclear paradigm" has tended to dominate our understanding of the scope of the regulations on the transnational circulation of people, knowledge, and things. This chapter has emphasized that, on the contrary, this was just one concern of the national security state, a state that realized the centrality of economic predominance to military strength, and that took extensive measures to control the circulation of unclassified technical data and know-how along a multitude of channels to the Soviet bloc, targeting both corporate and academic actors.

Secondly, this chapter has described the dilemmas surrounding policies to restrict knowledge circulation in a free society, the groping in the dark for acceptable solutions, and the accompanying intense controversies surrounding the scope of export controls on technical data. There were disagreements not only between different departments and agencies of the executive branch, but also between the United States and its closest allies. These disagreements involved far more than a debate over technical details. Regulating the circulation of unclassified knowledge in peacetime challenged a number of hallowed values, from openness and freedom of speech to scientific internationalism and academic freedom. The very fact that they persisted, and, as we shall see in the rest of this book, were expanded to meet new challenges attests to the dramatic transformation of the role of the state in American society during the Cold War in the name of national security.

PART TWO

CHAPTER FOUR

The Recalibration of American Power, the Bucy Report, and the Reshaping of Export Controls in the 1970s

In February 1976 the Department of Defense released a major report that called for a reassessment of the scope of export controls on American technology.[1] It was produced by a task force established in July 1974 by Malcolm Currie, the director of defense research and engineering. Currie was an applied physicist who had been a research vice president at Beckman Instruments Incorporated before joining the Pentagon in 1973. He and others in the administration, with the support of conservative politicians in Congress, were concerned that the Soviets were taking advantage of the opportunities for US-USSR trade made possible by the policy of détente "to acquire US technological know-how that [had] important military applications under what [were] supposed to be commercial agreements."[2] It was particularly distressing that Soviet negotiators spent a good deal of time learning all they could about the operation of sophisticated production processes in firms like Boeing and Lockheed without completing the deal, so "freely" acquiring technical understanding of how US firms manufactured some of their most advanced technologies.

Malcolm Currie asked J. Fred Bucy to help define policies to deal with Soviet acquisition of American technology. Bucy was executive vice president of Texas Instruments Incorporated with strong ties to the government.[3] He put together a fifteen-man task force that included a roughly equal number of corporate leaders and administration officials (from the Departments of Commerce and Defense, the CIA, and the White House) to look into the matter. To support it in its work, the task force designated subcommittees to investigate technology transfer in four industrial sectors. Each was chaired by a senior industry executive from the core membership and included participation from firms deemed "broadly representative of all 'high-technology' industries" (airframes, aircraft jet engines, instrumentation, and solid-state devices). The result of their deliberations was a report—colloquially known as the Bucy Report—that, in the words of political scientist Michael Mastanduno, "proved to be among the most influential documents produced on US export control policy."[4] Richard Perle, then on the congressional staff of Senator Henry "Scoop" Jackson (D-WA), and an expert on arms control, agreed: the Bucy Report demanded a "conceptual shift in the overall approach of US officials." It was an "important breakthrough" that "reshaped American thinking about the definition of technology" and about what needed to be protected.[5]

In his report and related writings Bucy introduced a novel and somewhat idiosyncratic conception of technology into the debate on export controls. As he explained in one of his published articles, "Technology is not science and it is not products. Technology is the application of science to the manufacture of products and services. It is the specific know-how required to define a product that fulfills a need, to design the product, and to manufacture it. The product is the end-product of this technology, but it is not technology."[6] This distinction between technology-as-know-how and technology-as-end-product was central to the Bucy Report. It argued that export controls should target the transfer of technology, defined in his sense, rather than the export of commodities, or end products, as was usually the case at the time. The group's leading industrialists thus broadly endorsed the policy initiatives being promoted by the Pentagon and their supporters in Congress. They also crystallized anxieties about the liberalization of trade authorized by the détente-inspired Export Administration Act of 1969 and insisted that the potential loss of know-how be a key criterion when licensing high-technology exports to Communist countries.

Legally speaking, the call for export controls on the circulation of know-how was not strictly necessary. As we pointed out in the previous chapter, sensitive know-how was covered by the Export Control Act of 1949, though its status was initially unclear.[7] The control of technical data allowed for in the act was associated with a material object, like a model, a design, a photograph, or a document. The status of intangible know-how, knowledge carried in one's head and shared orally, was however, somewhat ambiguous.[8] As described in chapter 1, it was clarified in 1962 in the case brought against Hydrocarbon Research Incorporated and its US president, Percival Keith. The Department of Commerce sanctioned them for not seeking a license to share unpublished "United States–origin know-how and experience," including an engineer's general technical knowledge acquired during a lifetime of work and while building a plant in Rumania.[9] Existing law did thus provide for controls on the circulation of know-how abroad, though, in the view of legal scholars Berman and Garson, enforcing compliance required an "expansive interpretation" of the law.[10] The legal significance of the Bucy Report was that it took decisive steps to clear up the existing ambiguities and moved controls over know-how from the periphery to front and center of the regulatory apparatus. It provided some support from industry for the view, expressed by an official in the Department of Commerce, that a potential adversary's "possession of technological know-how is often of greater strategic significance than finished items produced from that technology"—and that it was therefore necessary to control transfers of technical know-how more restrictively than commodities.[11]

The emphasis on export controls of technology and know-how, rather than end products, was called for by structural transformations in the world economy and the geopolitical balance of power in the late 1960s and the 1970s. These transformations were implicitly reflected in the Bucy Report and threaded through its conceptual core, giving its recommendations immense historical weight. The novelty of the Bucy panel's report did not lie simply in its emphasis on the need to control know-how. It lay in its developing a language to think about the role of export controls in a new global context. The concept of know-how was just one of its elements. It was embedded in a cluster of related terms that defined the goal of export controls as being to maintain US *strategic lead time* by restricting the sharing of *know-how* via *active mechanisms* of *technology transfer*.[12] The Bucy Report revolutionized export control policy by recognizing that the "technological gap" that separated the US economy from the rest of the industrialized world, including the Soviet Union, was closing rapidly, and could be maintained in key strategic areas only by targeting *technology transfer and the circulation of tacit knowledge* and not just the acquisition of new products.

This chapter is divided into three major sections. The first briefly describes the challenges to American technological, industrial, and political leadership in the 1970s brought about by state- and corporate-driven technological change in the Western alliance, and the associated recalibration of American global power. It then goes on to analyze the conceptual apparatus for export controls developed in the Bucy Report in more detail. The overall aim of this first section is to embed the cluster of terms "technology transfer / know-how / lead time" in a global economic and political context that called for the "reshaping of American thinking" spoken of by Perle and that generated the "conceptual shift in the overall approach of US officials" mentioned by Mastanduno, and captured in the language of the Bucy Report.

The second major section will delve more deeply into the Bucy Report itself. It will describe the detailed measures proposed to regulate technology transfer—rather than trade in commodities as such—that was seen as posing a major threat to US national security. The task force's report did not simply propose a new policy for export controls. It was aligned with a domestic political agenda that pitted the Department of Defense and conservative forces in Congress against the process of détente and the trade liberalization that it favored. Implementing its principles, which immediately received widespread support in the Department of Defense, proved contentious in practice, as we explain in the next subsection. Indeed Bucy himself was summoned back to Congress in summer 1978 to assess the highly controversial license application by Dresser Industries to provide deep-well oil and gas drilling equipment to the Soviet Union. His head-on clash with the Department of Defense's protrade interpretation of his policy guidelines not only provides valuable insight into the making of export control policy. It makes abundantly clear that, in practice, export controls are no more than guidelines that are negotiated in an intensely political process engaging multiple stakeholders with very different interests.

The Bucy Report remained embroiled in ongoing disputes over how to strike a balance between protecting national security and penetrating global high-tech markets that became even more intense as the United States' annual commodities trade deficit climbed to about $30 billion in 1979. The third and final section of the chapter explores the impact of these debates on the institutionalization of its recommendations and tracks its legislative trail into the 1979 Export Administration Act and the all-important Militarily Critical Technologies List (MCTL).

Interdependence in a Multipolar World: The Global Context of the Bucy Report

Although the United States was still the major global power in the 1970s, America's relative position in world markets was declining. With postwar recovery behind them, the economies of Europe and Japan had entered a period of sustained growth during the late 1950s. Overall the US share of developed-country export volume to the world fell steadily from 26 percent in 1960 to 19 percent in 1971.[13] The US share of OECD exports of manufactured products declined from 25 percent to 16 percent from 1954 to 1980, and from 35.5 percent to 20 percent for technology-intensive ones.[14]

The electronics, aircraft, and aerospace industries played a major role in the refashioning of the global distribution of productive capacity in the advanced industrialized countries. For the first two decades after World War II the United States was the undisputed technological leader in the world. American firms were "significantly ahead in developing and employing the leading edge technologies, their exports accounted for the largest share of world trade in their product fields and their overseas branches often were dominant firms in their home countries."[15] This was possible because the nation emerged relatively unscathed from World War II and because, drawing on the technological lessons of that conflict, both the federal government and corporations invested extensively in R&D, mobilizing industrial, government, and academic laboratories (under the contract system) to push the cutting edge of research across a wide spectrum of "science based" or high-technology industries. These firms exploited a rapidly expanding pool of scientists and engineers that grew from some fifty thousand in 1946 to about three hundred thousand in 1962.[16] And they acquired the organizational and management skills needed to bring new products and processes to market, opening a "technological gap" that separated the United States from other OECD countries in the 1960s.

In 1967 Jean-Jacques Servan-Schreiber famously described the contours of "the American challenge" with a mixture of admiration for US achievements and despair at the European lag.[17] Europe was doomed to become of a vassal of the United States, he insisted, if it did not act to close the technological and managerial gap between European and American high-tech industry. Massive foreign direct investment by US multinationals in the newly formed Common Market had reached about $14 billion in 1967, increasing by as much as 40 percent in just one year (from 1965 to 1966). Servan-Schreiber's cry of alarm was as much a plea for deepening Western Europe's political integration as it was for increased state-sponsored investment in high-tech industry, in research and development, and in the education of scientists, engineers, and innovation-savvy managers.

Even as Servan-Schreiber wrote, European firms were exploiting what they had learned from US multinationals to launch a period of sustained "modernization" later known as Les Trentes Glorieuses in France.[18] By 1971 R&D expenditure as a percentage of GNP was 2.4 percent in the United States. But it was also 2.2 percent in the UK, 1.8 percent in Japan, and 2.3 percent in Germany. America's share of the world's high-technology exports, while far in excess of its rivals, began to decrease from a high of 30 percent in 1970, while Germany climbed to about 18 percent in 1972, and Japan surged from 8 percent in 1965 to 13 percent in 1972.[19] This reorganization of market share was uneven between countries and industries but was frequently thanks to joint ventures and transnational licensing agreements with American firms. The internationalization of business, along with the training of strong national scientific and engineering communities that were linked in transnational networks and able to selectively adapt what they learned from one another, led to a convergence in the productive capacities of advanced Western societies. Richard Nelson and Gavin Wright pointed out that scientists from IBM, from Philips, and from Fujitsu met at conferences, exchanged papers, and built networks that produced technologies "that no longer have geographic roots."[20] A US trade balance in (select) electronic and communications equipment had plummeted from a surplus of about $300 million in 1965 to a deficit of some $2,100 million in 1976—even though US firms had made seventeen out of eighteen technological breakthroughs in semiconductor electronics.[21] Bucy's own Texas Instruments Incorporated had contributed to the shift: in the 1960s it exported to West Germany two turnkey production plants, including the technological know-how to operate, maintain, and improve the manufacture of plastic transistors and integrated circuits.[22]

The fields that Bucy's panel chose as representative of US high-tech industry in the mid-1970s were very much part of this process of international reorganization. Between the mid-1960s and the mid-1980s the export share of two closely related areas (aircraft and aircraft engines and turbines), along with computing and other office machinery had "held up well" in the face of rising international competition. What bothered Bucy is that the US lead in two other key areas (identified in the Bucy Report as instrumentation and solid-state devices) was being rapidly eroded.

Historians today see the 1970s as the onset of a new phase of globalization or, more precisely perhaps, international interdependence in which national politics and policies and transnational economic and financial considerations became inextricably entangled with one another in the capitalist world. The Vietnam War, as Paul Kennedy emphasizes, did show, in all its horror, that "vast superiority in military hardware and economic productivity [do] not automatically translate into military effectiveness."[23] In 1971 the United States registered a trade deficit for the first time in its history since 1893, the combined effects of a decline in exports and a voracious domestic appetite for imports. The country's share of the world's gold reserves shrunk precipitously. Massive transnational capital flows, the proliferation of off-shore havens for accumulating capital beyond the reach of national economic policy, and the growing power of multinational corporations led President Nixon, in summer 1971, to unilaterally withdraw from the Bretton Woods agreement. He ended the dollar's link to gold and then allowed its value to float in private markets, sacrificing currency stability to the need to deregulate capital flows.[24] Other countries, with the exception of Japan, soon followed suit.[25] US export market shares began to climb as the dollar plunged to new lows—to the consternation of the major oil-producing countries, many of them in the Middle East. The decrease in their revenues (and, for some, US support for Israel in the 1973 Yom Kippur War) led the Organization of the Petroleum Exporting Countries (OPEC) to quadruple the price of crude from $3 to $12 a barrel, precipitating a global recession. US share of exports fell back again as the demand for dollars to pay oil bills lifted the value of the dollar against other currencies. Paul Kennedy attributes these oscillations in the value of the dollar to underlying long-term trends: decreasing productivity growth in the private sector (from 2.4 percent between 1965 and 1972, to 1.6 percent over the next five years, to 0.2 percent from 1977 to 1982), large inflows of foreign cash attracted by high interest rates, and increasingly effective competition in the domestic market by imported household goods and other manufactured commodities.[26] The United States was being transformed from an "empire of production" into an "empire of consumption," in the words of economic historian Charles Maier, sending manufacturing jobs and social capital abroad, while "foreigners in turn bought Treasury bills to finance the United States' mounting debt and prodigal consumption."[27]

New modes of international economic cooperation were spawned by the turmoil in financial markets (e.g., the efforts made by European governments to limit fluctuations in the values of their currencies relative to one another to a narrow band), by what scholar Daniel Sargent calls "the power of petroleum as a political weapon," and by the debacle in Vietnam. They exposed the limits of the nation-state as an autonomous economic actor. As Henry Kissinger, Nixon's national security adviser, put it in 1975, "Old international patterns are crumbling; the world has become interdependent in economics, in communications, and in human aspirations."[28] In his eyes the era of Cold War competition between the two major power blocs had been overtaken by the emergence of a multipolar world and a geopolitical system "marked less by intense polarization than by fluidity and a blurring of alignments."[29] Détente, the liberalization of trade with the Communist world (including China) and Congress's replacement of the Export Control Act of 1949 with the Export Administration Act of 1969 are to be seen in this changing geopolitical context.

Détente and Its Discontents

The Soviet Union was largely excluded from the transformations wrenching the capitalist system, relying on restricted trade with Western Europe and its East European satellites to expand a lopsided economy.[30] By concentrating its resources in the military sector, the Soviet Union had reached "strategic equivalence" with the United States by the early 1970s, though at the expense of its civilian sector. The time had come to redress the imbalance, to improve the everyday lives of Soviet citizens, which had been neglected in the name of defense. The "main task" of the Soviet Ninth Five-Year Plan adopted in 1972 was "to ensure a substantial rise in the material and cultural standard of living on the basis of high rates of development of socialist production, a rise in production efficiency, scientific and technical progress and a faster growth of labor productivity."[31] The Central Committee of the Communist Party confirmed in 1973 that to achieve this goal it was necessary to "normalize" relations with the West and make a "radical turn towards détente and peace on the European continent."[32]

Détente is a general term used to signal a shift in US-USSR relations from a strategy of "containment" and confrontation between rival superpowers to one of "friendship and cooperation" in the framework of "peaceful coexistence and equal security."[33] It expressed General Secretary Leonid Brezhnev's determination to move beyond autarky and to improve commercial, scientific, and technological relationships with Western powers. It was coherent with Kissinger's vision of a new world order based on interdependence.[34] This fusion of interests led Nixon and Brezhnev to agree to expand contacts between American firms and their Soviet counterparts, setting a target of $2 billion to $3 billion in trade between them over the next three years. This was an ambitious goal. Between 1966 and 1969 French exports to the Soviet Union increased in value from $76 million to $265 million, and West Germany's from $135 million to $406 million, while the United States' only grew from about $42 million to some $106 million.[35]

These initiatives bore fruit. By 1974 about forty protocol agreements had been signed between the American private sector and the Soviet Union, establishing a framework for commercial sales.[36] Eleven knowledge-exchange agreements were also in place. An Agreement on Cooperation in the Fields of Science and Technology included an agreement on space that led to the famous handshake between crew members in the *Apollo* and *Soyuz* spacecrafts in 1975.[37]

Functionalist arguments were used to justify these cooperative activities. Both sides agreed to expand bilateral contacts and cooperation "in the mutual belief that it will further promote better understanding between the peoples of the United States and the Soviet Union and will help to improve the general state of relations between the two countries." The creation "of a permanent foundation in economic relations" would have the same positive effects.[38] Peter G. Petersen, Nixon's secretary of commerce, claimed that economic cooperation would "build in both countries a vested economic interest in the maintenance of a harmonious and enduring relationship."[39] At the microlevel, commercial interactions and bilateral scientific and technological exchanges would produce "a series of international and interpersonal relationships which, over time, could contribute to a lasting structure of peace."[40]

The optimism that accompanied these agreements during the first years of détente soon began to wither away. Looking back over a decade of developments in 1978, Samuel P. Huntington, professor of government at Harvard University, distinguished two "Eras" of US-Soviet relationships with a transition period in between. In Era I, the early Cold War period, the competitive aspect typified the Soviet-US relationship. It was followed by a period of cooperation known as détente that lasted from the mid-1960s to the early 1970s. Cooperation and competition were the hallmarks of Era II, which began around fall 1973.[41]

During détente the Soviets benefited substantially from trade liberalization, importing about $1.8 billion worth of machinery and equipment between 1965 and 1973. Much of it was bought with credit offered by Western banks at below-market rates.[42] The chemical, petrochemical, automotive, and energy industries benefited considerably from foreign trade—the annual average growth in various sectors of Soviet chemicals and petrochemicals, for example, varied from 8.3 percent to 11.2 percent from 1971 to 1975.[43] At the same time the Soviets intensified their military buildup and expanded their overseas military deployments and arms transfers to less developed countries (including supporting the involvement of their Cuban allies in civil wars in Angola and in Ethiopia). In fact the CIA estimated that from 1965 to 1978 defense spending in the Soviet Union absorbed a relatively constant 11–13 percent of GNP.[44] It also estimated that by 1979 the Soviets were investing $20 billion annually in Defense R&D, as compared to $13 billion for the US government.[45] Economic cooperation had not led to Soviet moderation. On the contrary, it had, as the hardliners in the United States predicted, been exploited by the Soviets to fuel what Huntington called "military adventurism."[46]

Bucy shared Huntington's analysis. In embarking on détente, he said in 1977, optimists had deluded themselves into thinking that the Soviet Union would change its military or political behavior. In fact it had "continued its buildup of strategic and conventional forces to the point where 'rough equivalence' [was] giving away to Soviet military superiority."[47] Soviet enthusiasm for détente indicated that "the USSR [had] finally dropped its long-time pretense that the Communist world was achieving self-sufficiency by its own devices."[48] This however did not lead to interdependence and compromise, as the optimists had hoped. On the contrary, the Soviet's fundamental motivation for détente, said Bucy, was to gain self-sufficiency in their military-industrial base by openly and avidly acquiring Western technology—especially electronics and computers—to compete for power and influence with the West.[49] One aerospace contractor angrily remarked that "what the Russians want from us is the whole damn plant."[50] The concerns expressed by Malcolm Currie in 1974 that led to the establishment of the Bucy task force occurred, then, at a turning point in US-Soviet relations. New instruments were needed to control the flow of high technology to the Soviet bloc in an increasingly globalized economy.

Fred Bucy and the Reshaping of American Thinking on Export Controls

The trade liberalization enthusiastically espoused at the level of high politics required Congress to agree to a major revision of export control policy.

For two decades, as we saw in previous chapters, the Soviet Union had been largely denied access to the increasingly knowledge-based products driving the growth of Western economies and the expansion of world trade. The Export Control Act of 1949, we remember, restricted trade in any materials or technology, including technical data, that could strengthen the military potential of any country that posed a national security threat to the United States. It also allowed for the control of items with only indirect military utility.[51] In 1962 its already considerable reach was extended, under congressional pressure, to restrict trade in materials whose "potential military *and economic* significance [and not only, as before, their potential military significance] may adversely affect the national security of the United States."[52] Export controls had explicitly become instruments of economic warfare.[53] This extension was becoming untenable by the late 1960s. "National weariness with the cold war and changing perceptions of superpower relations, a burgeoning balance of payments deficit, and recognition of the growing commercial value of an East-West trade in which the United States was not participating" enabled a majority in the Senate supported by a minority in the House to adopt the 1969 Export Administration Act (EAA).[54]

The export control criterion of economic significance was abolished in the EAA. Rather than seeing commercial transactions as a threat, the EAA emphasized the importance of trade as a means to further the sound growth and stability of the American economy. The door to closer collaboration with the Soviet Union was opened by stipulating that it was US policy "to encourage trade with all countries with which we have diplomatic or trading relations," restricting the export only of goods and technology "which would make a significant contribution to the military potential of any other nation . . . detrimental to the national security of the United States."[55] As a result, during the first year of the EAA's existence no fewer than 1,550 commodities listed in the 750 entries in the Commodity Control List were made available for export to Soviet bloc countries.[56]

Protecting Lead Time

The determination to encourage trade with the Soviet Union (and the People's Republic of China) not only removed restrictions on US access to Communist markets. It also made it far more difficult for the United States to constrain Western Europe and Japan from trading with these

TABLE 2. **Amount ($US Million) and Percentage of High Technology Exports to the USSR by Select Western Trading Partners between 1970 and 1980**

Year	1970	1970	1971	1971	1972	1972	1980	1980
Country	Exports	%	Exports	%	Exports	%	Exports	%
US	12.5	3.1	26.8	8.1	154.7	6.5	84.7	3.6
Japan	43.5	10.8	41.8	12.7	398.9	16.8	400.2	17.2
France	58.5	14.5	38.0	11.5	376.8	15.9	341.3	14.8
FRG	92.9	23.0	79.4	24.1	668.3	28.2	737.2	31.6

countries. During the first two decades of the Cold War the United States tried to impose a virtual embargo on trade with the Communist bloc and could secure assent to its export policies inside Cocom by virtue of its undisputed technological superiority and its willingness to tolerate East-West trade as long as it did not involve items of direct military significance.[57] Those days were drawing to a close. Western European countries were far more heavily dependent on export markets than was the United States and, along with Japan, had more extensive trade links in high technology with the Soviet Union (see table 2).[58]

A number of changes in the globalizing economy thus obliged American policy makers to redefine the goal of export controls: the increased foreign (i.e., non-US) availability of high-technology goods sought after by the Soviet Union and its "satellites," the importance of East-West trade to the growth of European economies, and growing allied determination not to be hemmed in by multilateral agreements in Cocom. The rationale for export controls was reshaped accordingly. As Dr. Edith Martin, one of the deputy undersecretaries for defense put it to Congress, "Recognizing that it is not possible to protect any given technology forever, US policymakers have concluded that it is essential to protect our technological lead times."[59] Denying technology to the Soviet bloc was no longer an end in itself: its specific aim was to ensure that America was always ahead of its rivals in the race to define, develop, and field new technologies.

This appeal to lead time was not novel, of course. It was simply given more weight in shaping policy in the Bucy Report. In chapter 1 we pointed out that the need to protect US scientific and technological preeminence, or "lead time," was one of the founding goals of the national security state, and intrinsic to the policy of deterrence. In the 1950s it was expressed as the need to develop increasingly sophisticated weapons systems that maintained the balance of military power in America's favor. The emphasis on an already-entrenched concept in the export control debate in the late 1970s was a response to the growing economic strength of America's allies in high-tech markets and a relaxation of trade restrictions with Communist countries that called for a more discriminating response than blanket denial.

Lead Time and Dual Use

The invocation of "lead time" also responded to the changing nature of the civil/military relationship in an increasingly globalized economy. Expanding on Edith Martin's strategy, Bucy explained that the point of export controls was "not to interdict trade, but to delay an adversary's acquisition of *commercial technology of military significance* for as long as possible."[60] Denying licenses for exports "that would make a significant contribution to [a rival's] military potential" now had to take into account the emergence of market-driven, so-called dual-use technologies that increasingly blurred the civil/military divide. Indeed all the subcommittees set up to advise Bucy's task force dealt with the development of civilian technologies that had military applications.

During the first two decades of the Cold War the US government accounted for most of the nation's R&D expenditure, most of it in the defense and space industries. Military use preceded civil application (typically, in the nuclear field). The situation was reversed over the next decade: from 1963 to 1975 private industry's share of industrial R&D rose from 42 percent to 68 percent. The military sector became increasingly dependent on technological advances made in the civilian domain that were embedded in weapons systems. Typically the time taken to develop industrial and consumer markets for dual-use technologies was from one to three years, and significant sales were achieved worldwide. Application to military programs often took from four to six years, and the number of units produced was rather limited.[61] The DoD's undersecretary for research and engineering, William Perry, told a congressional subcommittee in 1979 that the market for integrated circuits, "probably the single most critical technology which gives us a technological superiority over the Soviets today," was shared 7 percent by the Department of Defense, 93 percent by commercial industry.[62] From this perspective, a "protrade" Export Administration Act served as a loophole through which the Communist world could acquire dual-use civilian technology that could be used to strengthen their military posture. What Bucy wanted to stress was that export controls should be targeted to protecting US lead time in the development and commercialization of dual-use technologies.

The Emphasis on Know-How

As we would expect given his definition of technology, it was specifically the portability of know-how from the civilian to the military domains that was of grave concern to Bucy (and to the Pentagon). Consider computers, widely exported products that straddled the civil-military divide. Historically the dangers of their diversion had been managed by demanding that clients signed end-use agreements undertaking to restrict their use to the civilian domain. It was common knowledge that these agreements were not always respected: the French, for example, openly violated them by using powerful IBM computers in their nuclear weapons program.[63] End-use assurances or even the most elaborate safeguards developed in the early 1970s to control high-performance computers *after* they had been exported to the Soviet Union were not just ineffective but irrelevant as far as the Bucy task force was concerned, however.[64] For them it was not the "misuse" of products that mattered, but the acquisition of dual-use intangible know-how that posed a threat to national security. As the report put it, "the widespread use of computers, even in commercial applications, enhances the 'cultural' preparedness of the Soviets to exploit advance[d] technology. It gives them vital experience in the use of advanced computers and software in the management of large and complex systems. The mere presence of large computer installations transfers know-how in software, and develops trained programmers, technicians, and other computer personnel. All of this can be redirected to strategic applications."[65] Computers were generic platforms on which skilled personnel could acquire programming skills, and run software with multiple applications in all sectors of the economy, both civil and military. As the Bucy Report put it, "The receiver of know-how gains a competence which serves as a base for many subsequent gains," so threatening the technological lead time on which US national security depended.[66]

The DoD was deeply concerned by the Soviet quest for know-how. The Soviets sought manufacturing technology, rather than the products themselves, in multiple dual-use domains like aircraft engines, computers, integrated circuitry, telecommunications equipment, navigation systems and avionics, and specialized instruments. It was "the processing know-how, the whole flow of work and equipment in manufacturing plants: that's the technology they're after," said Malcolm Currie.[67] As an official from the Pentagon explained in 1974, what the Soviets wanted was "not plants but the knowledge that would allow them to build their own production facilities—complete with all the systems and quality control that are the hallmarks of US defense plants."[68] The shift in emphasis from controlling the shipment of products to controlling the sharing of knowledge and know-how introduced a new term into the language of export controls: technology transfer.

The Stress on Technology Transfer

The Bucy Report's findings and recommendations were arrived at by subcommittees who investigated technology transfer in four specific industrial sectors.[69] This attention to technology transfer was a defining mark of the "conceptual shift" in the Bucy Report, and it permeates all the congressional investigations from the mid-1970s onward.[70]

Technology transfer is a term of art that distinguishes the process whereby technology that is produced in one location is taken up and utilized in another. The haphazard yet effective global circulation of technology over centuries is conventionally thought of as occurring gradually, by "diffusion." It is a slow process that one scholar describes as "largely unplanned, unpremeditated, and barely noticed," "like the ooze of liquid seepage."[71] Discussions in the 1960s about the need to close the technological gap between the advanced industrial economies themselves (e.g., the United States and Western Europe) called for more deliberate efforts to *transfer* technology, understood as "a planned, predetermined effort to make technology available to those who lack it."[72] In this context transferring technology involved the "transfer of ideas and practices couched in licenses, drawings, manuals, know-how contracts, training, etc." from a technologically "advanced" country or firm to one that was lagging behind it.[73] Most fundamentally, as opposed to the notion of diffusion, technology transfer emphasized the role of human agency; highlighted the importance of practice and the sharing of tacit skills, know-how and "intangible" knowledge; and drew attention to the need for structured knowledge acquisition and learning by the recipient to close a "technological gap." Herbert Fusfeld, a research director at Kennecott Copper, explained that technology transfer was not "something like a pass from a thrower to a receiver." It was rather like an "organ transplant with all the attendant requirements of compatibility with the environment, plus the surgical (i.e. managerial) skills necessary to establish the intimate working relationships between transplant and the connecting parts of the system."[74]

Technology transfer was most successful when the host had the indigenous capability and infrastructure to exploit its potential. Thane Gustafson pointed out in a RAND report released in April 1981 that "the most important question about technology transfer in the long run is whether the receiving side is able to absorb the technology it imports, to diffuse it beyond one or two showcase locations, and to build upon it to *generate further technological advances of its own*."[75] This had major implications for export policy. The transfer of know-how mattered more than the sale of products (even if they could be reverse engineered to extract novel technological content from them) because it gave individuals a skill set that could be deployed to improve existing products and processes and create new ones. What particularly bothered Bucy about technology transfer was that, if successfully absorbed by the host, know-how enabled the receiving nation to move along unanticipated technological trajectories that were difficult to foresee by the donor, and difficult to control in advance. As he put it, "once a technology transfer is made, there can be no effective control of either the flow of products or the future applications of the technology. Once released, technology can neither be taken back nor controlled. Its release is an irreversible decision."[76]

Bucy was concerned not only with what a rival nation purchased but with how trade enhanced its technical "capability" in the long term. Speaking of technical capability directed attention to the potential of the manufacturing process to generate entirely new technologies, rather than simply looking at existing performance specifications to assess the military significance of an item. As one congressional panel put it, "The transfer of products stimulates trade; the transfer of technology results in a permanent transfer of production capability and a loss of future trade in products,"[77] as well as generating unexpected and unanticipated challenges to America's technological lead time. Lead time was a "perishable asset," dissipated as quickly as "the basic concepts and know-how [became] widely known and exploited."[78]

Bucy's historical significance lay in his grasping the interconnectedness of technology transfer, technological capability, know-how, and lead time, of their implications for trade and national security in the changed global political economy of the 1970s, and in having access to the political power needed to frame export controls to deal with their implications. It was no coincidence that the DoD, along with Bucy of Texas Instruments, invited senior people from Boeing, Lockheed, and McDonnell Douglas, from General Electric and Pratt and Whitney, from Fairchild, Intel, and Motorola, from Xerox, Perkin-Elmer, and Hewlett Packard to define the parameters of a new export policy in the mid-1970s. As leaders of powerful multinational high-tech industries in the military-industrial complex, they had direct experience of the deregulation of financial capital, of the relative decline of US technological leadership in key high-tech export markets, of the blurring of the boundary between civil and military technologies, of the narrowing technological gap between the United States and the advanced industrialized economies, and of the new trading opportunities with the Soviet Union and the Eastern bloc opened up by the policy of détente. They accepted there was a need for a "conceptual shift" in managing trade relationships. That shift assumed that the technological gap with the Soviet Union was rapidly closing, it highlighted technology transfer as the key site for export controls, and it insisted that the sharing of intangible know-how (in addition to technical data in tangible form and certain types of end products) posed a grave threat to American technological lead time and so to its national security. The Bucy Report and the language it used to frame its analyses were a "local" response to the increasing interdependence of the world's major economies that historians now see as inaugurating the new phase of globalization that has grown ever more expansive in recent decades. This chapter, and indeed the rest of this book, will reckon with the challenges that these seismic changes posed to US export controls in the name of national security.

Policy Recommendations of the Bucy Report

The Bucy Report made no pretense at neutrality. It was subtitled "A DoD Perspective." Produced by some of the country's leading corporate executives, it made findings and recommendations that added credibility to the successful lobbying by the Pentagon and by conservative members of Congress who had just secured changes in the procedures for granting export licenses to firms trading with Communist countries. Normally these licenses were awarded by the Department of Commerce after interagency consultation. An amendment to the EAA in 1974 empowered the DoD to overrule the Commerce Department. It authorized the secretary of defense to review all dual-use license applications for exports to controlled countries that were submitted to the Department of Commerce, and, if deemed necessary, to recommend denial to the president on the grounds of military significance.[79] The Bucy Report thus confirmed and legitimated a reorientation of export controls sought after by the secretary of defense and a powerful contingency in Congress that was critical of détente. It secured industry's assent to Currie's conviction that the flow of know-how should be controlled, and it legitimated a more muscular role for the Department of Defense in the licensing of dual-use exports.

The Bucy Report comprised six sections and a conclusion along with an "Executive Summary." It broke new ground in recognizing that, in the new world order that was emerging in the 1970s, America's technological "leadership" was no longer assured, and in proposing export policies to meet that situation. As pointed out earlier Bucy himself had a rather narrow definition of the term technology: "the specific know-how required to define a product that fulfills a need, to design the product, and to manufacture it."[80] His distinction between technology and science,[81] and between technology and products, was central to the reorientation of export policy being sought by Bucy and his industrial partners. It led them to specify three categories of export [that] should receive primary emphasis:

1. Arrays of design and manufacturing know-how;
2. Keystone manufacturing, inspection and test equipment;
3. Products accompanied by sophisticated operation, application or maintenance know-how.[82]

The first of these categories of technology transfer was the most critical. If accompanied by teaching assistance, it provided "the basis on which the receiving nation can build further advances in technology."[83] Computer-controlled process, inspection, and test equipment was a prime example of the second. It enabled the receiver to improve throughput and precision, in order to fulfill diverse manufacturing requirements, and it provided a "growth capability on which advanced new production skills can be built."[84] The third was important because it improved the overall performance of a manufacturing system.

While the four industrial sectors represented in the membership of Bucy's task force (airframes, aircraft jet engines, instrumentation, and solid-state devices) placed different weights on these several elements, all agreed that "the *detail of how to do things* is the essence of the technolo-

THE RECALIBRATION OF AMERICAN POWER   119

gies. This body of detail is hard earned and hard learned. It is not likely to be transferred inadvertently. But it can be taught and learned."[85]

How was know-how learned? Or, to use the report's terminology, how was technology transferred from the donor to the receiver? The Bucy Report distinguished active from passive modes of technology transfer depending on the degree of personal contact involved and on the intensity of face-to-face exchanges between the partners in the learning process.[86] Active relationships involved frequent and detailed communication between donor and receiver, usually aiming to improve the technical capability of the latter. They were iterative and could last for several years, the receiver requesting precise information, applying it, and seeking further (often proprietary) information until the desired capability was achieved. Active transfers worked best when the goal of the learning was well defined, when the receivers were technically competent, and when a national infrastructure was already in place and able to fully exploit the technology transferred (all reminiscent of the remarks by Fusfeld and Gustafson quoted earlier). By contrast, trade publications and trade shows were at the passive end of the spectrum, since they rarely communicated enough know-how to transfer the "essence" of the technology involved.[87]

The scope of export controls depended on the kind of trajectory the technology followed in the United States and in its trading partners. It could be revolutionary, so involving major breakthroughs, or it could be evolutionary, so moving ahead in small, incremental steps. The Bucy Report was emphatic that technologies that made a revolutionary contribution to lead time should be carefully controlled so as maximize the advantage that the United States could draw from them. By contrast, technologies on an evolutionary trajectory could be exported after weighing the immediate gain to the receiver and comparing it with the target state's ability to acquire the same technology by combining its indigenous efforts with the general diffusion of technology.

The overall aim of the Bucy panel was to streamline the export control process by getting it to focus on technology transfers that could deplete American technological lead time. It listed a sequence of questions to ask when screening license applications that brought together the findings of the four subcommittees. To answer them one had to have at least a working knowledge of the scientific and technological core of the device in question. For a product or material, one asked whether it had "a significant military utility in itself, based on performance capabilities." For production items, like a "turnkey plant" one asked if the manufacturing process supported strategic products or technologies, if it contributed technology useful for manufacturing or design, if the technology itself was changing with high velocity, that is, was on a revolutionary or evolutionary trajectory, and if the transaction engaged active means of technology transfer (i.e., involved extensive interpersonal exchanges of technical data and know-how).[88]

This approach called for the construction of a whole new database by knowledgeable individuals in the government and the private sector. In the past officials in the Department of Commerce who regulated the sale of products would use performance specifications and end use to decide whether or not to grant a license (or if one was needed at all). The decisions called for by Bucy's task force required officials to be technically qualified in the workings of the devices themselves and their process of production. Only senior executives and technical personnel in the Department of Defense's Division of Research and Engineering could do this properly, and avoid making conservative recommendations that would do little to improve the "unrealistically restrictive and overly cumbersome" control lists that then defined the terms of East-West trade.[89]

The DoD's expertise was also needed to manage the risks of know-how being lost at other active transfer sites. The Bucy Report noted that new modes of face-to-face interaction with Communist partners were being encouraged by the climate of détente. US citizens were being used as consultants by Communist countries advising on the use of key technologies. They were also being employed abroad as principals in firms that transferred embargoed technologies to Communist nations. Educational exchanges were also facilitating the circulation of expertise. As mentioned earlier, no fewer than eleven formal government-to-government scientific exchange agreements had been signed between various American and Soviet research bodies in the early 1970s. These were complemented by the training of citizens from Communist countries "at the more significant laboratories of US technical institutes and universities."[90] In an article published in 1977 Bucy specifically called on university administrators to evaluate the training of graduate students from Comecon countries in electronic and aerospace technologies with possible military applications. "A high percentage of these technical 'students,'" he wrote, "may have missions well beyond the area of academic enquiry."[91] The Bucy Report did not try to assess the significance of these recent "active" modes of technology transfer. But it did call on the Department of Defense to look into active mechanisms, like these, that had been sporadically monitored by the export control administration in the early Cold War, and it asked the DoD to make recommendations for controlling them.[92]

The antipathy to active technology transfer mechanisms, and particularly to the sharing of technical data and intangible know-how in face-to-face interactions, amounted to the wholesale rejection of one major noneconomic argument in favor of détente. That (functional) argument emphasized the role of personal contact, discussion, and knowledge sharing as ways of bringing individuals in the opposed camps together, of increasing mutual understanding, and of reducing tensions. Behind this lay the unstated hope held by some in the Western democracies that a greater awareness of the individual freedoms and opportunities available in open societies would slowly corrode the Soviet system from within. Bucy and his colleagues took the opposite position. Interpersonal contact and the sharing of know-how with trained people from the Soviet Union and Eastern Europe reduced US technological lead time, enhanced Soviet military capability, and posed a dire threat to national security. The trade liberalization encouraged by détente, if not managed properly, endangered the free world. The very same arguments are front and center in the debate over Sino-American trade and academic collaboration at the present time and will be addressed later in this book.

The Bucy Report and Intra-Western Trade

The Bucy Report looked beyond the risks of technology transfer from the United States to the Communist world to embrace the perils of intra-Western trade. Specifically it addressed the danger of the retransfer of strategic know-how through allied and neutral countries to Communist nations.[93] In chapter 3 we described the friction with allies in the 1950s caused by US attempts to impose extraterritorial constraints on trade in commodities that included a component made in America. The United States demanded that any US item or technology subject to export licensing had to be relicensed each time it was further exported, even as a component of a foreign-built device. For many Cocom members this extension of US export controls into the trading practices of its allies was seen as a violation of their national sovereignty.[94] This did not seem to bother Bucy. Instead his task force criticized them for constructing a leaky export control regime. Admitting that no better alternative to Cocom existed, the task force argued that it should be reformed to secure better compliance with American regulations. As a first step the Bucy Report recommended that, in future, the United States should impose sanctions on any Cocom country that failed to control a specific technology, by restricting the flow of know-how in that technology to the offending country. As for nonallied, non-Communist countries, they were not to be trusted at all, and should be allowed to acquire only technology that the United States would be willing to transfer to Communist countries directly.[95]

This punitive approach was deeply resented of course. The United States' own behavior in Cocom amplified the frustration. The United States liked to present itself as the "conscience of Cocom."[96] Yet between 1971 and 1975 it alone accounted for about 50 percent of the volume of Cocom exception requests (2,178 out of 4,423).[97] "Exception requests" were "petitions from firms in member countries to exempt from Cocom control, on a one-time basis, an item that appears on the Cocom embargo list and would otherwise be prohibited from sale."[98] Ironically these exceptions mostly concerned computers or computer-related technologies, where the United States dominated the world market and where the Soviets were actively seeking to enhance their indigenous strengths. Computers were precisely those technologies that the Bucy Report signaled out as posing risks to national security.

Mastanduno has pointed out that many American exporters suspected that, despite the emphasis placed on East-West trade and Soviet military capability, intra-Western trade was actually the primary target of the Bucy Report.[99] One reason for this was that existing legislation already required a validated license for exports of technical data (that included know-how) to the Soviet Union and Eastern bloc; exports to the West were less restrictive and simply required a general license and written assurances that there would be no unauthorized reexport.[100] Thus, for Mastanduno, the general claim in the Bucy Report that the existing export regime "overemphasized product controls and underemphasized technology applied more appropriately to intra-Western trade than to East-West trade." Effectively it authorized "looser *product* controls on trade with the East and tighter *technology* controls on trade with the West."[101]

Whether or not Bucy intended this to become policy—the report never addressed legal matters as such—he was certainly frustrated by liberal US trade policies with Western allies. In an article in the *New York Times* in September 1976, just seven months after the task force had released its report, he pointed out that design and manufacturing know-how was being exported to competitors who could beat the United States in international markets. "The threat is therefore twofold," Bucy wrote. "Exporting design and manufacturing know-how to potential enemies strengthens them militarily. And exporting that same know-how to potential economic competitors—friends or foes—strengthens them to compete against us for world markets."[102] The same frustration was expressed in an article in *Fortune* early in 1978 that gave examples of a French and a Japanese firm acquiring "front-end" American technology to produce respectively a civilian jet engine (SNECMA) and computers (Fujitsu). Bucy was quoted as saying that "today our toughest competition is coming from foreign companies whose ability to compete with us rests in part on their acquisition of US technology. . . . The time has come to stop selling our latest technologies, which is the most valuable thing we have got."[103] Conservatives in Congress shared Bucy's sentiments and made several efforts to amend the EAA to include punitive measures against Cocom nations that did not respect US export control policies. The Carter administration settled the issue temporarily in 1978 by rejecting tighter US controls on Western technology trade.[104] The question of the dangers of technology sharing with the allies would return with a vengeance in the mid-1980s with a focus on Japan, as we will show in chapter 7.

The Bucy Report also tried to turn the clock back on the reforms made in the 1969 Export Administration Act. That act, we will remember, specifically removed economic significance as a criterion for needing a validated export license. The Bucy Report reintroduced it for certain kinds of technology. By replacing the criterion of military *use* with that of military *significance*, and highlighting the need to control "design and manufacturing know-how," it pushed controls back into the production process itself. This potentially opened the door to denying licenses to sell manufacturing systems that shared know-how and produced goods for the civilian market that could also be applied or integrated into weapons systems. Guidelines for a new export control policy specifically called for controlling key technologies and products "on the basis of the capabilities they confer, rather than on the basis of commercial applications."[105] In short the Bucy Report, while putatively simplifying the export control system, added a new level of complication by targeting the techno-industrial base of dual-use production. It was a proponent of economic warfare undertaken in the name of national security.

DoD Implementation of the Bucy Report's Recommendations

The Bucy Report called on the Department of Defense to "develop policy objectives and strategies for the control of high-technology fields," and to "identify principal technologies that require export control."[106] In response, the new secretary of defense, Harold Brown, used it extensively to shape a major policy statement in August 1977 that provided interim internal guidance to the DoD on the export control policies it should follow.[107]

The statement confirmed the purpose of export controls as being to "protect the United States' lead time with respect to its principal adversaries in the application of technology to military capabilities." These controls were to interfere as little as possible with the normal practice of commercial trade. The point in having them was to provide time for the replenishment of technology through new R&D.

Brown's memo adopted the so-called critical technology approach in an effort to streamline policy. It defined critical technology as "classified and unclassified nuclear and non-nuclear unpublished technical *data*, whose acquisition by a potential adversary could make a significant contribution, which would prove detrimental to the national security of the United States, to the military potential of such country."[108] The security challenges posed by the export of dual-use technologies led Brown to specifically eliminate the "declared intended end-use by the recipient" as a basis for export control decisions.

What was "technical data"? Brown combined existing legal definitions of the phrase with an extension to know-how required by the Bucy Report. Thus technical data meant "information of any kind that can be used, or adapted for use, in the design, production, manufacture, utilization, testing, maintenance, or reconstruction of articles or materials." They could "take a tangible form, such as a model, prototype, blueprint, or an operating manual," or an intangible form such as a technical service.[109] This original definition of critical technology as data was then extended to include "certain associated end-products defined as 'keystone' that can contribute significantly in and of themselves to the transfer of critical technology because they 1) embody extractable critical technology and/or 2) are equipment that completes a process line and allows it to be fully utilized."[110]

The new guidelines made no distinction between Communist countries and Western allies: to protect strategic lead time it was necessary to control defense-related technology to *all* foreign countries. An exception was made for allies with whom the United States had a major security interest if the export could be shown to strengthen security or enhance interoperability. Even here transactions involving a revolutionary advance in defense-related technology would normally be denied: a "presumption for recommending disapproval" would guide any license application. Brown's memo reiterated the Bucy panel's frustration with allies whose export control regimes were less comprehensive than those put in place in the United States. It instructed the executive to work along with the intelligence and the security communities to identify unauthorized third-party transfers or other actions that compromised critical technology. Violations would normally lead to sanctions.

The secretary of defense's memo also proposed measures to deal with mechanisms of technology transfer. He targeted situations in which there was a high risk of the inadvertent transfer of critical technology. He called for improved measures to control visitors within the Department of Defense. He instructed the DoD to formulate guidelines for the appropriate agencies on "restrictions on the amount, extent or kind of interpersonal exchange in a given transaction." Brown's list of sites for technology transfer "transactions" included both "active" and "passive" modes of knowledge sharing. It mentioned "foreign liaison activities, scientific and technical exchanges, commercial visits, trade fairs, training programs, sales proposals and consulting agreements" and dealings in specific technology export cases.[111]

Brown's guidelines were embodied in the 1977 amendments to the Export Administration Act. If previously the act had spoken of restricting exports that made a significant contribution to a country's military capability, it now required the secretary of commerce to show explicitly that that enhanced capability "would prove detrimental to the national security of the United States." Instead of making a sharp distinction between Communist and non-Communist countries in administering national security export controls Brown prioritized a "country's present and potential relationship to the United States, [and] it's ability and willingness to control retransfers of United State exports in accordance with United States policy."[112] The trend then was to enhance East-West trade and to place more emphasis on the threat posed by the technology itself rather than its destination.

The concept of "critical technology" was central to the DoD's export control strategy. As William Perry, the undersecretary of defense for research and engineering, explained, "we have adopted the concept of strongly enforcing control on the export of selected critical technologies while simultaneously relaxing many existing product controls."[113] Perry accepted that "the primary objective" of controlling exports was to protect the US lead time and the application of technology to military capability. But he insisted that these goals depended on maintaining the health of US industry, whose profits were invested in further research that was crucial to sustain the technological edge that provided military superiority. There was no risk-free scenario in export controls for dual-use technology. The challenge was to balance the risk of allowing trade in a technology that would enhance the military capability of a rival against the risk of losing sales to the detriment of US industry, and ultimately to national security itself. The DoD had to move away from the conservative, risk-avoidance posture that was the hallmark of its current approach to export controls. Perry himself would take a "more critical view" of arguments that sales to adversaries gave them a military advantage. Brown's definition of a "critical technology" as one that could make a "significant contribution" to the "military potential" of a rival was too restrictive, in his view. "We expect a clear indication of a specific and a direct military advantage before we will reject a sale for that reason," Perry said.[114]

By sharpening the criteria for a "critical technology" as being one that *clearly* made a *specific* and a *direct* contribution to military advantage Perry hoped to protect a space for the sale of commercially significant dual-use technologies from conservative opponents who prioritized national security above all else. He also sought to set limits on the DoD's role when providing input to the decision-making process for export licenses to other countries. As he put it, "we attach considerable importance to a proper understanding of the extent of our responsibility and of our credibility with industry and government alike in recommending approval or disapproval of export license applications."[115] The DoD would assess a request for an export license on the basis of unambiguous "military significance" of the technology no more.

Perry's "narrow" definition of the impact of technology transfer on national security was soon tested. The occasion was the very public dispute over the license to export deep well drilling technology to the Soviet Union that was awarded to the Houston-based firm Dresser Industries in May 1978. Bucy argued that this contract would boost the Soviet energy exports, providing necessary foreign exchange to strengthen its military. Perry refused to stretch the role of the DoD in applying the Export Administration Act to include economic warfare. As far as he was concerned the DoD need not be consulted on the Dresser deal since deep well drilling technology was not of "direct military significance." Perry stood his ground, to the distress of conservative members of Congress and of Bucy, who expected the Department of Defense to take a more pugnacious approach in dealing with a rival superpower.

In September 1978 these congressional "hawks" tabled a bill for a Technology Transfer Ban Act. It rehearsed the usual set of arguments against détente, emphasizing that the Soviets had not changed their ways, and were using trade liberalization to secure US strategic technology. The bill also proposed to restrict the exports of goods and technology that "could make any contribution to the military *or economic* potential" of any nation, and which would prove detrimental to the national security of the United States.[116] This attempt to turn the clock back to a decade earlier failed, and the bill died in committee.

Notwithstanding their setbacks, conservatives in Congress remained convinced that the Department of Defense should have the main role in defining export policy to the Soviet bloc. Richard Ichord (D-MO), chairman of the House Un-American Activities Committee until 1975, and chair of the R&D Subcommittee of the House Armed Services Committee, dismissed the "whole system of export controls as operated by the Department of Commerce [as] properly characterized as a shambles and a can of worms."[117] Hopes were pinned on the "critical technologies" approach, using a list built using the Bucy Report, and adopted as policy by Defense Secretary Brown in August 1977, to put order in the system. It would establish the authority of the DoD as having "primary responsibility" for defining export control policy to the Soviet bloc. It would also define clear criteria to distinguish a small subset of critical technologies that needed to be protected in the name of national security from products that could be freely exported. The task proved far more difficult than anyone had imagined.

The Militarily Critical Technologies List (MCTL)

It will be remembered that Harold Brown's memo of August 1977 recommended controlling critical technology understood as technical data and "certain associated end-products defined as 'keystone' that could contribute significantly in and of themselves to the transfer of critical technology because they (1) embody extractable critical technology and/or (2) are equipment that completes a process line and allows it to be fully utilized." These concepts were translated into controllable items in the Export Administration Act of 1979 by defining a Militarily Critical Technologies

List (MCTL) as the basis for national security export controls.[118] The secretary of defense bore "primary responsibility" for developing such a list. It would eventually "become a part" of the Department of Commerce's Commodity Control List (CCL). The latter, for its part, would be revised to ensure that export controls were limited, to the maximum extent possible, to militarily critical goods and technologies *and* to the mechanisms through which such goods and technologies could be effectively transferred. Following the Bucy Report almost to the letter, it required that the secretary of defense give "primary emphasis" to "(A) arrays of design and manufacturing know-how, (B) keystone manufacturing, inspection, and test equipment, and (C) goods accompanied by sophisticated operation, application or maintenance know-how" if they were not possessed by countries to which national security controls applied, and if they "would permit a significant advance in a military system of any such country."

The drafters of the 1979 act hoped that it would reduce the number of controlled items by removing products that did not transfer militarily critical technology while tightening controls on a subset of items containing key technological know-how.[119] Congress specifically mandated that the list be "sufficiently narrow to constitute an improvement over the present system."[120] It would be reviewed every three years to harmonize it with changes in the Cocom list, and annually in cases where unilateral control was adopted to protect sensitive technologies in which the United States had considerable lead time. An indexing system would be established to enable items to be removed regularly as they became obsolete by US standards.

The act specified that an initial version of the MCTL had to be published in the *Federal Register* no later than October 1, 1980, and that its items should be "sufficiently specific" to guide the judgment of any official who used it to make an export license determination. These requirements were responses to concerns that it was proving more difficult to establish an operational MCTL than had been hoped.

The House of Representatives was alerted to difficulties of defining a small subset of critical technologies during the hearings and markup of the draft bill. These were held before a subcommittee of the Congressional Committee on Foreign Affairs spread over sixteen sessions from February to April 1979.[121] The DoD was represented by Ellen Frost, deputy assistant secretary for economic affairs, and by Ruth Davis, deputy undersecretary for research and development. Frost made an upbeat statement on progress. A Critical Technology Implementation Interagency

Task Group had been established under DoD leadership, and there were "extensive efforts underway to identify and continuously update a list of specific critical technologies and keystone equipment."[122] Davis was more circumspect. The DoD had not yet institutionalized the MCTL as an evaluation tool: they were still in "the formative stages" of the program. The department had drawn on "ad-hoc or interim actions, one-time studies and reliance on volunteer participatory groups from industry and government."[123] In fact she had only obtained official approval in the current fiscal year for an Office of Technology Export, charged with developing and implementing the critical technologies approach. Her first formal budgetary request would not be filed until the next budgetary opportunity, FY1981.[124]

Bucy got wind of this. Speaking at a symposium in Washington in January 1979, he deplored what he called the "intransigence of the administrative bureaucracy." In the five years since his task force had been convened, he said, there had "been minimal progress by the Executive Branch in either specifically defining critical technologies or making any discernible changes in the implementation of present controls."[125] To rebut such charges Davis and Frost shared a preliminary list of fifteen critical technologies that they had identified early in 1979 during the hearings on the draft bill.[126] The number was increased to eighteen by Perry in October 1979.[127] The complete list initially read:

Computer network technology;

Larger computer system technology;

Software technology;

Automated real-time control technology;

Composite materials processing and manufacturing technology;

LSI (Large Scale Integration)-VLSI (Very Large Scale Integration) design and manufacturing technology in microelectronics;

Directed energy technology;

(Military) Instrumentation technology;

Telecommunications technology;

Guidance and control technology;

Microwave componentry technology;

(Military) Vehicular engine technology;

Advanced optic technology (including fiber optics);

Sensor technology;

Undersea systems technology;

and, by October 1979, was extended to include

Cryptography;

Chemical technology;

Nuclear-specific technology.

These broad categories had to be filled out by identifying militarily significant component technologies relevant to the "design, manufacturing, utilization, testing and maintenance functions which can be subjected to export controls."[128] Davis pointed out that nine of the original fifteen main areas had been covered, and she provided illustrative sublists for three of them (guidance and control technology, microwave component technology, undersea systems technology). She also provided an illustrative listing of militarily significant component technologies for composite materials processing and manufacturing technology. It included both militarily critical and noncritical technology items.[129] Davis's lists were still far removed from anything that was envisaged in the Bucy Report.

Some of the practical problems implementing Bucy's recommendations were highlighted in the comptroller general's report in March 1979 on the need to clarify export controls. One DoD official warned him that a large number of products would have to be controlled because they had considerable intrinsic military value, or could readily be reverse engineered. Bucy had played down the importance of reverse engineering and had "seriously underemphasized the importance of controls for strategic items."[130] His emphasis on controlling technical data also seemed to miss the mark. Out of 50,737 applications for export licenses in 1977, only 299 were for the release of technical data (of which just three were denied).

The EAA deadline of October 1980 was met with the publication of an Initial Military Critical Technologies List (MCTL) in the *Federal Register*. It was accompanied by a list of military critical technologies produced by the Department of Energy.[131] Only the table of contents for the MCTL was published. It comprised technologies that could contribute to the development, production, or utilization of items currently controlled for national security purposes on the Commodity Control List. It also included technologies that contributed to items on the Munitions List that had present or future civil applications.[132] Detailed specifications and supportive documentation were classified by the administration on the grounds that it would provide a "shopping list" for America's enemies and even reveal the existence of technologies that they were unaware of.[133]

Although corporate leaders generally supported the MCTL concept in the hope that it would streamline export control policy, they resented the fact that it was classified and could be applied unilaterally. When they had access, they objected that it was far too long and far too broad. In 1982 the vice chairman of Control Data Corporation described it as "an exercise in futility." He explained that "in an analysis that my company has made of the list only 125 of its 700 technologies were found to be possible candidates for restrictive exporting and in many cases the restriction would have protected a proprietary process of particular companies rather than a technology that had any military significance."[134] Testifying to the government's Technology Transfer Panel in 1983, Larry Hansen, the executive vice president of Varian Associates, speaking on behalf of the semiconductor industry, complained that "practically all semiconductor equipment is presently set forth on the militarily critical technologies list. In fact," he went on, "my review of that list as much as I was able to see it, indicated that any piece of semiconductor equipment which had been developed in the last 20 years" was on the MCTL.[135] To emphasize its irrelevance, another participant representing the Semiconductor Industry Association suggested that one way to shrink the MCTL was to remove items that were included simply because they incorporated a commodity-level semiconductor, as found in the "Speak and Spell" toy and in many handheld calculators.[136]

The business sector not only complained that the MCTL was far too long and too broad in coverage. They also were emphatic that it had to be used in combination with an assessment of foreign availability and in consultation with other Western countries, especially the members of Cocom. In fact the EAA of 1979 went further than the 1977 amendment of the 1969 act in its stress on the need to assess foreign availability in implementing export control policy. Each government department, in consultation with the intelligence community, was instructed to assist in the determination of foreign availability, along with technical advisory committees comprising members of industry and the government. The act also took measures to preclude businesses using self-serving or uncorroborated evidence to exploit the appeal to foreign availability as a loophole through which to export sensitive technology. All claims to foreign availability had to be "made in writing, and supported by reliable evidence, including scientific or physical examination and expert opinion based on adequate factual or intelligence information."[137]

The tools were there, then: their implementation was another matter. "Our major concern with respect to administration," said Hansen, "is the handling of foreign availability. It has been abysmal."[138] Business put it down to the ineptness of the government. One corporate leader said it was time to abandon the misconception that "the United States is the originator of virtually all discoveries and advancement in technology.... One need only to attend the international trade shows, or the many other expositions worldwide to be aware of the foreign availability of the critical advanced technologies."[139] Speaking for the semiconductor industry Hansen pointed out that "right now it is possible to build a production line, the most modern production line that can manufacture a 64K RAM or the most modern microprocessors that we know of, without buying a single solitary piece of semiconductor manufacturing equipment from the United States." He was frustrated that US firms were denied licenses for exports to the Eastern bloc when the same technologies were sold to them by the Cocom allies and non-Cocom sources.[140]

The government may have been inept. However it also had to face the skepticism of its allies in Cocom. In fact, it was futile to hope that the members of Cocom would be supportive of American policies. Mastanduno, in interviews with British officials, found little support for either the Bucy Report or the "critical technologies" approach. Technology, defined as know-how transferred interpersonally, not only was too elusive to control effectively. It would require the government "to open every letter and put a guard on every scientist who left the country" said one Western official.[141] What is more, "active" mechanisms of technology transfer were valuable precisely because they enabled Western governments to monitor end-user compliance by Eastern bloc clients. Officials in Britain, France, and Germany were also convinced that domestic political pressure would make it impossible for the US government to seriously decontrol products of military significance to the Soviet Union, even if they were not "critical technologies." As far as Mastanduno was concerned, "in the alliance, as well as in the United States, the critical technologies approach was essentially a non-starter."[142]

This was too quick a dismissal. Eight hundred pages of classified information on critical technologies and thousands of hours of work were not by any means wasted. Already in his 1979 report the comptroller general had suggested that the emphasis on critical technology "could result in a data-rich, systematically developed set of guidelines which can serve as the basis for export license decision-making."[143] A panel convened by the National Academies eight years later agreed almost verbatim. It recognized that Bucy's proposals relied on distinctions like "critical," "revolutionary," and "keystone" that were open to widely differing interpretations that were difficult to reconcile. All the same it suggested that, for all its faults, the MCTL "serves a useful but limited purpose as a reference document for developing control proposals and making informed licensing decisions."[144] The MCTL would survive as a classified encyclopedia of advanced technologies to be referred to as and when needed.

The Bucy Report is widely accepted as having had a major impact on export control policy. It was a local response to a specific problem at a specific moment in the history of the twentieth century: the first stirrings of a global knowledge-based economy, in which America's military dominance had been matched by its Soviet rival, and its market dominance was being whittled away by its major Western competitors, especially in dual-use technologies. By placing the transfer of advanced technologies and know-how at the heart of the regulatory regimes of the national security state a corporate giant in the semiconductor industry hoped to simplify the system of export controls, and to deny the Soviet bloc access to the new knowledges and practices on which American technological lead time depended. In the event, Bucy did not dislodge the existing criteria for denying trade in a dual-use high-tech manufacturing process or commodities with the Soviet bloc. Instead he led the executive branch to pay closer attention to controlling critical new technologies and to incorporate a new criterion, tacit knowledge or know-how, "the detail of how to do things," alongside technical data in the scope of export controls.

Bucy did not streamline the export control system in practice, as he had hoped. He complicated it technically and administratively in two ways. Firstly, he placed controls over the circulation of militarily critical technologies in the scope of the Export Administration Act, so expanding the role of the intelligence community and the Department of Defense back into the realm of commercial activity that was traditionally managed by the Department of Commerce. Secondly, by highlighting the dangers of transfer of "technology" or "know-how" in face-to-face interactions he brought to the fore the risks to national security posed by the "active" sharing of skills and tacit knowledge in learning contexts in general, and in American research universities in particular. In chapter 3 we described the efforts made to maintain the international circulation of scientific and technical knowledge in the interests of scientific progress, academic freedom, and the values of an open society when export controls on technical data were first implemented. The extension of export controls to incorporate knowledge more generally, and tacit skills and intangible knowledge above all, reopened a debate during the Reagan years on the role of the state in restricting international scientific and technological collaboration that we shall turn to in the next two chapters.

It makes sense for us to follow the implications of the Bucy Report into the Reagan years, when détente gave way to the "Second Cold War," and the full implications of Soviet access to American advanced technology were revealed to the shock of the new administration. We should not forget, however, that Bucy was just as concerned about the growing economic strength of America's allies as he was about the military advantages the Soviets had gained from détente. Indeed during the 1970s and into the 1980s, American global leadership was challenged economically by the technological development of its allies, above all Japan, while its military preeminence was challenged by the Soviet Union. Japan's aggressive inroads into the semiconductor and related industries exposed the fragility of American high-tech industry, raised questions about whether the country could lead the free world if it persisted with a laissez-faire approach to its core techno-industrial base, and caused a fundamental reassessment of what was needed to protect national security. A new concept entered the debate on the role of export controls in maintaining national security: the concept of economic security. This concept was progressively embedded by successive administrations into their framing of export policies to secure a global Pax Americana. The history of these developments, which also had their origins in the events defined in this chapter, follows in chapter 7. It provides the bedrock for the rest of volume, which concludes with an epilogue showing the centrality of the concept of economic security to the Trump administration's policies concerning the transnational transfer of corporate and academic critical technology, knowledge, and know-how between the United States and the People's Republic of China.

CHAPTER FIVE

The Reagan Administration's Attempts to Control Soviet Knowledge Acquisition in Academia

Détente died with a succession of events that dispelled any hope that the opening to the West would "democratize" the Soviet regime. The invasion of Afghanistan in December 1979, the internal exile to Gorky of leading Soviet physicist and dissident Andrei Sakharov in January 1980, and Moscow's support for the declaration of martial law in Poland in December 1981, which banned Solidarity and other prodemocracy movements, led a new conservative president to dub the Soviet Union the "evil empire" in March 1983. Coexistence yielded to confrontation in what historians have dubbed the "Second Cold War."

The changed geopolitical climate had a major effect on US-Soviet trade and scientific exchange relations. Two major CIA reports confirmed anecdotal evidence that the Soviets were exploiting détente to strengthen their military industrial base through the legal and illegal acquisition of Western technology.[1] The legal channels ranged from commercial transactions, including purchasing turnkey plants and brokering joint ventures, to gathering information and know-how at trade shows, at conferences, and during academic visits as well as by exploiting thousands of open-source technical publications distributed annually in the United States. Combining legal strategies with espionage and illegal deals to bypass US and allied restrictions on technology transfer, the Soviet Union and the Eastern bloc had embarked on a vast, orchestrated program to close the technological gap with the United States and its Western allies.

Retaliation was rapid and increasingly comprehensive. President Carter, working closely with the CIA, immediately took a number of ad hoc measures against the Soviets. He imposed sanctions, cancelled several trade deals, and called for a boycott of the Moscow Olympics in 1980. The policies adopted by President Reagan, who took office in January 1981, took a far more confrontational and wide-ranging approach to the Communist threat. Reagan was convinced that the Soviet economy was doomed to fail and believed that economic pressure could hasten its demise.[2] Immediate measures were taken to enforce existing trade regulations with the Soviet Union and, to a lesser extent, the Eastern bloc. The Departments of Commerce and State put immense pressure on America's allies in Western Europe and Japan to follow suit, claiming that "the USSR has created a veritable 'Soviet Lobby' in Western business and government circles" to exploit Western dependency on its markets.[3] The United States convened a meeting of Cocom at the ministerial level in January 1982—the first since the 1950s—to coordinate multilateral action that was essential to deterring Soviet acquisition efforts.[4] It also took a range of measures to control the flow of technology and know-how to the Communist enemy. This was inspired by the conviction that the Soviets had made giant strides in enhancing their offensive capabilities by successfully transferring technology from the advanced industrialized West. This view not shared by all, though it was loudly proclaimed by conservative forces in Congress and in other parts of the government.[5]

The alarm raised by the so-called hemorrhage of sensitive technical data to the Soviet bloc led to calls for tighter export controls and increased classification powers in the name of national security to restrict the access of foreign nationals to the production and dissemination of knowledge. In this chapter we describe the multiple loci of contestation over the open circulation of new knowledge in the late 1970s and into the 1980s: the visits of foreign nationals to university campuses and manufacturing plants; their participation in "international" conferences on US soil; the public dissemination of advances in cryptography; the constraints on divulging (unclassified) information acquired while working for the government; the terms of the contractual relations between the Department of Defense and the academic research community. Many of these measures were contested by a research community that was deeply suspicious of government intrusion into the open circulation of the results of research that fostered the growth of knowledge and secured American preeminence at the research frontier. They drew inspiration from the First Amendment of the US Constitution, which, they argued, was being undermined by an administration described by Harvard's vice president in 1984 as being "particularly active in making claims of national security to curtail civil liberties."[6] Then-current interpretations of the scope of the First Amendment's protection of the freedom of "technical speech" are briefly summarized in a concluding coda to this chapter. They played a crucial role in formulating policies that created a nonregulated space for basic and applied research on university campuses that was formalized in 1985 after intensive discussions and negotiations described in chapter 6.

Early 1980s: Soviet Legal and Illegal Acquisition of Western Technology and Know-How

The trucks ferrying Soviets troops and materiel into Afghanistan in December 1979 confirmed, for conservative critics, the futility of trying to make the Soviets more peaceful. The trucks were built at the Kama River Motor Vehicle Plant (KamAZ), a giant facility covering forty-eight square kilometers, and planned to employ eighty thousand workers and to produce 150,000 trucks and 250,000 diesel engines annually by the early 1980s.[7] KamAZ was equipped with $1.3 billion worth of Western manufacturing technology and engineering assistance, of which some $500 million was acquired in the United States. The Soviets reinforced a strong, independent in-house capability in truck building with shrewd foreign acquisitions, including advanced Western machine tools, highly sophisticated and automated foundries provided by Swindell-Dressler, and an engine assembly line provided by Ingersoll-Rand of a size, production rate, and type of automation that made it unique in the world.[8] Hardware was complemented by know-how. A rotating team of seventy Soviet engineers was assigned to the United States to oversee engineering and design work of the three foundries that they bought there.[9]

The Reagan administration's worst fears were confirmed by a file that French president François Mitterrand handed to the new American president at a summit meeting in Ottawa in April 1981. It contained thousands of KGB (Soviet Committee for State Security) documents, the so-called Farewell papers. Farewell was the codename given by the French administration to Colonel Vetrov, a fifty-three-year-old engineer who had recently defected to the West. To ensure that the public and America's allies were aware of the threat, the CIA put one of its analyses of the Farewell dossier in the congressional record within a year. Another analysis dating from June 1982 was released only in 1999.[10] These two CIA documents gave a remarkably detailed account of the organization of the Soviet acquisitions system and claimed to reveal the full extent of the Soviet intelligence collection efforts for science and technology.[11]

The central claim made by the CIA was that the Soviet Union was making a concerted effort worldwide to secure Western technology, including manufacturing technology, to meet both its defense and its "defense-industrial" needs. While there was a Soviet tradition going back to the 1930s of acquiring technologies from others, there were two major new trends. Firstly, these acquisitions were focused on militarily significant technology, meaning "equipment, material and technology having direct and immediate impact on Soviet military research, development, and production." Secondly the Soviets were targeting new emerging technologies being developed in the West, items like adaptive optics, very high speed integrative circuits, superconductive systems, state-of-the-art computer devices, and genetic engineering and recombinant DNA—items one memorandum characterized as the "most advanced and least protected" in Western countries. It is perhaps not a coincidence that the CIA's characterization of the technology acquired illegally by the Soviets—it was "emergent" and "militarily significant"—was precisely that which was being used by the DoD to identify the types of advanced American technology and know-how that had to be protected by export controls (see chapter 4).

The documents reviewed the range of military hardware acquired by Soviet and East European agents. The list included copies of US air-to-air and surface-to-air missiles, and data on the guidance subsystem of the Minuteman ICBM, on gyros, on solid-propellant missiles, on radars, on tank engines, and on antitank weapons. A table of "notable successes" of legal and illegal technology acquisitions included computer hardware and software; automated and precision manufacturing equipment for electronics, materials and perhaps optical and laser weapons components; complete industrial processes and semiconducting manufacturing equipment to produce high-quality microelectronics; precision machinery for grinding ball-bearings for inertial guidance systems up to a tolerance of twenty-five millionths of an inch (purchased with Nixon's approval from a US supplier);[12] and advanced jet-engine fabrication technology and jet-engine design information. The CIA claimed that the "multifaceted" Soviet acquisition program had saved the country hundreds of millions of dollars in R&D costs, and years in R&D development time. It had helped them modernize critical sectors of their military and reduce engineering risks, achieve greater weapons performance than otherwise, and incorporate countermeasures to Western weapons.

The hawks in Congress were appalled. The president had proposed that the country spend $1.6 trillion over the next few years to counter the Soviet threat. As Bill Armstrong (R-CO) saw it, the United States was "groaning under the strain of financing two military budgets—our own and a significant portion of the Soviet Union's." To drive the point home he surmised that "much of the additional money that must be spent on defense is required to offset Soviet weapons that probably could not have been built without our assistance."[13]

To achieve these objectives the Soviets had built a vast technology acquisitions system that was "well-organized, centrally directed, and growing."[14] The CIA estimated that in the early 1980s the system comprised an extensive complex of "information departments affiliated with Soviet Research institutes, design bureaus and production facilities."[15] They exploited every possible legal and illegal avenue to secure advanced Western information and technology. Unclassified, academic, commercial, and official science and technology agreements and exchange programs, even those arranged with the Soviet Academy of Sciences, along with trade shows and company visits, as well as the published technical literature from over one hundred countries, was used to keep Soviet scientists, engineers, and technicians abreast of foreign developments. Assistant Secretary of Commerce Brady remarked in March 1982 that "operating out of embassies, consulates, and so-called 'business delegations' KGB operatives have blanketed the developed capitalist countries with a network that operates like a gigantic vacuum cleaner sucking up formulas, patents, blueprints and know-how with frightening precision."[16] The legal and illegal collection effort was allegedly so effective that, according to one member of the intelligence community, by the early 1980s Line X, the operating arm of the KGB tasked with collecting what they could from the R&D programs in Western economies, had fulfilled two-thirds to three-fourths of its collection requirements for radar, computers, machine tools, and semiconductors.[17] Table 3 is an extract from an early 1970s official Soviet Requirements List that ran from four hundred to five hundred pages.[18] It gives one an idea of how field agents were informed of the technical data that was needed, how important it was, and how much it was worth (in rubles). It covers university and corporate laboratories, and it targets data and documentation on space systems and weapons manufacturing in the military and in its contractors.

TABLE 3. Excerpts from Soviet Military-Industrial Commission Requirements List (Early 1970s); Three Items, Priorities B-I, B-II, and A-I

Reqmt. No. Priority Validity	Subject of Requirement and Quantity	Max. Ruble Cost per Unit	Known Characteristics and Features	Country/Firm (Date of Production)
T062-0467 B-I 1971	Automatic assembly line for manufacturing noncontact fuse XM596 with integrated circuits, for 40 mm grenade fired from M75 rapid-fire recoilless launchers 1 set of technical documentation	2,000	Folders of working blueprints of the assembly line Technological process of assembling the fuse on the automatic assembly line Source: *Electronic News*, no. 676 (1968), 4–5	USA Harry Diamond Laboratory Universal Instruments and Federal Tool Company
Kh039-0311 B-II 1970	Technical Report No. 6915506 One copy	2,500	Ultraviolet and infrared spectra of free radicals in irradiated polyethylene Source: *Star* 7, no. 5 (1969), 783	USA Northwestern University, Evanston, IL (1969)
A010-0001h A-I 1975	Artificial satellites used for photo- and radiotechnical reconnaissance of ground targets 1 set	5,000	Scientific information and technical documentation: -Photographic equipment— resolution of the camera in relation to altitude and distance of the object: - precision matching of the pictures received to the terrain - film supply - design of the recovery capsule containing the exposed film	USA Customer: AF SAMSO, Los Angeles AF Station, California Firm: Lockheed Aircraft Corporation

The intelligence community deplored the imbalance in the flow of sensitive information and know-how to Soviet visitors and exchange students in US laboratories and universities (a view not shared by the State Department).[19] As Deputy Secretary of Defense Frank Carlucci put it, "in our considered view . . . the [scientific] exchanges to date, in the main, have not been reciprocal. Rather, it is quite apparent the Soviets exploit scientific exchanges as well as a variety of other means in a highly orchestrated, centrally directed effort aimed at gathering the technical information required to enhance their military posture."[20] CIA director William Casey regarded US-Soviet scientific exchange as "a big hole. We send scholars or young people to the Soviet Union to study Pushkin poetry: they send a 45-year old man out of their KGB or defense establishment to exactly the schools and professors who are working on sensitive technologies."[21] In May 1982 Senator William Cohen (R-ME) presented a comparative list, constructed to highlight the stark difference in topics pursued in international exchange visits. US students in the Soviet Union studied topics like musical genres in Russian music from the last half of the sixteenth century to the first half of the eighteenth century; the administration of the Russian Empire under Catherine the Great, 1762 to 1796; and a study of the linguistic basis of Pushkin's iambic tetrameter.[22] Carlucci gave some examples of topics being studied by Soviets in the senior scholar exchange program administered by the International Research and Exchange Board: "Preparation of micro-tunnel diodes in gallium arsenide by annealing and/or molecular beam epitaxy"; "Theory of computer science and programming methodology"; "Thin-film metals in semiconductor technology"; and "Machinability of difficult to machine materials."[23]

Time and again the 1982 memoranda emphasized that, in the past, most Soviet participants in exchange programs "had conducted very basic scientific research. Now almost all of them proposed to study in the emerging scientific fields, with most of these fields having direct and immediate military application."[24] Cutting-edge knowledge gained in face-to-face interactions was especially highly prized. Visits to companies were "among the most important sources of technology loss because of the 'hands-on' experience and collegial working relations with US counterparts gained by Soviet participants."[25] Universities and research centers were targeted by hostile intelligence services that, since the late 1970s, had used "an increasingly large and highly qualified pool of scientific officers with enhanced technical collection capabilities."[26] Research areas were not chosen randomly. Scientists and engineers who participated in academic, commercial, and official science and technology exchanges were briefed by Soviet authorities before they visited the West on what was required and debriefed on their return home.

The restructuration of academic research under way in the 1980s played into the Soviet intelligence services' hands and triggered government action. The 1980 Bayh-Dole Act, and related legislation that authorized university researchers who had government grants to patent the results of their work, pushed them closer to the "D" end of the R&D spectrum.[27] The commercialization of research may have released the entrepreneurial energies of academic researchers. It also enabled increasingly large numbers of graduate students from abroad to do (unclassified) research that enabled them to delve "deeper into applications-oriented research" and to work at "technological frontiers" in dual-use domains.[28] In fact in 1979 foreign nationals obtained about twelve hundred (almost 50 percent) of the doctorate degrees awarded in engineering in US universities.[29] Export controls limiting foreign access to sensitive knowledge were one instrument invoked during the Reagan administration to deal with this situation.

Early Attempts to Control the Dissemination of Knowledge

The CIA reports were part of a broad and ambitious campaign by the Reagan administration to "upgrade its efforts to control the dissemination of unclassified scientific and technical information to foreign nationals."[30] These efforts were essential in the early 1980s for three reasons that are worth recalling here. First, the Bucy Report's emphasis on the need to regulate the circulation of knowledge and know-how highlighted the vulnerability of these exchanges. Second, basic research in dual-use technologies was now central both to securing the US share of global markets and to national security. And thirdly, the transformation of the practice of university research and teaching engaged an increasing number of foreign nationals that worked at the interface between research and development in emerging technologies. As officials from the Department of Commerce put it, government controls were needed because of "the increased level of university involvement in applied research activities over the past several years and the growing numbers of foreign scholars at US academic institutions."[31]

In the last year of the Carter administration, and accelerating after Reagan took office in January 1981, the government made a number of different but interrelated challenges to the free circulation of knowledge. They followed distinct tracks depending on the agency involved. One track was that followed by the Department of State and the Department of Commerce, which used export controls on technical data to intervene in academic programs and conferences in which sensitive information was shared with foreign nationals, especially those from Communist countries. The other track was explored by the National Security Agency (NSA) and the Department of Defense, the latter having a long tradition of sponsoring basic academic research via contracts (and grants) with the universities. The NSA sought acceptable ways to limit the dissemination of knowledge by imposing precirculation constraints on the publication of cryptographic research, while the DoD was focused on militarily sensitive technology. This section will follow each of these tracks in detail, highlighting the often-uncompromising response by academia to the government's sometimes clumsy efforts to control research. The First Amendment protection of "free speech" was an ever-present concern in the background of these conflicts; its pertinence will be briefly explained in a coda to this chapter.

The first startling signs that the regulatory system was being tightened up was the widely publicized last-minute intervention by the Department of Commerce in the First International Conference on Bubble Memory Materials and Process Technology held in February 1980 in Santa Barbara. It was sponsored by the American Vacuum Society (AVS) and was attended primarily by representatives of firms making bubble memories for computers. The technology was on the verge of moving from the prototype to the mass-production stage, and the companies present hoped to agree on standards, specifications, and the means of ensuring reliability of supply.[32]

A few days before the meeting the conference organizers phoned AVS president John Vossen at the RCA Laboratories in Princeton. They had been told by the Department of Commerce, probably at the instigation of the CIA, to disinvite Soviet, Eastern European, and Chinese delegates. All foreign nationals admitted to the meeting would also be obliged to sign a nondisclosure agreement. Just before the conference began Vossen received a letter from the Department of Commerce telling him that, according to section 379 of the Export Administration Regulations, "oral exchanges of information in the US with foreign nationals constitute the export of technical data." A license would be needed to authorize the export of such data at the Bubble Memory Conference, and "foreign importers," that is, those who attended the meeting, would have to agree in writing that such technical data would not be "reexported to proscribed technical destinations." Vossen agreed to disinvite the visitors from the Communist countries only when he was told that he personally, and the AVS, could be fined $10,000 or five times the value of the export, whichever was greater, and/or spend ten years in prison if they did not comply with the law.[33]

The delegates from Hungary, Poland, and the Soviet Union did not come. The Chinese were already in transit and got news that they had been disinvited when they arrived in Santa Barbara. At this point the State Department, who had no objection to the Chinese attending the gathering, overruled the Department of Commerce and authorized their attendance. Roughly thirty foreign nationals attended the conference, about half of them Japanese, as well as the three Chinese visitors. They all signed a letter stating that information obtained at the gathering would not be "divulged to nationals of [eighteen] countries unless prior authorization [was] obtained from the Office of Export Administration."[34] Their attendance had to be licensed and the letter of assurance signed because the Department of Commerce determined that this was not an "open conference," that is, that "the technical data were not generally available to the public and the technical information was scientific data directly and significantly related to design, production or utilization in industrial processes."[35]

There is no question that the constraints imposed by the Export Administration Act and its earlier variants were applicable to this meeting (see chapter 3 for similar practices in the 1950s). There were sessions on crystal growth, grinding technology, and wafer slicing. Manufacturing procedures were discussed at an associated workshop. Frank Press, President Carter's science adviser, rightly insisted that "there are certain laws and regulations that have been on the books for years." These laws had been invoked because bubble technology "was very high technology and was on the embargoed list for export," meaning that both the shipment of devices and oral communication were regulated. It is to be noted that bubble memory computers were also on the MCTL published in the *Federal Register* in October 1980 (see chapter 4).

The research community was dismayed. Herbert Feshbach, the president of the American Physical Society, objected to Secretary of State

Cyrus Vance that the conference dealt exclusively with "scientific information that is either published, or about to be published, in the open literature."[36] Industry representatives pointed out that sensitive manufacturing data were already amply safeguarded by companies' own concern to protect proprietary information.[37] Allan Bromley, the president-elect of the American Association for the Advancement of Science (AAAS), insisted that it was unacceptable for the United States to erect barriers to free circulation when they objected to other countries doing so.[38] Vossen himself described the behavior of the administration as an "unprecedented, frivolous, and foolish exercise of hamhanded bureaucratic power over a technical meeting."[39]

Attempts to change the exercise of bureaucratic power failed, however. The AAAS's Committee on Scientific Freedom and Responsibility called a meeting with officials from the Departments of State and Commerce in April 1980 to have them clarify their policies. This led to "several rounds of good-natured and repetitious arguing," notably over the invocation of export controls to restrict the circulation of information at a scientific meeting—all to little avail.[40] Both parties held their ground, and the government representatives were asked to clarify their positions and to improve communications with the scientific community. Frank Press, for his part, was at pains to emphasize that the scientific community had nothing to fear. The administration had no intention of suppressing publication in scientific publications like the *Journal of Applied Physics*. The government had intervened in the bubble memory conference because the information shared would "enable somebody to build something," and it concerned "actual technical processes" dealing with manufacturing know-how. In addition he had learned at the last minute that one of the foreign delegates was not a bona fide scientist.[41]

In defending the administration's actions, Press disingenuously claimed that "there's no change in Administration policy." There was of course. Intervention at Santa Barbara highlighted the administration's determination to tighten up technology export controls with the Soviet Union in the wake of the invasion of Afghanistan, and of the internal exile of Andrei Sakharov. Press himself announced that the State Department's détente-era formal US-Soviet bilateral scientific agreements were being "deferred" if they were in "highly visible areas." The new administration of President Reagan maintained the pressure. In mid-January 1981 the Department of Energy instructed the directors of the US national laboratories, and Department of Energy contractors, to clear in advance all anticipated discussions with Soviet citizens. In late February the State Department refused to allow six Soviet bloc nationals—five coming from abroad, one a postdoc at the University of Texas—to attend a conference in San Diego on lasers and electro-optical systems held jointly with one on inertial confinement fusion because they included a large exhibit of scientific equipment.[42] These were sensitive fields, and a license would be needed before allowing the "visual" export of information about the instruments to foreign nationals from the Soviet Union.

Research trips by scholars in Communist countries to American campuses were also targeted. In June 1981 the several universities that had been asked to host a Soviet visitor, Mikhail Gololobov, in the framework of a US-Soviet interacademy exchange program were advised by the State Department that his professional activities had to focus on fundamental research that had been published in the open literature. Gololobov was not to be given visual, documentary, or verbal access to activities funded by DoD contracts or grants. One of his host institutions, MIT, was told in November that, in addition, he should be denied access to research in genetic engineering, as well as to work in nutritional research and the production of food supplements. One senior faculty member found the latter to be "so outrageous as to be incredible," all the more so as Gololobov was invited by the Commission of International Relations of the National Academy of Sciences. The provost filed a protest with Frank Press, now president of the National Academy of Sciences.[43]

The scenario was repeated in December 1981. Institutions hosting Dr. N. V. Umnov, an expert on robotics, were told to restrict his activities to theoretical research, and to stop him visiting industrial facilities.[44] The activities of a Chinese visitor, Qi Yulu, assigned to professor W. R. Franta in the Department of Computer Science at the University of Minnesota, were also severely restricted.[45] Franta was collaborating at the time with industrial partners at Honeywell in the design and implementation of hardware and software for real-time distributed computer systems. Keith Powell, in the Office of Chinese Affairs in the State Department, was concerned that Qi Yulu would have access to technology covered by an export license, notably "computer software technology," this being "an area with military applications." The letter also wanted more detail on Qi Yulu's "planned program of study and research," which was also subject to export control regulations.

Powell was at pains to explain that Qi Yulu should not break laws regulating the export of technical data.[46] The Chinese visitor, he warned, should have no access to "unpublished or classified government-funded work"; his program was directed to "emphasize coursework with minimum involvement in applied research," and he should have "no access to the design, construction or maintenance data relevant to individual items of computer hardware [or] to source codes or their development." Powell authorized the university to provide Qi Yulu with "as full an academic program as possible" within this framework, even while restricting what he could see and use to "the published software for operating systems subroutines."[47]

The Departments of State and Commerce did not act alone in impeding foreign nationals exporting sensitive knowledge. In 1982 the Reagan administration formally authorized the Commerce Department to work closely with the US Customs Service to implement export controls in a program called Operation Exodus. The primary objectives of Exodus were "(1) to stop the outflow of critical technology to the Soviets and to enforce compliance with export laws by intercepting or seizing shipments of items and of data that were being exported illegally; and (2) to disrupt groups and individuals responsible for these illegal exports by arrests, prosecutions and other legal sanctions."[48] In May 1982, Operation Exodus provided the official backing for the action of federal customs officials who summoned five Chinese scholars and engineering students from their seats just before their plane left New York, bound for China. They were told that the authorities were confiscating some suspicious-looking items in their luggage. These included articles from scientific journals, classroom notebooks, theses and lecture materials, slides, computer software, and tapes of rock music—all of which were later returned to the visitors.[49]

Leading members of the academic community protested vehemently against the use of export controls to restrict the activities of foreign visitors. On February 27, 1981, the presidents of five major American research universities (David Saxon, University of California; Marvin Goldberger, Caltech; Frank Rhodes, Cornell; Paul Gray, MIT; and Donald Kennedy, Stanford) addressed a joint letter to the secretaries of the Departments of Commerce, Defense, and State.[50] Their goal was to persuade the authorities that export controls were "not intended to limit academic exchange arising from unclassified research and teaching." If export controls on knowledge circulation became policy, they wrote, it could mean that "faculty could not conduct classroom lectures when foreign students were present, engage in the exchange of information with foreign visitors, present papers or participate in discussions at symposia or conferences where foreign nationals were present, employ foreign nationals to work in their laboratories, or publish research findings in the open literature." Universities would also be barred from admitting foreign nationals into many graduate programs.[51]

The five presidents opposed export controls on both pragmatic and principled grounds. They noted that securing compliance was impossible given the "necessarily decentralized and fluid nature of most campuses." It was also incompatible with the organization of universities, which were "neither structured nor staffed to police the flow of legitimate visitors to a given laboratory or the dissemination of information by the faculty at international conferences, or, indeed even in the campus classroom where foreign students happened to be present." They would also "conflict with the fundamental precepts that define the role and operation of this nation's universities" by restricting publication and discourse among scholars, and by discriminating on the basis of nationality against the employment of faculty and the admission of students and visiting scholars.[52] In October 1981 Peter McGrath, the president of the University of Minnesota, echoed their sentiments. He objected strongly to the State Department's attempt to limit the activities of Qi Yulu on the basis of his Chinese citizenship. The mission of an American research university, he wrote to Powell, was "teaching, research and public service, and neither our faculty nor our administrators were hired to implement government security restrictions." The restrictions advocated by the State Department could only have "a chilling effect upon the academic enterprise" and threatened the "integrity of academic principles, traditions and obligations that [were] the foundation of education in a democratic society."[53] Francis Low, the provost of MIT, also objected that the restrictions imposed on Gololobov's visit were "inconsistent with the spirit and practice of a university as an open community of scholars, teachers, and students."[54]

The DoD ran into the same difficulties as did the Departments of Commerce and State when it proposed constraints on access and publication in a new federally sponsored research program in very high speed integrated circuits (VHSIC). In March 1980 nine contractors were funded to explore the technical requirements for developing and manufacturing 1.25-micron silicon chips. On December 12, 1980, the program director informed scientists in participating universities that, although basic research and its results were not generally controlled, it was also "the preference of the Program Office that only US citizens and immigrant aliens who [had] declared their intention of becoming citizens participate."[55] The problem arose because the line between basic and applied was difficult to draw in this case, and the devices and technical data related to process or utilization technology (as opposed to basic research) would be subject to ITAR and EAR. In similar vein, at about the same time at least one university was told by the Department of Commerce that some foreign nationals should not be allowed to participate in its sponsored research programs on account of their citizenship.[56]

Stanford president Donald Kennedy and his four colleagues singled out the VHSIC program to reinforce their opposition to invoking export controls on academic research. To get around them the director of the VHSIC program office suggested limiting the dissemination only of applied research. The five presidents insisted that the basic/applied distinction was unworkable in areas such as device design and fabrication techniques, process equipment, and software—areas where the government wanted to limit publication or the presentation of results at conferences. For them the operative distinction was classified/unclassified, not basic/applied. Unclassified parts of the research should be conducted in universities. The most sensitive questions should be dealt with in the government's dedicated classified research facilities. The five presidents reiterated their view that "restrictive and virtually unenforceable" export control regulations had no place in academia.[57]

The letter from the five presidents was published in the scientific and educational press to the consternation of the government.[58] Two DoD officials visited each of the five to get a better sense of their concerns and to reassure them that the department had no intention of damaging their research activities. The managers of the VHSIC program also called a meeting of investigators to discuss how to prevent the transfer of technology to the Soviet Union. Even though no conclusions were reached, it was clear that severe differences of opinion over the use of export controls to restrict the circulation of technical data could seriously damage the DoD's plans to work closely with university scientists and engineers in their R&D programs. An alternative approach to restricting the dissemination of basic research in cryptography was then being explored by the National Security Agency, though it too quickly ran into opposition on both pragmatic and constitutional grounds.

The NSA and Cryptography

In the mid-1970s university studies in cryptography and related areas of mathematics came to the attention of the National Security Agency.

Cryptography is a system of information protection. It consists of methods for transforming data by a process called encryption, which renders it unintelligible to someone not authorized to have it.[59] This was a dual-use technology that was invaluable both to the intelligence community and to civilian entities like banks. There was a long history of coding sensitive information in wartime and in diplomacy. At the dawn of the information age there was also a growing demand for data protection by the private sector. Unclassified sensitive information contained in personal records, in commercial and financial exchanges, in electronic mail, in electronic funds transfer, and the like also had to be protected from unauthorized monitoring or use. Restrictions impeding the sharing of research on new methods of encryption, imposed in the name of national security, could thus undermine the ability of US industry to compete effectively in world telecommunications and data-processing markets.

In October 1977 the University of Wisconsin at Milwaukee filed a patent application for an encryption device developed by George Davida, a professor in electrical engineering and computer science. Six months later he was informed by the Patent and Trademark Office that the Invention Secrecy Act of 1951 had been invoked, and that he would be subject to a fine of $10,000 and two years in prison if he divulged the principles of his invention to anyone but federal agents. The same fate befell three engineers in Seattle soon thereafter who filed an application for a patent on a simple voice scrambler, and received a secrecy order instead. The subsequent outcry led to the removal of both restrictions within a few months.[60] The director of the NSA subsequently claimed that the issuance of the first order was a bureaucratic error. He also acknowledged that he had personally authorized the imposition of the order on the voice scrambler but declined to explain why the order had been lifted.[61]

In 1978 Admiral Bobby Inman, director of the NSA since July 1977, had spoken openly of his fears that the uncontrolled dissemination of basic research in cryptography undermined the agency's mission and was a threat to national security.[62] It would not only enhance the capacity of foreign governments to protect their data more effectively, making it more difficult for the NSA to access them. It would also enable foreign governments to penetrate more easily into the secure telecommunications of the US government.

Two steps were taken to deal with Inman's concerns. First, the National Science Foundation (NSF) agreed to send any proposals for funding research in cryptography to the NSA for review, reserving the right to fund such research at its own discretion.[63] Secondly, the American Council of Education, with NSF funding, established the Public Cryptography Study Group, consisting of nine people, including representatives from the NSA and from the academic community. It first met on March 31, 1980, to deal with "the dilemma of reconciling important First Amendment rights with the NSA's concern for the nation's communications security and intelligence-gathering activities."[64]

The study group limited its deliberations to controls on the domestic dissemination of nongovernmental technical information relating to cryptology. (Foreign dissemination of cryptographic equipment and related technical information were subject to ITAR.)[65] Inman suggested that authors should be *obliged* to submit their work for prepublication review; publishing without prior clearance by a designated agency, like the NSA, would be a criminal offense. To reconcile prior review with the First Amendment, Inman stipulated that only technical advances that would clearly put national security at risk would be restricted. The burden of proof that this was the case would lie with the government, not the researcher as proposed by the NSF.[66]

When Inman made this proposal some authors, institutions, and publishers were already voluntarily submitting proposed publications for review and comment by the NSA as to the sensitivity of the information involved (see chapter 3 for early Cold War attempts to use voluntary cooperation as a means of export control). MIT researchers, for example, routinely sent all their cryptography-related research results to the NSA for information at the same time as they sent them out for peer review to others in the field.[67] However, the NSA had no statutory authority to require submission of proposed publications for review, or to require changes in publications prepared outside the agency and not under an NSA contract or grant. This was a loophole that Inman wanted to close.

The study group's report was released in February 1981. It was opposed to Inman's suggestions. The group noted that, on the basis of past decisions, the Supreme Court was more than likely to oppose prepublication constraints, and would require the government to make an extremely strong case if it wanted to uphold a mandatory policy in the face of the First Amendment. All the same the majority admitted that Inman had a legitimate concern. To resolve the dilemma the group recommended that the agency adopt an alternative nonstatutory system, on a trial basis, in which authors voluntarily submitted their papers for prepublication review to the NSA. They suspected that many people in the field would welcome being told in advance that their findings would substantially risk compromising US national security interests. They were emphatic, all the same, that "there would be a clear understanding that submission to the [prepublication review] process is voluntary and neither authors nor publishers will be required to comply with suggestions or restrictions urged by NSA."[68]

Even this "soft" compromise was controversial. George Davida, who was one of the nine members of the study group, submitted a minority report opposing prior publication review by the government on the grounds that it was likely to be found unconstitutional by the Supreme Court. He also believed that the procedure would diminish the quality and direction of basic research in computer science, engineering, and mathematics. He was particularly concerned that, if the trial period produced results deemed unsatisfactory by the NSA, the government would call for new legislation, using the study group's recommendation as expert testimony in favor of restraints. In Davida's view the NSA's attempt to control cryptography was "unnecessary, divisive, wasteful and chilling," and he urged the agency to perform its mission in "the old-fashioned way: Stay ahead of others."[69] A few months later his opposition was implicitly endorsed by the National Science Foundation's Advisory Committee for Mathematical and Computer Sciences on the grounds that the NSA policy would discourage US researchers from staying at the forefront in public sector uses of cryptology. Like Davida, they were opposed to the agency vetting papers prior to publication: scientists need send their manuscripts to the National Security Agency only for information purposes. Another voice was added to the growing chorus of opposition toward the end of 1981, when the Association of American University Professors (AAUP) suggested that the Cryptography Study Group's proposals would impugn academic freedom.[70] There the matter rested. The problem of how to control the loss of sensitive knowledge in the face of strong academic opposition remained, however. The Department of Defense tried another approach that would at least get around the constitutional objection.

The DoD Woos Back the Research Community

In 1980 and 1981 the DoD's Defense Science Board sponsored summer studies that dealt with the declining health of the nation's technological base. This was an issue of great concern to the House Armed Services Committee, whose hearings in April 1981 asked the DoD to explore ways of harnessing university research to defense needs without compromising national security or antagonizing the research community. In October 1981 Richard DeLauer, the undersecretary of defense for research and engineering formally constituted the Defense Science Board Task Force on University Responsiveness to National Security Requirements.[71] It made a number of suggestions that were to influence the debate on the role of government controls in academia for the next five years.

The immediate aim of the task force was to propose ways to rebuild links with the research community that had been strained a decade earlier. The relationship had been damaged politically by opposition to DoD-sponsored research on campus during the Vietnam War. It had been impacted financially by the Mansfield Amendment, adopted by Congress in 1969, discouraging the DoD from using its funds "to carry out any research project or study unless [it had] a direct and apparent relationship to a specific military function," even though these tight constraints were subsequently loosened. [72] The first question discussed by DeLauer's task force embodied these concerns. It asked, "Is there real university interest in performing classified and unclassified research with clear-cut DoD application and sponsorship? If so, are the conditions under which this research will be performed compatible with national security interests? If not, what steps can be taken (by either the DoD or the universities) to improve the situation?"[73] The task force was convinced that, with the Vietnam War a fading memory, the political climate had changed such that there was little principled opposition to accepting funding for projects that had a clear-cut DoD application. But the task force was concerned that neither industry nor academia was any longer able to "keep pace in their abilities to support the nation's increasingly complex and sophisticated defense needs."[74] A steady decline in DoD funding, critical shortages in university equipment and facilities, a shortage of highly skilled "manpower"—these were some of the factors eroding the US technological base. In fact, the task force found that "the US lags far behind the western democracies and Japan, no less the Soviet Union, in general science, in mathematics and in engineering education."[75]

In their wide-ranging analysis of the situation, the task force identified two main issues requiring attention (in addition to the need for sustained financial assistance by the DoD and other federal agencies). Firstly, as we mentioned earlier, there was the high percentage of foreign students in graduate science and, especially, graduate engineering programs. For example, in 1969 one out of ten doctoral students in engineering was a foreign national; in 1979 the ratio had grown to two out of three.[76] Export controls were one way of controlling their access to sensitive unclassified research.

This sensitivity was the second major, urgent, and highly visible issue. There were several reasons why the DoD argued that it had no option but to demand compliance with ITAR and EAR in sponsored research on campuses. There was Bucy's shift from controlling hardware to controlling know-how. There was the shift in emphasis toward dual-use applied research in university laboratories. And there were the constraints imposed by the DoD's MCTL, which contained "over 620 technology titles with literally thousands of critical elements specified under those titles," many of which were found on campuses.[77] The task force knew that this would be unpopular. But they hoped that the university community would come to understand that some "information, technologies and critical elements were important militarily and should be subject to some form of review and ultimately of control." They might then be more willing to accept export controls, enabling the two parties together to "work out mutually acceptable terms for safeguarding critical information."[78]

What domain of research needed controlling? The task force "assiduously reject[ed] any control guidelines that restrain the development and dissemination of the fruits of basic research."[79] It identified one broad category of research that was subject to ITAR and to EAR for which a license was needed: "manufacturing and process-oriented research (as opposed to basic research)" in DoD-funded unclassified research projects. Though they did not use this term, we will call this a "gray zone" following the well-established Cold War notion later used again by the Corson panel and discussed in chapter 6.

The DoD had one instrument it could use to restrict the circulation of research in the gray zone using ITAR and EAR: the *sponsored research contract*. As the task force pointed out, "if DoD is funding the research, it is reasonable that DoD could, in turn, monitor for national security purposes the flow of information and technical data emanating from the research."[80] Why not then build the constraints imposed by export controls and the MCTL into the contract itself? As the task force put it, "The focal point for control is the DoD contract; the government negotiates the terms of the release of information with the contractor. The Project Office or Contract Monitor within the DoD thus becomes the interpreter of militarily criticality and the extent to which ITAR or EAR is applicable.

The system is voluntary in the sense that the contract does not have to be accepted. If guidelines for release of information are accepted as part of the contract, then there would be little room for misunderstanding later."[81]

This procedure respected First Amendment considerations, if only implicitly, and should satisfy academia. It thus allowed for a healthy dialogue within the scientific community. The task force saw no difference in principle from the situation that arose in industrially funded university research. Corporations did not object to professors teaching basic science and technology. And in both cases the researcher was not obliged to accept the contract offered by the sponsor.

The proposed system was bureaucratically lean. Since the terms of the contract were negotiated in advance between the DoD and the university administration, in consultation with the Departments of State and Commerce, both parties would know what was expected in terms of information release. The DoD would apply the filter of military criticality and assess the relevance of ITAR and EAR to the proposed research. The Departments of State and Commerce would not be flooded with unnecessary license applications, and those that did cross their desks would be more focused and so easily resolved.

What could be done to regulate the access of *foreign nationals* to gray zones of research? The task force rejected the use of visas as a regulatory instrument, quoting verbatim the objections raised by the five presidents in their February 1981 letter to using universities "to police the flow of visitors." Instead the research contract would be used to stipulate the restrictions to be respected by the university. If it seemed that some parts of the work could produce ITAR-protected data, the university could be asked to assign only US citizens and immigrant aliens (green card holders) to continue working on it. Foreign nationals doing research in the other parts of the project could be asked to declare that "they do not intend to expatriate their acquired knowledge."[82]

The DoD saw prepublication review as an essential component of this control system, authorized by virtue of its sponsorship of the research. It was presented as an instrument to streamline knowledge production in sensitive domains that might be subject to export controls. As the task force put it, "In the course of the contract work, pre-publication review would allow for a contractor to change or modify the presentation of technical data so that it could be releasable to the public without going through the licensing process. Pre-review could in a sense be the DoD's mechanism for interpreting ITAR for the universities and may be less onerous than requiring universities to submit formal license requests to State (ITAR) or Commerce (EAR)."[83]

Prepublication review was also "the most sensitive area" from a constitutional point of view, "and one which the DoD should approach with the utmost caution." As the task force emphasized, "an overly ambitious program of information control could easily end up in the courts." To avoid First Amendment challenges the DoD contract should emphasize the voluntary nature of the program, stress the principle of peer review, explain that there was a mechanism for appeal if researchers felt that the restricting guidelines had been unfairly imposed, and point out that there was a time limit for the government to decide if there was a problem with the results (thirty to sixty days). Absent indications to the contrary, researchers were free to publish their work once the specified period had run its course.[84]

On January 27, 1982, the chairman of the task force submitted his report to Norman Augustine, who chaired the Defense Science Board in the office of the secretary of defense. He urged the DoD to act quickly on its recommendations, notably in the area of export control. "The public at large and the university community in particular," he wrote, "are becoming increasingly alarmed by the prospect of government overreaction which in the end may harm the Department's chances of establishing a strong and healthy relationship with the universities."[85] That "overreaction" was already underway at the time and precipitated an ill-tempered response from a senior representative of the scientific community.

The CIA Muddies the Waters

Early in January 1982 Admiral Bobby Inman, now the deputy director of the CIA, participated in a panel discussion entitled "Striking a Balance: Scientific Freedom and National Security" at the annual meeting of the AAAS.[86] Inman was palpably irritated by what he saw as exaggerated and dangerous demands for scientific freedom. The tension between it and national security, in his eyes, resulted from "the scientist's desire for unconstrained research and publication, on the one hand, and the federal government's need to protect certain information from foreign adversaries who might use this information against this nation," on the other. Inman made a point of emphasizing the "spectacular" service scientists had rendered to the nation in war and in peace. He recognized that restrictions on science and technology should be considered only "for the most serious reasons." But he expressed disdain for the scientific community's "blanket claims of scientific freedom." It was "hollow" to suggest that "national security should not have an impact on 'scientific freedom,'" granted the long history of a symbiotic relationship between them. "Scientists," Inman said, "do not immunize themselves from social responsibility simply because they are engaged in a scientific pursuit." He was particularly irritated by their attitude to the federal government. Inman resented scientists' willingness to accept restraints on publication imposed by corporate sponsors while objecting to restrictions imposed in the name of national security just because it was the federal government that applied them. The frequently heard objection that the government had not made its case for restrictions stemmed from "a basic attitude that the government and its public servants cannot be trusted."

Inman stressed that every major foreign intelligence service was collecting data on technical subjects in the United States. He identified areas like "computer hardware and software, other electronic gear and techniques, lasers, crop projections and manufacturing procedures," where publication of certain technical data could harm national security. Research in cryptography was another extremely sensitive dual-use field, in which the "indiscriminate publication" of results "could cause irreversible and unnecessary harm to US national security interests." It was also one area of "special, long-standing concern" to him. The dialogue on voluntary constraint between scientists and public servants that he had initiated as director of the NSA had yielded "reasonable and fair results" to date.

Winding up his speech Inman said that a "joint search" for "workable and just solutions" satisfying the needs of both science and national security was essential.

To start the conversation, Inman proposed shifting government oversight deep into the research process, and well beyond the voluntary level of constraint proposed on an experimental basis by the majority of the Public Cryptography Study Group in February 1981. The way forward, he said, "may lie in an agreement to include in the peer review process (prior to the start of research and prior to publication) the question of potential harm to the nation." This system should both protect national security and ensure that no "unreasonable restrictions" were imposed on research, publication, or the use of the results.

Inman provoked the research community even more in the discussion that followed. He remarked that the intelligence community had evidence of a "hemorrhage of the country's technology" to the Soviet military system—an implicit reference to the CIA reports we discussed above, one of which was to be released in April 1982. The scientific community's antipathy to regulation, Inman warned, "was about to be wiped away by a tidal wave" of public outrage when the "depth and degree of technology transfer" to the Soviets was revealed by this report. This could precipitate, before the decade was out, an expansion of the definition of technologies that needed to be controlled to include "economic competition terms, not just military" (as indeed happened; see chapter 7). He urged the AAAS to encourage its members to engage in dialogue with the pertinent government agencies at once, before the administration "overreacted" to a widespread "public outcry, How the hell did that happen?"[87]

Inman was certainly attuned to the deliberations of the DoD task force even if he had perhaps not seen their report (it had not been released when he spoke). There was a major difference in tone however: the DoD went out of its way to be sensitive to the fears of academia, while Inman was losing patience with them and deeply resented their antistatist posture. His refusal to consider academic opposition to a voluntary system of prepublication vetting of cryptography manuscripts, and to push government surveillance of research even more deeply into the federally sponsored research process, indicated his determination not to be deflected by academic sensitivities. Assistant Secretary of Commerce for Trade Administration Lawrence J. Brady took his message to its logical conclusion while speaking to the Association of Former Intelligence Officers in Washington, DC, in March 1982. The Soviets, Brady said, were able "to exploit the 'soft underbelly'" of American openness, including "the desire of academia to jealously preserve its prerogatives as a community of scholars unencumbered by government regulation." And he wondered aloud whether these freedoms, as basic as they were to the American way of life, had to be preserved. It was "time to ask what price we must pay if we are unable to protect our secrets," Brady said.[88] Brady, like Inman, sent a clear warning to universities that some people in the Reagan administration were frustrated by their principled appeals to academic freedom and were willing to launch a major challenge to it in the name of national security.

William Carey, the executive officer of the AAAS, replied to these threats, notably those made by Inman, with uncompromising hostility.[89] He implied that CIA had taken a leaf out of the Soviet KGB's book, showing a similar penchant for secrecy and censorship. He insisted that American scientists working at the leading edge of their fields were "better equipped than CIA functionaries" to judge when their research touched on national security interests. He objected strongly to the idea that American scientists should submit their work to intelligence agencies "prior to the start of their research and prior to publication," as suggested by Inman, so "censoring scientific research at its points of origin." Carey expressed the same antagonism to government intervention as Inman had expressed to scientific freedom, going so far as to say that it was humiliating for scientists to accept CIA restrictions on their research. Scientists, Carey said, would not "easily accept the shame of prostrating their minds and their work at the doors of the intelligence community." He repeated the widespread view in the scientific community that security lay in achievement, which in turn depended on the open circulation of ideas, whereas for the government, and the CIA in particular, security lay in secrecy that stifled scientific exchange.

Mutual antagonism and suspicion of this kind were no basis for framing consensus. But they clarified once again that the issues at stake touched deep-seated values for partisans on both sides of the divide. There was common ground: both agreed that both national security and academic freedom had to be protected. The challenge lay in coming up with arrangements that satisfied both parties and that respected the First Amendment. This challenge was taken on by a panel set up by the National Academies whose task was to strike a balance between scientific communication and national security. Their deliberations, and the solution they proposed to resolve the thorny issues discussed in this chapter, will be the subject of the next.

Coda: The First Amendment, Export Controls, and the Freedom to Publish Scientific Results

Scholars in academia may not be aware of the significance of the First Amendment to the US Constitution to their publication practices. Yet, as we have just seen, it was appealed to time and again to limit the government's ability to control the publication of certain categories of research findings. For those unfamiliar with this issue, we will here briefly describe the debate among some legal experts over the First Amendment right to freedom of publication, which was a hot-button issue in the 1980s, and which had an important impact on framing policies to regulate the circulation of scientific knowledge and know-how.[90]

In the 1980s legal scholars were emphasizing that the courts typically resolved cases involving the First Amendment by "balancing" the interests of freedom of speech and the press against other important interests at stake in the case at hand.[91] For much of its history the Supreme Court had used the criterion whether the speech at issue posed a "clear and present danger" in order to decide whether it was protected by the First Amendment or warranted some type of government regulation. In the mid-1980s it concluded that this one standard alone could not deal with the variety of modes of speech posing First Amendment problems. This led authorities to adopt what attorney James Ferguson described as "a hierarchical view of the First Amendment—a view that assigns different levels of constitutional protection to different categories of speech."[92] Thus, in a series of decisions around the time of his writing (1985), the court held that "political speech" warranted full protection, and "commercial speech" (e.g., in advertising specific products) warranted only an intermediate level of protection, while "sexually-explicit" speech warranted no protection at all from government restraints. Where did "scientific and technological" speech fit on this sliding scale of First Amendment protection, in Ferguson's view?

The First Amendment divides the entire range of human behavior into two broad categories: protected "expression" and unprotected "conduct." This distinction between protected "speech" and government-restricted "action" was blurred in the case of technical knowledge. This is because there is a causal link between "knowledge" and "action" in the case of "technological speech." Ferguson explains it thus: "Typically a body of technical knowledge will enable a nation or enterprise to pursue a course of action in ways not previously possible for that nation or enterprise. More broadly, a major advance in technological knowledge will often introduce a new form of material power, and thereby enlarge the potential range of human action."[93] The most dramatic example of this arose in 1979 in *United States v. Progressive Inc*. In this case a district court ruled against the publication of a magazine article by freelance journalist Howard Morland on how to design and manufacture a hydrogen bomb. The court granted a preliminary injunction barring publication as a breach of national security, even though Morland's article contained no officially classified information at all. As the court put it, *prior restraint* was acceptable because "what is involved here is information dealing with the most destructive weapon in the history of mankind, information of sufficient destructive potential to nullify the right to free speech, and to endanger the right to life itself."[94] Of course, the extreme nature of the threat limited the significance of the ruling as creating a legal precedent. However it clearly illustrated the government's growing ambitions to regulate scientific and technological expression in specific cases.

With this in mind, Ferguson appealed to a series of other Supreme Court decisions for guidance on where the boundary lay between the intangible realm of ideas and the material world of action on which the First Amendment rested. He concluded that "the broad category of technological knowledge appears to warrant a full measure of constitutional protection," except for the "narrow class of technical data on the design of military weapons," as happened with Morland's article.[95] This did not mean that the government could not impose restrictions on the dissemination of all other technological knowledge in the name of national security. It meant only that it carried a heavy burden of proof that such restrictions were not unconstitutional. This burden was all the more onerous in the case of prior or prepublication restraint, where the government had to demonstrate a "compelling" interest in regulation. As Thomas Emerson puts it, "The doctrine forbidding prior restraint is one of the major underpinnings of the system of freedom of expression [whose] roots go back to the English censorship laws against which John Milton protested. [The doctrine] holds that the government may not . . . prohibit or restrict expression in advance of publication, even though the material published may be subject to subsequent punishment."[96]

Did the use of export controls to oblige foreign nationals to withdraw papers from conferences, or to deny a foreign student access to sensitive unclassified information on campus, amount to cases of prior restraint warranting protection by the First Amendment? We cite three different views to give one an idea of the complexity of the issue and to foreshadow the discussion on export controls in the next chapter.

David Wilson thought that the application of export controls to research contexts would be constitutional only in quite specific situations.[97] Wilson, a political scientist and executive assistant to the president, University of California in Los Angeles, served on a DoD-university forum on export controls.[98] He discussed the legality of using export controls to regulate teaching foreign nationals, to stop the presentation of papers at conferences attended by foreign nationals, and to control the publication of articles in journals accessible at home and abroad by foreign nationals. Wilson concluded that "there seems little doubt in the opinion of jurists that licensing requirements for any such activities would constitute prior restraint on freedom of speech and publication, a governmental action of extreme dubiety under the doctrines of the First Amendment tradition." Without a clear and well-defined standard for immediate and intentional damage to national security, Wilson wrote, there seemed to be "a widely held legal view" that the application of EAR as well as ITAR "to scientific and engineering research for the most part would be unconstitutional and unlawful."[99]

Edward Gerjuoy, a physicist and editor in chief of the journal *Jurimetrics*, was less certain, at least as regards the publication of results. He argued that even basic science did not warrant full First Amendment protection. To make his case he drew on his knowledge of physics to argue that the distinction between "basic" and "applied" science "lies in the eye of the beholder." Gerjuoy noted, for example, that an abstruse mathematical problem—how to determine the prime factors of very large numbers—had major implications for encryption systems.[100] He also devised a hypothetical, yet plausible, example in which basic research into the absorption properties of water could have major implications for the vulnerability of the United States' nuclear submarine fleet without the researcher even realizing it. Gerjuoy surmised that, if the Supreme Court was convinced that export controls were wise government policy, the government could produce enough examples of this kind to convince the court to uphold "even those features of export controls which impose prior restraints on basic research publication" in the name of national security.[101] Universities would also have great difficulty convincing the court that prepublication review of "basic" research—which would involve only brief delays in publication—was unduly restrictive since they willingly accepted such delays in research collaborations with private industry.[102]

United States v. Edler was one of the rare cases that established judiciary precedent.[103] In 1976 an American corporation and its president were convicted of exporting technical data related to items on the United States Munitions List without securing a license.[104] Edler had offered to provide technical assistance to French firms on tape-wrapping techniques for durable lightweight materials and carbon composites that had both commercial and military applications (e.g., in the production of golf club shafts and of missile casings, respectively). Edler pleaded that the export of technical data was protected by the First Amendment even though its application for an export license on the hardware itself had been denied. The court did not agree. However, in doing so, it severely restricted the scope of the ITAR as regards regulating technical data. It stipulated that, to be subject to restrictions, technical data had to be "significantly and directly related to specific articles on the Munitions List."[105] It added that, when dual-use technology was concerned, as in this case, the exporter also had to know, or have reason to know, that the data was for a prohibited military use. This narrowing of the reach of the ITAR was intended to protect First Amendment rights, and the court noted that a broad interpretation of controlled "technical data" would "seriously impede scientific research and publishing and international scientific exchange."[106] This case lies between Wilson's view that export controls would not apply to most academic activities, and Gerjuoy's view that, if the Supreme Court favored the use of export controls, it could decide to restrict the circulation of basic research under certain tight conditions.

The political context in which the legal debate over the constitutionality of prior restraint took place bears mentioning. John Shattuck provided a particularly graphic analysis for the congressional subcommittee that discussed civil liberties and the national security state in 1983–84. Shattuck had served in various legal capacities at the American Civil Liberties Union from 1971 until 1984, when he became vice president of Harvard University. Harking back to the early 1970s, he argued that the failed attempt by the Nixon administration to block the publication of the Pentagon Papers in 1971, while celebrated as a victory for the freedom of the press, also "set in motion the development of a formal law of national security secrecy." This was because in its ruling "the Supreme Court abandoned the longstanding limitation of prior restraints on publication to narrow wartime circumstances." It also enabled Congress to pass statutes authorizing prior restraint, lowering further the standards for imposing government controls over the circulation of information. "The cat was out of the bag," Shattuck argued, and a succession of post-Watergate cases "transformed it into a tiger with a ravenous appetite for the first amendment."[107] *Snepp v. United States* was just one example of this.[108] By establishing a new principle based on the law of contract it forestalled these ambiguities as regards federally sponsored research.[109]

When Franck Snepp joined the CIA in 1968 he signed an agreement that he would not publish any information concerning its activities without the agency's approval. Two years after leaving the CIA he published an account of its activities in Vietnam. It contained no classified information, and he did not seek the agency's approval to release his book. Snepp was accused by the government of violating his contract. In response Snepp argued, inter alia, that the prepublication review clauses in his contract with the CIA violated his right of free speech under the First Amendment.

The dispute between Snepp and the CIA was brought before the Supreme Court. It upheld the government's position in its entirety. The court held that since Snepp had voluntarily signed a prepublication review when he accepted employment with the CIA, he had forfeited his First Amendment right to publish freely. This ruling probably encouraged the Reagan administration to include prior review clauses in any contracts signed with the government, including those sponsoring research. Ferguson suggests that it was on solid ground in doing so. As he put it, "the Supreme Court has consistently upheld the validity of the agreements in which individuals have accepted restraints on a constitutional right in exchange for some benefit from the government."

Once individuals preferred the benefits of an agreement with the government to the unfettered exercise of a constitutional right, they lost considerable power to challenge governmental restrictions of that right. In this case the government would no longer have to demonstrate a "compelling" or even a "substantial" reason for including prepublication review provisions in sponsored research in militarily sensitive areas. It was enough for the government to argue that the restriction used was a "reasonable means" of furthering the government's aims, here to protect national security. The state's position was further strengthened because, as in any prepublication review system, the court simply assumed that the government was in a far better position than an individual to assess *in advance* the national security significance of sensitive information, as is needed in prepublication review. As Ferguson puts it, "where competency is lacking and the stakes are high . . . the Court would be strongly inclined to pay deference to the expert opinions of government representatives."[110]

To conclude, "technical speech" warranted a considerable, but not unconditional protection from government intrusion under the sliding scale in place for assessing one's First Amendment rights in the early 1980s. For this book the issue of prior restraint was the core constitutional question surrounding the government's attempts to restrict access to, or the circulation of, technical data by the research community in the early 1980s. Censorship of unclassified research publications by the government was inadmissible. Legal experts, however, differed over whether the ITAR and the EAR could be used to restrict the participation of foreign nationals in research activities, and to impose less drastic (than classification) prepublication restraints on members of the research community—though most thought that the government would have considerable difficulty doing so. There was a consensus though that the government could constitutionally restrict access to research and constrain the publication of its results by including these conditions into contracts for federally sponsored research. Here the researcher diluted his or her First Amendment protection in return for federal sponsorship of his or her research. This provided a legal path for establishing controls over the publication of unclassified research findings.

CHAPTER SIX

Academia Fights Back

The Corson Panel and the Fundamental Research Exclusion

The administration's "broad, clumsy and unexplained attempt to impose restrictions on the flow of scientific information," which we described in chapter 5, and the hostile response from the academic community, "started a complicated discussion between the scientific community and the government."[1] The DoD was the driving force of this discussion. As we saw in the previous chapter, its Defense Science Board task force was anxious to engage universities in a dialogue on the national security implications of their research, and to increase their funding to strengthen the national technological base. In fact, between 1980 and 1986 the federal contribution to the national R&D effort increased from $91.3 billion to $112.5 billion constant 2020 dollars. Of that, the DoD's share soared from 47 percent in 1980 to 69 percent in 1986. This major increase in investment in R&D gave the DoD considerable leverage to shape the terms of the dialogue with the research community on the risks of knowledge dissemination.

That leverage was used in two main institutional settings. As mentioned above, early in 1982, Richard DeLauer, the undersecretary for defense research and engineering, established a DoD-University Forum in consultation with the heads of three associations of higher education.[2] It also included university presidents and DoD officials. Its cochairs were Donald Kennedy, the president of Stanford University, and DeLauer himself. It first met in February 1982. It established a Working Group on Export Controls, chaired by David Wilson, executive assistant to the president, University of California in Los Angeles, along with Dr. Edith

Martin, deputy undersecretary of defense, research and advanced technology. It first met in April 1982.

At about the same time discussions between DeLauer in the DoD and Frank Press, the president of the National Academy of Sciences, led to the creation of a panel on scientific communication and national security under the auspices of the National Academies complex. It was chaired by Dale Corson, president emeritus of Cornell University, and populated with eighteen other distinguished members from high-tech industry, former federal agency officials, and senior members of university administrations and faculties. The panel was cosponsored by the DoD and maintained close liaison throughout the summer with the Forum Working Group on Export Controls, which commented on its report for the DoD-University Forum and for the Department of Defense itself. The DoD, then, was a main target audience for the report produced by the Corson panel that was published in October 1982.[3]

This chapter describes the deliberations of the Corson panel in some detail. Its historic importance lay in the acceptance, by the academic community, that provision should be made to restrict the transnational circulation of a class of knowledge that fell in a "gray area" that was neither sensitive enough to be classified nor so remote from national security concerns as to be allowed to circulate freely. After several years of heated discussion inside the administration between control hawks and those advocating for the widest circulation of knowledge possible, it was decided to abandon the notion of a "gray area" altogether. A new politico-epistemological category of "fundamental research" was created to protect basic and applied research that was published openly from government regulation, including export controls. This Fundamental Research Exclusion (FRE) remains in force today, even though its generosity has been challenged on several occasions. The chapter concludes by describing one particularly strong attack on its provisions in the mid-2000s through the attempt to strengthen the so-called Deemed Export controls on sharing knowledge with foreign nationals in the United States.

The Corson Panel and Its Recommendations

The Corson panel met against the backdrop of the continuing use of export controls and other instruments to constrain the presentation of papers at international conferences, described in chapter 5. In mid-August

1982 the DoD obliged the Society of Photo-optical Instrumentation Engineers (SPIE) to withdraw over 150 out of 626 scheduled papers being presented at its annual international technical symposium in San Diego.[4] The research in many of them had been sponsored by the DoD, and cleared for public release by program managers. Higher officials in the Pentagon restricted them shortly before the meeting began, arguing that they were subject to ITAR and that there were foreign nationals at the meeting. An even more contradictory situation arose the next month, in September 1982, at a meeting of the Aerospace and Electrical Systems Society sponsored by the Institute of Electrical and Electronic Engineers (IEEE). The air force asked the conference organizers to destroy all conference records, and to restrict the publication of certain papers. The conference organizers agreed to do so if the air force paid the estimated cost of $25,000 to $50,000. A day later the air force representative withdrew his request. The scenario repeated itself in November 1982, shortly after the Corson Report was released. The air force instructed an official working at Texas Instruments that three papers by their engineers on very large scale integrated circuits, long since submitted for clearance, be withdrawn from an upcoming meeting in Philadelphia sponsored by the IEEE. Five days before the meeting began the conference organizers were asked to remove the papers from the already-printed conference proceedings, and to recall copies that had been distributed to reporters. The conference organizers objected strongly, and, after reviewing the papers again, the air force agreed that they could be presented as planned.

The members of the Corson panel shared the administration's determination to restrict the loss of sensitive knowledge to the Communist bloc. They realized that existing control mechanisms were not up to the task since they concentrated on hardware and technical data of obvious military significance. They did not address the risks of leakage via unclassified scientific communications and foreign visitors at research centers. Much of the recent controversy had arisen because of the government's sometimes makeshift use of export controls to intervene in research activities, conference participation, and the free movement of foreign scientists and foreign students on campuses.

The Corson panel focused on technology transfer from the United States to the Soviet Union. It did not consider specific issues that may have arisen with the Eastern bloc or with China.[5] It further limited its analysis of the tension between scientific communication and national security in two respects. Firstly, it concentrated on framing policies for federally sponsored (as opposed to proprietary) research since the government had greater leverage to control dissemination if it "owned" the research. Secondly, it limited itself to the impact of regulations on university research. Universities were unique in two ways. They integrated research and teaching. Negative impacts on the former would adversely affect the quality of education for the next generation of scientists and engineers. Also universities, unlike other research institutions, had never established broad controls to protect sensitive information. Faculty regarded such controls as an unwelcome and unfamiliar intrusion that both undermined the cardinal values of academic life and were actually detrimental to national security.

The Corson panel was emphatic that—contra the deeply held convictions of some government officials—security lay in accomplishment, not in secrecy. Policies to secure US technological lead time had to protect the openness and the free flow of information that fostered creativity, which weeded out mistaken ideas, and which stimulated competition. This was the kind of research environment that had laid the foundations for American technological leadership, stimulating economic and military strength. If controls were needed they had to be directed at well-defined targets and be limited to what was necessary to achieve specific goals. The panel thus defined its task as being to "assess how much harm to our national security... could be attributed to information losses from members of the scientific community, including university scientists," and "to identify and evaluate alternative approaches [to controls] that cause the least damage to the capacity of the research community to make its many contributions to American life."[6] Thus the Corson panel took it for granted that some controls on the open dissemination of scientific research (other than classification) were necessary. Their aim was to establish what needed to be controlled, and to suggest the least offensive way to do so.

The members of the panel consulted extensively with officials in the Departments of Commerce, Defense, and State. All had security clearance and could access classified information shared with them by the intelligence community. They were fully briefed on the extent of the legal and illegal acquisition of American science and technology by the Soviet Union and the Eastern bloc, and of the multiple paths for "technology transfer" through the Iron Curtain. They were particularly concerned by evidence showing that Soviet visitors and students were less interested now in doing basic science and were targeting new, emerging technologies that were close to application.

The Corson panel quickly narrowed down the extent of the loss of sensitive information from university sources. Their thinking was guided by none other than Admiral Bobby Inman himself, who stated publicly in

May 1982 that about 70 percent of the militarily significant technology acquired by the Soviet Union had been acquired by Soviet bloc intelligence organizations using clandestine, technical, and overt collection methods. Most of the rest was acquired by legal purchases and open-source publications by organizations like the Soviet Ministry of Trade. "Only a small percentage comes from the direct technical exchanges conducted by scientists and students," Inman said, retreating from his earlier hawkish position.[7] This was confirmed by the panel's own discussions with representatives of all the US intelligence agencies, which "failed to reveal specific evidence of damage to US national security caused by information obtained from US academic sources."[8]

Having exonerated the universities as major sources of technology loss, the Corson panel then focused on the specific "active mechanisms" of sharing know-how (reminiscent of Bucy's concerns) that distinguished universities from other knowledge transfer agents. As the report put it, "the transfer of know-how involves information that is generally not captured in scientific papers. The transfer mechanism for such detailed information involves neither documents nor equipment, but more typically is the 'apprenticing' experience that takes place, among other means, through long-term term scientific exchanges that involve actual participation in ongoing research."[9]

Thus the panel stressed that "the danger to national security lies in the immersion of the suspect visitor in a research program over an extended period, not in casual observation of equipment or research data."[10] This mode of transfer was intrinsic to the university's mission to combine teaching with research. It was also a leading concern of the intelligence community, the report said.

The Corson panel did not elaborate. However, Lara Baker, an official in the International Technology Office at the Los Alamos weapons laboratory, gave a very clear illustration of what they were getting at. In May 1982 he explained to a Senate subcommittee how this "apprenticing experience" occurred in a typical electrical engineering program at MIT or Stanford. After a year in such programs, he said, a student starting with a "blank notebook" will hold a microprocessor chip in his hand. He will have "used computer-aided design to design the micro-processor, he will have used computer-aided layout to lay out the processor on silicon, manufactured the chip either in the laboratory or in collaboration with a manufacturer, tested the circuit, packaged the circuit, mounted the microcomputer on a printed circuit board, and made the resulting computer work." All this took place "under the supervision of experts, with careful hand-holding throughout the program, to make sure that the student understood his activities." Seen from this angle, US-Soviet exchange programs, Baker warned, were "a particular coup on the part of the Soviets, since the best technology transfer organization in the world is the United States university system."[11]

The main finding of the Corson panel followed readily from this analysis. If the main asset of a university education was a skill set whose benefits were acquired after sustained immersion in a research project, which would take time to bear fruit, it followed that *"in comparison with other channels of technology transfer, open scientific communication involving the research community does not present a material danger from near-term military implications."*[12] Or as Corson himself put it to a congressional hearing in May 1984, his panel was predominantly concerned about visitors using "products of basic research to influence production of military equipment or systems by an adversary in one so-called production cycle which may mean 10 years, 12 years."[13] The time from research to application was thus a cardinal variable for the Corson panel. Unless there was clear evidence that the gap between basic and applied research was narrowing, and the time from application through development to deployment was short, there was no need to control basic research. The dynamism of the American research system would sustain the nation's technological lead time.

That said, the panel recognized that the situation was evolving rapidly. They did not dispute Inman's claim that, looking ahead, the Soviets were determined to acquire component and manufacturing technologies from defense contractors and research-intensive small- and medium-sized firms and—in line with an increasing trend since the late 1970s—"new Western technologies emerging from universities and research centers."[14] This was additionally sensitive because the "development of equipment and processes for the manufacture of various items [was] often only an extension of the equipment and processes developed to conduct the basic research."[15]

With the specific danger identified, the panel laid out its scheme for balancing scientific communication with national security. It defined three categories of university research. The first and by far the largest was that in which the benefit of total openness outweighed the possible near-time military benefits to the Soviet Union. There should be no restriction of any kind limiting access or participation in basic or applied research in this category. The second was those areas of government-supported research that demonstrably would lead to military products in a short time, when classification should be considered. And then came the panel's key proposal: "Between the two lies a small 'gray area' of research activities for which limited restrictions short of classification are appropriate." Basic and applied research fell in this "gray area" if they involved a technology that met four criteria simultaneously:

- It was developing rapidly, and the time from basic science to application was short;
- It had identifiable direct military applications or was dual use and involved process or production-related technologies;
- The transfer of the technology would give the Soviet Union a significant near-term military benefit; and
- The US (or foreign countries with similarly strict security systems) was the only source of information about the technology.

An example of the gray area that embodied these four criteria was "large-scale integrated circuit work in which on campus research merges directly into process technology with possible military application."[16]

Let us compare this recommendation with the proposal made by DoD's Defense Science Board Task Force, which had also set apart a zone of unclassified research that fell in a gray area, to use the Corson panel's terminology. Table 4 compares the characterization of the gray areas that warranted restriction in the reports by the task force released in January 1982 and by the Corson panel released nine months later. It has three striking features. First, both limit the gray area to manufacturing and process-oriented research. Second, to avoid running afoul of the First Amendment, both locate controls in the research contract between the federal sponsor and the university. Third, while the DoD task force used export controls extensively as an instrument of control, the Corson panel avoided using them whenever possible.

The Corson Report's guiding principle was that the "national welfare, including national security, is best served by allowing the free flow of all scientific and technical information that is not directly and significantly connected with technology critical to national security." In its view, using export controls to impose prior constraints on communication (i.e., for prepublication controls) was the "most serious and least tolerable limitations of First Amendment freedoms."[17] The law preferred to punish the few who abused their right to free speech after they had broken the law,

TABLE 4. Comparison of Controls on Gray Areas in Federally Funded Research by the Defense Science Board Task Force and the Corson Panel

Issue	DoD Task Force, 01/82	Corson Panel, 09/82
Area of concern	Unclassified manufacturing and process-oriented research, as opposed to basic research	Manufacturing technology developed or employed in the research itself. Process technology with possible military application (Corson to congressional committee, 05/84)
Criteria for control	Militarily critical research that was subject to ITAR and EAR, as defined by the DoD using the MCTL	Simultaneously: an area that is rapidly developing, related to production processes, highly and immediately militarily significant as defined by a streamlined MCTL, and not available from non-US sources
Control instrument	Clauses in the federal contract	Clauses in the federal contract
Denied access to research programs in gray areas	All foreign nationals excluded in highly sensitive areas of research likely to develop ITAR-controlled data	Foreign nationals from designated countries
General access		Allowed physical access to university spaces and facilities, and to classroom instruction
Prepublication process	MSS submitted to contract officer; free to submit to a publisher after 30–60 days without a response	MSS submitted simultaneously to publisher and federal contract officer; maximum 60-day delay; the government had no right to order changes, but it could classify the work
Prepublication controls	Export controls used to distinguish what needs a license to be released	Classification; export controls should not be used

"rather than throttle them and all others beforehand." There were rare exceptions, as in *United States v. Edler*, where the Supreme Court had narrowed the application of ITAR down to technical data "significantly and directly related" to articles on the Munitions List (see chapter 5). But, the panel suggested, in the academic context, it "might well be unconstitutional to use ITAR or EAR to bar an American scientist either from informing his or her colleagues, some of whom might be foreign nationals, on the results of an experiment or from publishing the results in a domestic journal."[18] That said, it also admitted that in federally sponsored research that included a contractual clause allowing for controls on communication "the constitutional limitations on the government's authority under ITAR and EAR do not apply with the same force."[19]

The Corson panel's caution was due to the confused and ambiguous place of export controls in regulating flows of technical data that we touched on in discussing the constitutional limits of government restrictions on research at the end of chapter 5. The panel insisted, with frustration, that "the government has the responsibility of defining in concrete terms those technical areas in which controls on information flow are warranted."[20] The precedent set by *United States v. Edler*, though dealing with the private sphere, was a starting point for reflection. Borrowing the language of that ruling, the panel recommended that the circulation of unclassified information, as well as information that was not "directly and significantly" connected with technology that was "critical" to national security, should be granted a general license.[21] A streamlined version of the Militarily Critical Technologies List (see chapter 4) could be used to establish such criticality. This effectively exempted the domestic and international dissemination of these categories of information from the formal licensing process.

Debates, Disagreements, and Delays

The National Academy of Sciences officially released the report *Scientific Communication and National Security* (the Corson Report) on September 25, 1982. The initial responses in the *New York Times* and *Science* took comfort in the Corson panel's finding that open scientific communication, particularly by universities, played little or no role in the leakage of sensitive knowledge to the Soviet Union. More critically, Rosemary Chalk, program head of the AAAS's Committee on Scientific Freedom and Responsibility, regretted that the panel had not made an "open-ended review" of the complex choices required to protect both scientific openness and national security. Instead it had reduced its task to defining a gray area where "the information controls sought by government officials could be imposed with minimal damage to university and research interests." This was only to be expected, she said: after all the DoD was its "major client."[22] And indeed Edith Martin, the deputy undersecretary for research and advanced technology in the DoD, saw in it a "phenomenal" change of attitude by a section of the research community. Science journalist David Dickson explained why: "the compromise academic leaders told the Pentagon they were prepared to accept" by recognizing a gray area "represented endorsement of the principle that even in peacetime, the government is entitled to place controls on the conduct of unclassified research and the dissemination of its results on grounds of national security." This was in "direct conflict with the traditional notion that no restrictions of any kind should be placed on basic research."[23]

The intelligence community did not like the report. The chairman of the Technology Transfer Intelligence Committee (TTIC) objected that the panel had not recognized just how essential it was to have controls on the dissemination of sensitive information. In his view, the panel had completely underestimated the nature and scope of the Soviet acquisition effort. He regretted that the members of the intelligence community had had no "'smoking gun' examples involving universities [that] completed the cycle from US university into Soviet weapons system." He insisted that the panel was "naïve" to think that the Soviet system would have difficulty absorbing new Western technology because of institutional impediments to technology transfer in its secretive, compartmentalized procurement system. He was frustrated that the panel "rejected our analysis and forecast that Soviet bloc acquisition efforts against universities are likely to increase in the future."[24]

The first official response to the Corson Report was National Security Study Directive (NSSD) 14-82, entitled "Scientific Communication and National Security," signed by President Reagan and issued by his national security adviser William P. Clark. It was released on December 23, 1982. The directive established an interagency panel, chaired by the Office of Science and Technology Policy (OSTP), and including representatives from NASA and the NSF, the intelligence community, and the government departments concerned with technology transfer policy and implementation. It called for a report for the National Security Council by March 1, 1983, to "review the issue of protecting sensitive, but unclassified scientific research information" taking account of the Corson panel's findings.[25] Its aim was to devise controls on this category of information that were acceptable to the scientific research community. Its deliberations were classified. It had still not completed its study by May 1984. At that point the terms of its review had been altered twice, there had been multiple changes in personnel at the OSTP, and still under discussion was whether "some sort of unclassified document" should be released.[26]

This delay was caused by disputes inside the DoD, and between the DoD and the academic community over the acceptable scope of the government's intrusion into the production and circulation of knowledge. Two factors undermined achieving a quick consensus. Firstly, there was the challenge posed to the scope of Corson's gray area by President Reagan's Strategic Defense Initiative (SDI). Secondly, whereas the Corson panel wanted to limit government constraints on knowledge circulation to the barest minimum called for by national security, the Reagan administration was determined to expand the regulatory apparatus of the national security state by increasing the breadth and depth of classification. Any policy flowing from the Corson panel's recommendations had to respect these new administrative requirements.

President Reagan's Strategic Defense Initiative, announced in March 1983, posed a major challenge to the Corson panel's proposals. SDI, popularly known as "Star Wars," was intended to defend the United States from an attack by Soviet ICBMs by intercepting incoming missiles at various phases of their flight. It relied on complex technological systems that had not yet been researched and developed, and that engaged multiple fields of science and technology.[27] Control policies were needed to deal with mission-oriented basic research that might evolve in quite unexpected ways that could not be foreseen when the contract was signed between the federal sponsor and the university. One option favored by the DoD—prepublication clauses in the contract authorizing the government to classify the research if the need arose—was anathema to academia. No other viable option was evident.

The impact of the Corson panel on this particular debate was limited, even if it remained an ever-present benchmark expressing the universities' position. It was thwarted by the complexity of the changing R&D profile in university and corporate research and the emerging technologies they spawned. It was paralyzed by the deep divisions in the DoD between control hawks determined to isolate the Soviet Union and officials more attuned to the demands of academia. And then there was the danger of getting entangled in constitutional challenges posed by constraints on prepublication.

The second contextual factor cutting across implementation of the Corson panel's suggestions was the scope of the classification controls envisaged by the Reagan administration. Their ambition produced such a public outcry that they culminated in two major congressional hearings held in 1984. A subcommittee of the Committee on the Judiciary began meeting in November 1983 and continued on into September the year after, in hearings symbolically entitled *1984, Civil Liberties and the National Security State*. Its proceedings, almost thirteen hundred pages long, included debate, testimony, academic publications, newspaper clippings, and summaries of court cases. The other hearing was held in May 1984 by a subcommittee of the Committee on Science and Technology and was entitled *Scientific Communications and National Security*. It was addressed by Corson himself, by a representative from industry, and by a senior official in the DoD. It was also a turning point in the framing of policies to deal with sensitive, unclassified knowledge.

Expanding the Scope of Controls over Knowledge Circulation and Its Implications for the Corson Panel's Recommendations

The Reagan administration's Executive Order (EO) 12356,[28] dealing with "National Security Information," was published in the *Federal Register* on April 2, 1982, while the Corson panel was in session. It announced several major steps away from classification policies already in place and that had evolved from Dwight D. Eisenhower's presidency in the 1950s, through several presidents, to Jimmy Carter. While differing in their details, previous policies shared the view that the scope of classification should be limited in the public interest. As the Carter administration put it, the government had to "balance the public's interest in access to government information with the need to protect certain national security information from disclosure." Executive Order 12356 removed the balancing test: classifiers no longer needed to weigh the public's "need to know" in a decision of whether or not to classify information. Where Carter had specified that information was not to be classified unless "its unauthorized disclosure reasonably could be expected to cause at least *identifiable* damage to the national security" (emphasis added), EO 12356 required only that the government have "a reasonable expectation of damage" to restricted information, whether or not it could identify just what that damage would be. For Carter, if there was a "reasonable doubt" that information should be classified at all, it should remain in the public domain. For Reagan, "if there is a reasonable doubt about the need to classify information, it shall be safeguarded as if it were classified."[29] Thus where Carter treated borderline cases by raising the bar at which material was classified, Reagan lowered it, moving material on the borderline to a higher level of classification.

Executive Order 12356 also specified that "basic scientific research information not clearly related to the national security may not be classified" (sec. 1.6 [a]). However, it also allowed the government to "reclassify information previously declassified and disclosed" and specifically to classify or reclassify information "after an agency has received a request for it under the Freedom of Information Act" (sec. 1.6 [c], [d]). A committee of the American Association of University Professors painted a dire picture of what might lie ahead for their professional constituency. In their reading, EO 12356 gave "unprecedented authority to government officials to intrude at will in controlling academic research that depends on federal support. It allows classification to be imposed at whatever stage a research project has reached," they wrote, "and to be maintained for as long as government officials deem prudent." If a research program evolved from producing unclassified results into one that generated classified data, "academic research not born classified may, under this order, die classified."[30]

A year later, on March 11, 1983, the White House went further. It released National Security Decision Directive (NSDD) 84, entitled "Safeguarding National Security Information." NSDD 84 imposed a new round of constraints on the dissemination of knowledge. Characterized by Sissela Bok of Harvard University as "perhaps the most dramatic and far reaching of the new efforts to control information," it had two particularly startling provisions.[31] The first required government employees to sign nondisclosure agreements as a condition of gaining access to classified information and to Sensitive Compartmentalized Information (SCI), that is, classified information concerning or derived from sensitive intelligence sources, methods, or analytical processes. Those who were granted access to SCI agreed to submit information and materials for prepublication review "during the course of my access to SCI and to all times thereafter."[32] The *New York Times* reported that by mid-August 1984, 120,000 employees had signed such agreements. The second provision authorized the government to order polygraph tests of government employees under investigation for possible unauthorized disclosures of classified information.

The first of these directives, the lifetime prepublication requirement, was stoutly resisted by the AAAS Committee on Academic Freedom and Tenure. It objected that "the effect of the directive is that government officials may require anyone with current or lapsed access to high-level classified information to submit any writing intended for publication to the government agency for prior review. Those who have ever had access to classified information would accordingly, we take it, be placed indefinitely under the constraints of government censorship."[33] Thomas Ehrlich, provost of the University of Pennsylvania, warned a congressional hearing in May 1984 that NSDD 84's proposed lifetime prepublication requirement would have disastrous effects on the quality of governance that drew on expertise from a wide range of backgrounds, including academia.[34] It would be a serious disincentive to those who temporarily left academia to work for the government. Academic scholars would fear that they might later inadvertently divulge materials deemed classifiable in scholarly publications, in classroom lectures, and so forth. As Ehrlich put it, "Who could be sure that any particular piece of prose would not be found by some unidentified bureaucrat to contain some information that, if not classified, should be classified." NSDD 84 would cast more than simply a chill over academia's willingness to serve their government: "The result would be no less than a deep freeze."[35]

To deal with the currents flowing in favor of increased restrictions on knowledge circulation, the DoD reorganized its administration of technology transfer in December 1982, shortly after the Corson panel reported. The departing secretary of defense, Frank Carlucci, signed an interim departmental directive, 2040xx, that made two major changes. Firstly, it expanded the scope of technology controls beyond "critical" technologies, as defined by the MCTL, to include technologies that were "sensitive" and "significant." This was anathema to universities and industry alike.[36] Second, and related to this, draft directive 2040xx created a new Steering Committee on Technology Transfer. It focused on a range of aspects related to controlling knowledge circulation, from the instruments used (contracts, visas) to the sites for control (emerging technologies, scientific conferences, publications) to legislative impediments (rules for exemption to the Freedom of Information Act).[37] These issues had previously been dealt with in Richard DeLauer's Office of Research and Engineering. Carlucci shifted prime responsibility to the Office of Defense Policy, whose key personnel were hardliners Fred Ikle, Richard Perle, and Stephen Bryen.

This administrative reshuffle caused considerable consternation in the academic and scientific communities and infuriated DeLauer. DeLauer had a background in industry and had little taste for controls. Perle and his associates had been only minimally associated with the ongoing discussion between members of the Corson panel, and the DoD-University Forum's Export Control Working Group in 1982. In addition, Perle, who was responsible for international security policy, was a control "hawk" who was determined to do all he could to deny technology to the Soviet Union and the Eastern bloc.

Basic policy disagreements were inevitable. Indeed the committee spent much of 1983 debating the composition of a new intradepartmental committee to administer controls on research and development as such, including that in universities.[38] A simmering dispute became public in December that year, when Perle personally intervened to ask Defense Secretary Caspar Weinberger to put draft directive 2040xx into immediate effect, transferring most power for technology transfer issues to him, at DeLauer's expense. This was not simply a "turf-war" between DeLauer and Perle that, according to one DoD official, took on "a dimension of a personal nature that [was] terrible, deplorable and unfortunate."[39] More fundamentally its outcome would have a major effect on the scope given to "control hawks" to restrict the dissemination of knowledge.[40]

The dispute was settled by Weinberger on January 24, 1984. He signed DoD directive 2040.2, titled "International Transfers of Technology, Goods, Services and Munitions."[41] It established an International Technology Transfer Panel that had two subpanels. One dealt with export control policy (Subpanel A). Subpanel B, chaired by Edith Martin, dealt with research and development.[42] Martin had been and remained the cochair of the DoD-University Forum Working Panel on Export Controls. She described her new post as a vindication of her struggle throughout 1983 to convince her hawkish DoD colleagues that "security by accomplishment"—rather than by technological denial—had to be recognized in the DoD's technology transfer policy. The creation of Subpanel B was their acceptance of the idea that research and development needed separate consideration from technology acquisition.[43]

Martin's goodwill notwithstanding, prepublication constraints still haunted DoD-university relationships. In late February and early March 1984 the department informed universities of a new directive that, though still not formally issued, had apparently been accepted as de facto policy. It specified the conditions regarding the release of information that would be included in its research contracts with universities that produced unclassified information:

> In areas that were not deemed to be sensitive, papers would have to be sent to the DoD solely for its information at the same time as they were submitted to the journal for publication.
>
> In militarily sensitive areas of basic research, a researcher would be required to send papers to the DoD sixty days before submission for publication. The DoD's review would be purely advisory, and the final decision on publication would be left to the researcher's discretion.
>
> The third area involved papers derived from grants and contracts in sensitive areas of exploratory development. These would have to be sent to the DoD

ninety days before submission for publication. The DoD reserved the right to insist on changes or to withhold publication.

It was also rumored that the DoD may also "require that foreign nationals be barred from participation in some sensitive research projects, particularly those involving exploratory development" (the third category above). This was possibly a reference to SDI research. It was also in line with directive 2040.2's specific mention of the need to "give special attention to rapidly emerging and changing technologies" (sec. 4.4.5).

Prepublication restraints on unclassified university research in gray areas were immediately opposed by the presidents of three research universities—Caltech (Goldberger), MIT (Gray), and Stanford (Kennedy). Their letter to DeLauer and to Presidential Science Adviser George Keyworth warned that it "it would be impossible for our institutions—and for the majority of American universities—to accept a contract that . . . would require government approval of publication."[44] The issue was discussed at a meeting of the DoD-University Forum on March 22, 1984, and reported in *Science*.[45] Representatives from Caltech and Stanford who attended the meeting repeated that they would refuse DoD contracts that removed the university's control over publication. David Wilson from the University of California said that his institution too "would not yield to any sponsor authority over final approval for publication." Frustrated, Edith Martin asked just how much money the universities were willing to give up for their principles.[46]

Richard DeLauer took a major step toward resolving the dispute at the same meeting by abolishing its root cause. Listening to the heated exchanges between Martin and the university representatives, he remarked that "he saw no reason to make a distinction between sensitive and classified research. If information should be kept secret, then classify it; if not, it should be unclassified. 'I don't think we should add the burden of another category,' he said."[47] A major blow to the whole idea of gray areas had been struck.

The creation of gray areas had never won universal assent in the scientific community. We will remember that the AAAS's Rosemary Chalk had already seen it as a concession made to the DoD, whom she characterized as the panel's "major client." Stephen Unger from the Computer Science Department at Columbia University, who served on the AAAS's Committee on Academic Freedom and Tenure, was also against them. As he put it to the congressional subcommittee *1984, Civil Liberties and the National Security State*, Congress should state clearly in law that the various regulations such as ITAR, the Commerce Department regulations, the Invention Secrets Act, the Export Administration Act, etcetera, should be interpreted as not to permit restrictions on information that is not classified. I would make that a blanket, overall statement, not leaving any gray areas for officials to exploit. Our experience is, if you leave a loophole, they will drive elephants through that loophole. It is fine to say we can identify very narrow gray areas, but in practice I don't think that can be done. Therefore I would force the Defense Department and the administration in general explicitly to classify material they feel is of national security importance.[48]

The different views of the principle of carving out intermediate gray areas were dramatically revealed to the public during the meeting of the Congressional Subcommittee on Scientific Communications and National Security on May 24, 1984.[49] Corson was the first to speak. He explained why Cornell University had objected to restrictive language in a $450,000 contract offered by the air force that expanded restrictions in the gray area to demanding that "the contractor [i.e., Cornell] agrees that it will obtain prior written approval before assigning *any* foreign national to perform work under this contract or before granting access to foreign nationals to any technical data provided by the government, *or generated under this contract*."[50] Corson found it acceptable to restrict the participation of foreign nationals in a sponsored project. It was impossible to restrict all thirteen foreign nationals in Cornell's Department of Electrical Engineering from technical data generated under the contract. This was a symptom of what Corson called "creeping grayness." "There appears to be a growing interest in extending the concept of grayness to ever more areas defined as critical, or sensitive, or emerging," Corson said, insisting that "we must limit the concept of grayness to as few areas as possible." In a word, Corson confirmed Unger's objection that gray areas were a loophole that, once available, allowed the government to "drive elephants" through it—but he still defended their need.

Edith Martin turned the tables on him. Speaking last, and deviating from her official statement, she made an announcement that "caught Congressmen, witnesses, spectators and even a few Pentagon officials in the red-and-gold carpeted House hearings room by surprise."[51] She said that for seventeen months the DoD and OSTP had tried to implement the gray areas that had been carved out by the Corson panel. They had come to the conclusion that "the gray area concept is good in theory but not workable in practice"[52]—meaning, for Leo Young, the DoD's director of research and management, that it left too much discretion to military contract officers. Martin then read the text of a draft administration policy advocated by George Keyworth. It suggested that *gray areas of so-called sensitive unclassified research would be suppressed altogether.*

DeLauer had persuaded Martin that gray areas were unworkable: either fundamental research in science and engineering was classified, or it was not.[53] Shortly after Martin made her statement Keyworth met with DeLauer, Ikle, Perle, and Taft in the DoD. Their unanimous agreement with the proposal settled a problem "that has been blown out of all proportion," Keyworth wrote, doubtless a reference to the very public disagreement between DeLauer and Perle.[54]

Martin did not elaborate on the meaning of the phrase fundamental research. In a memorandum dated October 1, 1984, DeLauer explained that it meant research supported by DoD budget category 6.1 ("basic research"). He added that "unclassified research performed on campus at a university and supported by 6.2 funding (the budget category that corresponds generally to applied research) shall with rare exceptions be considered 'fundamental'" and therefore also exempt from restrictions. Those rare exceptions would arise when "there is a likelihood of disclosing performance characteristics of military systems, or of manufacturing technologies unique and critical to defense."[55] In these cases any restrictions on publication would have to be agreed to by the DoD and the university performing the work before a contract was signed. The White House was reported as wanting all federal agencies to follow the DoD's policy.

The Fundamental Research Exclusion (FRE)

DeLauer left government service and returned to private life in January 1985. His allies feared that those who still bitterly opposed the June 1984 policy would take the opportunity to impose their demand for stricter regulations on university research.[56] This was not to happen. A new baseline for regulating academic research in sensitive areas was established in 1985 and was embedded in the structure of the national security state: the Fundamental Research Exclusion (FRE).

The draft text for this had been presented by Edith Martin to Congress in May 1984. DeLauer's operationalization of fundamental research in terms of DoD budget categories 6.1 and 6.2 wound its way through the bureaucracy and underwent several revisions that produced language generalizing it beyond the specific needs of the DoD. The official version was presented in National Security Decision Directive (NSDD) 189, signed by the president on September 21, 1985, "National Policy on the Transfer of Scientific, Technical and Engineering Information."

National Security Decision Directive 189 dealt specifically with federally funded fundamental research at colleges, universities, and laboratories. It abolished gray areas: either research was unclassified, or it was classified. The research contract was used to stipulate the restrictions that would be applied to the conduct or reporting of unclassified research when that was required by law. As NSDD 189 put it,

> To the maximum extent possible, the products of fundamental research should remain unrestricted. It is also the policy of this Administration that, where the national security requires control, the mechanism for control of information generated during federally-funded fundamental research in science, technology and engineering at colleges, universities and laboratories is classification. Each federal government agency is responsible for: a) determining whether classification is appropriate prior to the award of a research grant, contract, or cooperative agreement and, if so, controlling the research results through standard classification procedures; b) periodically reviewing all research grants, contracts, or cooperative agreements for potential classification. No restrictions may be placed upon the conduct or reporting of federally-funded fundamental research that has not received national security classification, except as provided in applicable US Statutes.[57]

What counted as fundamental research? The DoD's budget categories were replaced by a general term that could be adapted to specific situations by different agencies of the federal government. Fundamental research was "basic and applied research in science and engineering, the results of which ordinarily are published and shared broadly in the research community, as distinguished from proprietary research and from industrial development, design, production and product utilization, the results of which are ordinarily restricted for proprietary or national security reasons."

This Fundamental Research Exclusion (FRE) was welcomed by academia as throwing a "loop" around academic research "within which loop there should be no restrictions on dissemination or participation."[58] The FRE was indifferent to the nationality of the researcher. It expanded the scope of free circulation beyond basic research—where it had been situated in 1958, as we saw earlier—to include basic *and applied* research in the definition of fundamental research.[59] The boundary between knowledge circulating freely and knowledge that had to be controlled now lay, not in the kind of unclassified research one did, basic or applied. What mattered was whether the researcher published the results (or intended to do so) in the open domain without any restrictions imposed by the contractor for proprietary or national security reasons.

What of export controls? These, and other legislative restrictions on the "conduct or reporting" of unclassified, federally sponsored fundamental research, were allowed for in the very last clause in the FRE (i.e., the phrase "except as provided in applicable US Statutes"). In 1986, revisions to the implementing regulations under the Export Administration Act went further, exempting "fundamental research" from export controls altogether unless prepublication restrictions were included in contract agreements with the sponsor.[60]

The Fundamental Research Exclusion obviously pleased those members of the research community who were opposed to applying constraints on unclassified research. It also pleased the business community. In March 1982 a group of university administrators and businesspeople met at Pajaro Dunes in California to discuss how to reconcile their mutual interest in commercializing "basic" university research at the frontiers of molecular biology with traditional academic values.[61] The meeting was called at the behest of Donald Kennedy, the president of Stanford University, and attended by representatives of eleven major US corporations and the presidents of the University of California, Caltech, MIT, and Harvard. Four of these presidents—Harvard's being the exception—were the signatories of the letter that had been sent to the Department of State in February 1981 strongly objecting to the imposition of export controls on unclassified university research as an instrument to restrict Soviet access to "basic" research that was increasingly interwoven with "application." The FRE recognized the shared interest of those who met at Pajaro Dunes in commercializing research and limiting government constraints on it as much as possible. As a report to the president in 1982 by the National Science Board explained, with the increased interdependence of university-industry research, "the distinction between basic and applied work disappears. Fundamental ideas and approaches become a necessity and they are used in both universities and industry."[62] The FRE accepted this fusion in its definition of fundamental research and kept the state at arm's length in its implementation, to the frustration of the intelligence community.

The pertinence of the FRE was confirmed by Secretary of State Condoleezza Rice in December 2001, after the terrorist attacks on the World Trade Center. The DoD reconfirmed its pertinence and liberalized it even more in 2008 in a widely circulated memorandum by John Young, the undersecretary of defense for acquisition, technology, and logistics. Young's approach was accepted and clarified in some areas by his successor in the Obama administration in 2010, Ashton B. Carter.[63] The labels for budget categories 6.1 (Research) and 6.2 (Exploratory Development) were replaced by Research, Development, Test and Evaluation Activity 1 (Basic Research) and 2 (Applied Research). To ensure that the products of fundamental research, as defined in NSDD 189, remained unrestricted to the maximum extent possible, the memos emphasized that awards for the performance of contracted fundamental research should not involve classified items, information or technology, other than in exceptional circumstances. Carter's directive, somewhat more explicitly than Young's, also insisted that a deliberate effort should be made to give the FRE free rein: as he put it "unclassified contracted fundamental research awards should not be structured, managed or executed in such a manner that they become subject to controls under US statutes and regulations, including US export controls and regulations." Nor should it be "managed in a way that it becomes subject to restrictions on the involvement of foreign researchers or publication restrictions," except in exceptional and rare cases. Finally, to deal with an annoying situation that sometimes arose when an entity performed subcontracted unclassified research that counted as fundamental research in terms of the FRE, Carter's memo stipulated that provisions be made to ensure that "DoD restrictions on the prime contract do not flow down to the performer(s) of such research," for example, by a university "subcontractor."

Many scholars in American academia are not aware of the significance of the regulatory regimes of export controls to the domestic and international circulation of their research. That does not mean that export controls are irrelevant to what they do. It is simply because the dissemination of their results is covered by a general license that exempts it from government control, as the Corson panel insisted. That exemption was secured in *political* negotiations and discussions between the national security state and the research community that began in 1982 and that have peaked several times since as circumstances have changed. In fact since the 1990s the liberal policies enshrined in the FRE have been an ongoing source of frustration for control hawks in the government and have been challenged in various ways, including by appealing to the so-called deemed export regulations.

Deemed Exports and the FRE

Put simply, a deemed export can be defined as the release of technology or source code, that is, knowledge, that has both military and civilian applications, to a foreign national within the United States.[64] It can occur at a technical meeting with a client in a corporate boardroom on American soil, or in an American university classroom or laboratory when "sensitive" knowledge is shared with certain categories of foreign nationals. Thus even though the transfer takes place within the confines of the country, the face-to-face exchange is "deemed" to be an export and potentially subject to government regulation since the recipient can take it back to his or her home country. This unusual site of government intervention was already targeted in the 1950s (see chapter 3). It was rarely if ever invoked over the ensuing decades. However it gained prominence in the 1990s with the increasing presence of foreign nationals in the American research ecosystem.

In 1994 a revised version of the Export Administration Act stipulated that the release of any controlled technology or source code to a foreign national was "deemed to be an export to the home country or countries of the foreign national" (clause 734 [b] [2]). The "home country" was defined as the individual's most recently established legal permanent residency or most recently established citizenship. What bothered the government was that universities were allowing foreign nationals in the United States (whom they abbreviated as FNUS) who satisfied these criteria, for example, who had a green card, to use controlled technologies in their research as long as it fell within the scope of the FRE, which was "blind" to national origins.

One way the government hoped to close this loophole on knowledge circulation in 2004 was to specify clearly what it meant to "use" controlled technology.[65] The Export Administration Regulations defined using equipment as meaning its "operation, installation (including its on-site installation), maintenance (checking), repair, overhaul and refurbishing." The universities interpreted this liberally: a deemed export license was needed if an FNUS used technology or software in all six of these ways, that is, they interpreted the word "and" at the end of this clause inclusively. Of course this situation arose very rarely. The Department of Commerce suggested that, on the contrary, the "and" was to be read as "and/or," that is, it wanted the specified modes of use to be treated as a list in which case a license was needed if an FNUS performed any one of the six operations. Their proposal was successfully resisted by the universities on several grounds: there was no evidence that the liberal interpretation was a threat to national security, the proposed interpretation was incompatible with the FRE, and it was quite impractical to implement on a large research campus that had an inventory of thousands of items of research equipment and that hosted thousands of FNUS.

In 2006 another attempt was made to invoke deemed export regulations to challenge the generous provisions of the FRE. The Department of Commerce established a Deemed Export Advisory Committee (DEAC), chaired by Norman Augustine, a former CEO of Lockheed Martin, which was asked to find ways to ensure that existing legislation was protecting national security while allowing US corporate and academic research to "continue at the leading edge of technological innovation."[66] Specifically, the government wanted access by FNUS to controlled technology and software to require an export license if the individual in question was *born* in a country to which technology transfer was restricted by the Export Administration Regulations. Their most recent permanent residence or citizenship status was irrelevant.

The DEAC grappled with the implications of this suggestion. As they wrote in their report, "in today's post–Cold War globalizing, Internet-connected world, knowledge is a commodity that is exceptionally difficult to control if for no other reason than that it can be stored in the human brain, and humans are becoming increasingly mobile."[67] They did not want to discourage foreign nationals from studying and staying in the United States. Many had made major contributions to the nation's research ecosystem. But they also realized that if controls were too lax universities and corporations would be exposed to industrial and defense espionage.

But how to strike the balance? The DEAC agreed that a foreign national's permanent residence was not reason enough to exempt him or her from deemed export regulations. But neither did it think that an individual's country of birth, as suggested by the Department of Commerce, was a sufficiently robust criterion to decide if deemed export legislation applied to his or her use of controlled research equipment. What mattered most was whether the individual's *loyalty* was tied to a country of concern. To establish what the Augustine Committee called a potential licensee's "national affiliation" the government should "include consideration of country of birth, residence, and current citizenship, as well as the character of the person's prior and present activities."⁶⁸

There is a decisive shift here. First, a researcher's loyalty is being invoked explicitly as one criterion to establish whether deemed export regulations apply to his or her research practices. Second, the criteria for invoking the law have shifted from an individual's country of birth, as suggested by the Department of Commerce, to include the "character of a person's prior and present activities," that is, from a formal bureaucratic consideration to a vague, culturally mediated one.

This move is reminiscent of the loyalty-security system established by the Truman administration to weed out disloyal, un-American citizens. Jessica Wang suggests that using a criterion like this to establish loyalty assumes "the existence of a certain kind of private and authentic self accessible through indirect means, in which one's reading material, organizational affiliations, political acquaintances, tendency toward dissent, rejection of authority and expressed political beliefs provide clues about reliability and allegiance."⁶⁹ This "loyalty test" has not been enshrined in law, as far as we know. What we can stress is that the export control regime is in as state of permanent flux, and that the scope of its restrictions are constantly being adapted and extended to meet new challenges to American global economic and military dominance. This is only too clear from the threat posed by the People's Republic of China that we will deal with in detail in the last chapter, where we will learn that there are a variety of other measures in place today to deny foreign nationals access to advanced knowledge in academia.

PART THREE

CHAPTER SEVEN

"Economic Security" and the Politics of Export Controls over Technology Transfers to Japan in the 1980s

On November 9, 1989, only a few hours before the border police of the German Democratic Republic, in a series of surprising, stunning moves, began to open the Berlin Wall, the US Congress Joint Economic Committee began a series of hearings entitled "American Economic Power: Redefining National Security for the 1990's." The hearings' convener, Representative Stephen J. Solarz (D-NY), could not know how quickly the Cold War was about to wind down. But he had no doubt that the old bipolar international order was coming to an end, marking a true watershed moment in history: "Future historians will identify the final decade of the present century as a transitional period between two historic epochs." Clearly, the post–Cold War world of the 1990s would look markedly different from the global conflict that had, as late as in the early 1980s, reached new heights of intensity. Solarz believed that it was "fair, and not presumptuous, to say that we are on the threshold of a period that may be as different from the bipolar superpower rivalries since 1945 as the 20th century has been from the Victorian Age." The hearings' leading assumption was that these recent developments demanded not only a reassessment of American national security *policies* but also a redefinition of the *concept* of national security itself.[1]

Despite the symbolic date, and even though the hearings began with a nod to Soviet military power and the nuclear stalemate, the end of the Cold War was not the main reason for convening the hearings. Their key question was—as the title suggested—to what extent the challenges of rapid economic and technological globalization that had picked up steam since the late 1970s had affected US global power and national security. The "nature and the composition of the threat we face is undergoing transformation," Solarz stated. "Clearly economic factors weigh more heavily than they used to, although there has always been an economic factor in the broad definition of national security. The accelerating pace of technological change and the challenge from abroad to our once unquestioned technological leadership represents two of many developments that require us to devote more attention and perhaps national resources to the economic side of the security equation."[2]

As we emphasized at the end of chapter 4, this threat to America's technological lead and security did not emanate from the Soviet Union. The real challenger was first and foremost Japan. The closest US ally in the Pacific had, by the end of the 1980s, become "an economic superpower"[3] second only to the United States. Propelled by high technology innovations, competitive advantages in manufacturing know-how, and trading policies that combined robust export strategies with a careful protection of domestic markets, Japan's meteoric rise caused deep concern in the US national security community of the Reagan and Bush administrations, with reverberations in academia and the US public. The fears of Japan's power complemented a pervasive perception of American economic and technological decline. The many variations of this declinist narrative — something like a national obsession, and not only in the 1980s — shared the perception that the United States was losing ground everywhere: it was losing global market shares, its ability to innovate, jobs, and national wealth, as well as its military capability to project power by means of superior weapons technology. Even worse, all these crisis phenomena, which caused US power and economic prowess to crumble, were closely linked. This was a systemic crisis, an existential threat to the very foundations that the United States was built on, and it was at least fueled and precipitated, probably even caused, by Japanese competition.

In the late 1980s, the notion that Japan posed a threat to US national security was not an extreme position that only some hardline Congress backbenchers nourished. Though never uncontested, it entered the political mainstream and played a crucial role in the intense, complex, and often contradictory debates about the course that American trade, security, alliance, and technology policies should take at a time of dizzying change of the international economic and political landscape.[4] The perceptions of the challenges posed by globalization, technological change, and Japanese-American competition gave rise to a term that was not entirely new, but derived great significance, weight, and meaning from its 1980s context: "economic security." This concept was a reaction to the widely shared perception of an economic, military, and political decline of the United States since the 1970s. "Economic security" was mainly an economic offspring of the "realist" school of international relations and was steeped in a century-long debate between the realist "mercantilists" and "liberals" about the "right" relationship between states and markets, between international commercial exchange and national power.[5] In the 1980s, the main driver of this discussion was the US confrontation with Japan.

As this chapter will show, the debates over Japan's economic and technological ascendancy and America's parallel decline had a deep and lasting impact on the export control system. The most important effect was an extension of the reach of export controls and the addition of new bureaucratic instruments. The United States already regulated the sharing of knowledge by controlling the movement of goods, people and information. In the late 1980s, facing the new phenomenon of Japanese foreign direct investment (FDI), which facilitated the transfer of high technology, especially in the semiconductor industry, the export control system was expanded to scrutinize the cross-border movement of money. In 1988 the Exon-Florio Amendment institutionalized FDI national security screenings and gave the US government the authority to block direct investment in any sector of the national economy.

Exon-Florio marked a significant departure from the principles of the Cold War export control system. The inclusion of FDI screenings targeted specifically knowledge sharing within the Western alliance—no Communist enemy invested in the United States. Before the amendment, the impact of export controls on inter-Western interaction was mainly a (sizeable and strong) side effect of the denial policy against the Soviet Union and its Communist allies, and the main stumbling block was usually reexport control regulations. To be sure, trade and knowledge sharing within the Western alliance were never entirely open and uninhibited. During the entire postwar period the United States constantly regulated the access to technological knowledge as a means to keep its technological lead and shape its relations to its allies.[6] But free-market principles prevailed within the alliance. Even though there is not much research available, it seems safe to say that if inter-West controls were used they mostly curtailed the sharing of military and nuclear technology. Export control interference in predominantly commercial civilian and dual-use technology transactions was the exception, and outright, open denial was a rare event.

This changed in the 1980s. The Exon-Florio Amendment blurred the lines between enemies and friends by transferring the export control logic to economic relations with US allies and opened the door for systematic interference in knowledge sharing within the Western free-trade sphere.[7] This development was fueled by larger economic, technological, and political trends, which converged to undermine the well-worn convictions and certainties of the Cold War era. Economically, the allies—especially Japan—had been catching up with the United States since 1945. The United States had promoted the postwar expansion of industrial and technological capacities within the alliance to buttress its friends against the Communist threat, but also to enlarge markets for American industry. In the 1980s it seemed that the allies had indeed caught up and began to breathe down America's neck. At the same time as the globalization of markets progressed with increasing speed, political allies were increasingly perceived as serious economic competitors—and even as a threat. These growing tensions within the alliance were exacerbated by the rapid changes in the Cold War confrontation. Whereas the 1980s began with an intensification of East-West confrontations, the rise of Gorbachev paved the way for a rapid redefinition and transformation of the bipolar order. How this would change the Western alliance was far from clear. But it appeared plausible to American policy makers and pundits that the alliance would lose its cohesion, making room for intensified political and economic competition. In short, the Cold War distinction between friend and foe became less clear-cut in the 1980s.

These trends became even more complex as they were intertwined with technological changes that also weakened the political distinction between the military and the commercial sphere. Dual-use technologies like semiconductors and computers played a central role in the economic globalization push of the late twentieth century. Not only was inter-West competition in these sectors especially fierce as they promised high growth rates and profits. Because cutting-edge weapons systems increasingly depended on the incorporation of dual-use technologies, commercial market shares appeared to be directly translatable into military power. Thus the reshuffling of economic weights within the Western world had direct security implications at a time when the international postwar security architecture began to falter.

Thus, during the 1980s the technological, military, and political leadership that the US export control system had been established to defend experienced massive pressures that weakened the binaries of East and West, military and civilian, commercial free trade and national security interventionism. Critics of the export control system pointed this out and pushed for its overhaul in order to adapt it to its shifting environment. A large and vocal faction demanded an end to most export controls. They argued that because the United States had, compared to its allies, the most stringent restrictions on technology trade, American companies suffered from a distinct competitive disadvantage. In their view controls should be limited to a very small number of critical military technologies whose circulation was beyond any doubt a danger to US national security. But there also was a loosely organized, yet politically powerful hardline faction in the US administration and among members of Congress, parts of the business community, academics, and journalists, who advocated stricter controls over the sharing of US technology with friendly countries in order to secure the American technological lead. They are the main actors in this chapter.

These advocates and detractors of export controls shared a common goal: they wanted to make the United States more competitive and internationally stronger. But their basic understanding of how markets, national security, technology, military power, and international relations were linked followed diametrically different philosophies, which roughly adhered to the realist/liberal divide. The 1980s saw a constant tug-of-war between these philosophies, and while, even in retrospect, it is not an easy task to decide which one of them prevailed at any given moment in history, the existing literature claims that the liberal, free-market principles won the day.[8]

We beg to differ and will show in this chapter that the sharing of high technology with Japan and other Western allies since the late 1980s was increasingly subject not only to intense scrutiny and but also to thoroughly enforced export controls. This becomes visible only by our going beyond the established narratives of the existing literature that pay only marginal attention to knowledge sharing as a central political issue. The literature on Exon-Florio has focused on the economic and political aspects of FDI regulations but has not analyzed its place in the US export control system. We will show, however, that Exon-Florio's main accomplishment was that it established a direct link between FDI and knowledge sharing and thus incorporated money flows firmly in the export control system. Moreover,

Exon-Florio was part of a wider reassessment of the dangers and benefits of sharing knowledge within the free-market sphere of the West. This chapter will show how and why this logic of denial of technology to friendly states became a prominent feature of the US export control system in the late 1980s and early 1990s.

The Rise of "Economic Security" Thinking: The Role of Japan

The hearings on economics and national security that Representative Solarz staged in 1989 were part and parcel of a broader revisionist debate about the very meaning and scope of "national security." Starting in the early 1980s and until well into the 1990s, an endless stream of government reports, newspaper articles, and academic papers argued that the traditional Cold War concept of national security, which focused on military power, had to be supplemented or even replaced by an understanding of security that emphasized the importance of the political economy of international markets and especially issues of technological and industrial competitiveness to US power in the international system. This discourse identified companies with their "home" nation-states, thus turning economic factors like company profits, and commercially innovated and marketed technologies, as well as the skills of the workforce of private firms, into extensions and sources of state power. Markets were meeting grounds for states, which manipulated them to win advantages over their competitors. Hence the definition and enforcement of market rules had to be understood as issues of national interest and security policy. The size of "national" market shares and control over market access, especially in the realm of dual-use high technology, thus came to be seen as the foundation, essential to the expression of military and political power. In sum, the debates about "economic security" led to an "economization" of national security thinking—and to a "securitization" of markets and commercial interactions.[9]

Academics played a crucial role in this intellectual shift toward what was often called "economic security." Political scientists, economists, and historians were not simply driven by the intense political debates about Japan's impact on US high technology. They also had an immense influence on the way politicians thought and talked about the "Japanese Challenge." In fact, conceptually, policy makers and academics went hand in hand. They were each other's readers, talked to each other in congressional hearings, and together developed and implemented the ideas, theories, and terms that shaped the perception of technological relations with Japan and paved the way toward a shift of export control policy against US allies. It was thus not by accident that Representative Solarz invited people like the Yale historian Paul Kennedy to testify to Congress. His book *The Rise and Fall of Great Powers* had just become one of the most widely discussed analyses of American economic and political decline.[10]

As academics were an integral part of the establishment of the economic security paradigm, it seems necessary to historicize their work, not least because it still wields strong influence over the way historians think and write about the US-Japanese conflicts of the 1980s—including in this chapter. More importantly, an analysis of core assumptions of "economic security" thinking will show how the conceptual and ideological environment of export control practice changed after 1980, and especially why and how Japan, a close ally, could become one of the control targets. An important side effect of such an intellectual history is to address a blind spot of the existing historical studies on the genealogy of the US concept of national security, which tend to overlook the shift toward economics in the security debates of the late Cold War and the early 1990s.[11]

Echoing Solarz, and summing up the security debates of more than a decade, the widely quoted edited collection *The Highest Stakes: The Economic Foundations of the Next Security System* (1992) announced the dawn of a "new security era—an era in which military threat to territory and society recede, only to be replaced by new, more sophisticated ones; an era in which 'security threat' no longer refers just to tanks and missiles but also to the control of markets, investment, and technology; an era that recycles old security vocabulary to fit new issues: market share, protectionism, relative gains from trade. An era, simply, that would reconceive the very character of security, redefine the international power game, and resituate its players." Even though these lines were written while the Soviet Union collapsed, they talked about Japan and Europe. "Well before the . . . collapse of the Soviet Union," their "industrial and technological initiative" had started a process of "redistribution of global economic capabilities," leading to "new arrangements of power" and a "wholly new system that minimizes US influence." The end of the Cold War was only a catalyst to this secular change. Without the Soviet Union's military threat, "the reconstruction of the international security system will depend more directly than ever on the relative economic power and influence of the United States, Europe, and Japan."[12]

In a similar vein, Edward N. Luttwak from the Center for Strategic and International Studies in Washington, DC, wrote in 1990: "Everyone, it appears, now agrees that the methods of commerce are displacing military methods—with disposable capital in lieu of firepower, civilian innovation in lieu of military-technical advancement, and market penetration in lieu of garrisons and bases." Eschewing the term "economic security," Luttwak advocated an analytical perspective of "geo-economics" to understand the economic changes of the 1980s. "This neologism is the best term I can think of to describe the admixture of the logic of conflict with the methods of commerce—or, as Clausewitz would have written, the logic of war in the grammar of commerce."[13]

Such hawkish conclusions were highly influential in the Japan debates and indeed often dominated the public discourse.[14] Yet even scholars who were much less aggressive in their outlook agreed that something had changed since the 1970s. James R. Golden, professor at West Point and author of a quite balanced study, *Economics and National Strategy in the Information Age* (1994), argued very much along liberal-internationalist lines against the implementation of "economic security" policy. And yet he stated, "In the emerging environment," after the Soviet Union's collapse, "the threats to national security are real, but they are more diffuse and less likely to provide a clear focus for standing alliances or to justify the subordination of economic issues to security concerns. Instead, national strategy will have to balance economic and security interests. . . . At the same time the United States must meet the economic challenge of sustaining high and rising levels of national income in the face of intense regional competition."[15]

Statements like these expressed a growing interest in the economic dimension of national security. The ideas that drove the economic security debate should not be mistaken for a coherent theory. But the discussions were dominated by a set of common arguments, concepts, and assumptions that pervaded US perceptions of Japan and informed the political actions taken to tackle the conflict with the Pacific ally. "Economic security" was first and foremost a brainchild of the realist school of international relations. Accordingly, advocates of this extension of national security centered their analysis on state interests, power, and security in an anarchic, competitive international system of zero-sum relations. From this perspective—often in open contradiction with what mainstream economic theory had established[16]—trade was always adversarial and not to the benefit of all sides involved. Trade always produced winners and losers. Samuel Huntington, for example, who since the late 1980s had frequently weighed in on the Japanese-American conflicts, declared, "The idea that economics is primarily a non-zero-sum game is a favorite conceit of tenured academics. In the course of economic competition that may produce economic growth, companies go bankrupt; ... managers and workers lose their jobs; money, wealth, well-being, and power are shifted from one industry, region, or country to another."[17]

Indeed, economic security advocates saw economic performance as a function of power, particularly the power of nation-states. In its simplest form this meant that security and the means to influence the international environment had a price tag. The US military and an activist foreign policy were affordable only if the national economy yielded sufficient revenues. A poor state simply did not have the wherewithal to project power. "Economic power," "a term used by security analysts but rarely by economists,"[18] thus became in fact interchangeable with military and political power. The decisive source of economic power was, in this view, "market control."[19] This entailed not only the size of the global market share a nation could secure for itself, but also "market structure and market function." Whoever made the rules of the game derived political and military power from it. Realists acknowledged that it was not states that conquered markets, but first and foremost multinational firms whose national identity became increasingly blurred. From a realist point of view, however, multinationals were mere extensions of their home state. In short, even if markets were populated by private actors, the "security issues do not disappear"; they only "become submerged and hidden by market relations."[20]

Moreover, power was understood as a relational concept. It could be successfully exerted only if there was a gap, a hierarchy in the means of its projection between the parties involved. If, therefore, economic interactions were not a win-win relationship, it mattered who gained more in comparison to the other party. The constant assessment of relative gains was at the core of economic security thinking.[21]

When the postwar gap in economic performance between the United States and its allies, especially Japan, was closing, realists interpreted this accordingly as an all-encompassing decline of US power. The causes and extent of this decline were as much subject to controversy as their consequences for US policy making. But there was at least a rough consensus that since 1945, but especially since the 1960s, the aggregate weight of the United States in the global economy had shrunk whereas Japan and

Western Europe had increased their shares of worldwide production and trade (see also chapter 4). The loss of US technological leadership and the catching up of the other states of the Cold War West was to most observers an important part of this development.[22] This could have been interpreted as a predictable reequilibration of a world capitalist system that had been shaped by the preponderance of American power at the end of World War II and in the early Cold War. Instead many US academics and also large swaths of the public saw it as an existential crisis and threat.

The growing US trade deficit became the main political symbol of this decline.[23] Since the mid-1970s, the gap between imports and exports had widened quickly, from $9.3 billion in 1976 to $36.4 billion in 1982—and it was about to triple in the following two years. As early as 1982, almost 50 percent of the deficit ($17 billion) was created in the trade relations with Japan. Even more disconcertingly, Japanese exports to the United States had grown disproportionately in the manufacturing sector, the former pride of the US economy.[24]

The most influential contribution to a veritable wave of "declinist" academic literature and public commentary was certainly Paul Kennedy's *Rise and Fall of the Great Powers*, published in 1987 at the height of the escalation of American-Japanese competition. Even though it did not use the term, Kennedy's study interpreted five hundred years of world history through the lens of "economic security," focusing on "the *interaction* between economics and strategy, as each of the leading states in the international system strove to enhance its wealth and its power, to become (and remain) both rich and strong." For Kennedy, too, relative gains were key: "The relative strengths of the leading nations in world affairs never remain constant, principally because of the uneven rate of growth among different societies and the technological and organizational breakthroughs which bring a greater advantage to one society than to another."[25] Kennedy argued that because the United States experienced relative economic decline in the 1980s, it had to reduce its military expenditures and its international obligations if it wanted to address its "imperial overstretch," which only accelerated its loss of power. His "overstretch" metaphor was nothing more than an appeal to tone down the focus on military security and put more emphasis on commercial activities in global markets. Kennedy stressed the dilemma that "a very heavy investment in armaments, while bringing greater security in the short term, may so erode the commercial competitiveness of the American economy that the nation will be *less* secure in the long term."[26] Clearly, Kennedy's main concern was not the Cold War any longer but the changes of the economy of the "Free World." This implied changing relations with Japan and the European Economic Community, who were still military allies but also "rival[s] in economic terms."[27]

Indeed, the term *competitiveness*—a relational concept to assess the US position vis-à-vis these "rivals"—became *the* buzzword of US economic policy for the entire 1980s and well into the 1990s.[28] Borrus and Zysman were not the only scholars who argued that the "debate on US competitiveness must become a debate about national security."[29] The rhetoric of competitiveness implied fears of a nation losing its grip and growing weak and impotent. In this narrative, infused with the ideology of American exceptionalism, other nations like Japan were about to fill the vacuum created by American power in decline and would ultimately push aside the United States as world leader. It appeared to be a bitter irony that the most threatening economic adversaries were political friends. Even worse, the United States itself had enabled their rise after 1945 as part of its Cold War strategy of shoring up the West economically to contain Communism militarily and ideologically. It seemed that after having given generously for decades, the United States was being taken advantage of by ungrateful allies, and had ended up with the short end of the stick.

Even worse, for some observers, the US economy was not just losing its grip—it was under attack. Numerous US pundits and politicians argued along economic security lines that the Japanese successes were not just based on a smart industrial policy, superior management techniques, or higher product quality alone. In their view, Japan made inroads into US markets because of unfair trade practices that directly harmed American power. Japan shielded its markets with protectionist measures, used price dumping to hurt US competitors, and doled out subsidies to its companies. In short, Japan took advantage of American open markets without reciprocating, thus enhancing its relative gains and tilting the economic playing field in its favor.[30] Huntington summed it up in especially stark terms, "It takes only one side . . . to produce a cold war. Japanese strategy is a strategy of economic warfare." And after quoting Japanese officials who stated as their main goal to overtake the United States as the leading economic power, Huntington added a stern warning, comparing Japan to America's main totalitarian enemies of the twentieth century: "In the 1930s Chamberlain and Daladier did not take seriously what Hitler said in *Mein Kampf*. Truman and his successors did take it seriously when Stalin and Khrushchev said, 'We will bury you.' Americans would do well to take equally seriously both Japanese declarations of their goal of achieving economic dominance and the strategy they are pursuing to achieve that goal."[31]

Huntington was not the only one to claim that the Japanese government and industry were the better realists and had implemented a determined economic security policy before the United States had even realized the changing relationship between economic, political, and military power. The developmental strategy the Japanese state had pursued since the 1950s was a policy to "maximize Japanese economic power," said Huntington, and was for him "totally consistent with the 'realist' theory of international relations." But whereas American realists had been preoccupied with military power, the Japanese had applied "all the assumptions" of realism "purely in the economic realm" while it had abjured military power.[32] Thus, Huntington put the fact that Japan demilitarized its political system and foreign policy—not least because of the pressure of the US occupation government—in a nefarious light. He also added an aggressive spin to academic analyses of Japanese security concepts that markedly differed from the US post-1945 political tradition.

Consider, for example, political scientist Richard J. Samuels's description of the core of Japan's strategies as "a fusion of industrial, technology, and national security policies" based on the "powerful belief that national security is enhanced . . . by the ability to design and to produce."[33] To Japan "national security was more a matter of economic advantage than of maintenance of a 'war potential.'"[34] Samuels understood Japan's catching-up policies as an expression of "commercial technonationalism" that used the US-led international order of the Cold War to transfer and indigenize Western dual-use technologies. "National learning" and the acquisition of technological knowledge that could be put to commercial use for the national collective was at the very heart of Japanese national security thinking.[35] Samuels refrained from political criticism, but his historical analysis was nevertheless part of a larger discussion about whether the American Cold War understanding of national security had not turned into a competitive disadvantage in its burgeoning rivalry with its allies.

The focus of the "economic security" discourse on the power of nation-states and the reflections on the role of the government in the planned development of Japan's economy, and especially its high-tech industry, led into a broader discussion about whether the US government should play a more active role in the economic sphere. Huntington, for example, insisted that the US government had "to become a *competitive state*, that is, a state whose purpose it is to enhance American economic competitiveness and economic strength in relation to other countries." He envisioned his competitive state as a combination of Franklin Roosevelt's "welfare state" and Harry Truman's "national security state" that would fuse the national well-being and strategic interests—in fact, an economic security state with a techno-nationalist agenda.[36]

Proposals for an activist state policy bore many names. "Industrial policy" was one term that agitated the US policy makers, journalists, and the public in the 1980s in the quest for new public-private alliances against increasingly successful Western competitors. But there were other concepts like "strategic trade theory," which argued for the targeted support of certain high-tech industries that promised to have huge potential on international markets and whose expansion would have wide-ranging positive ripple effects in the national economy.[37] Concerns about ways to maintain and protect the "defense industrial base" were also widely discussed as we will show in our case study below. Even though they often disagreed about the why and how, these and other strands of public and scholarly debate pushed for an economically more activist state.[38] They sometimes segued into outright protectionist strategies at the very same time as the Reagan administration aggressively used trade instruments, like tariffs or measures to open up markets, against Japan.

These interconnected "(neo-)mercantilist" debates revolved very much around high technology and dominated American trade and Japan policy in the 1980s and early 1990s. They went against the grain of, and often openly criticized, the Reagan administration's ideological emphasis on laissez-faire: a small state, trust in market solutions, free trade, and deregulation.[39] In fact, mercantilist ideas made considerable inroads into the executive branch, with the Department of Commerce and the US trade representative often (but not always) at the forefront of state intervention. The result of this intellectual and political shift was not the end of the still-dominant free-market ideology. But it came under pressure, was in all Japan-related matters constantly questioned, and was in practice time and again put on the back burner in favor of policies that protected industries, supported R&D in arguably "strategic" high-technology sectors, or tried (in the name of free trade) to pry open foreign markets shielded by even more activist states.

Whereas the understanding that US liberals had of free trade embraced globalization as a blessing, the concepts of economic security, strategic trade, and protectionism were critical of international integration and centered instead on the nation-state, emphasizing the risks and costs of the openness of global markets.[40] In the eyes of mercantilists, the leaps globalization had taken since the 1970s had weakened not only the US economy, but also, and more importantly the power of the US state.[41] To them, the international system that the United States had set up after 1945 had become a boomerang that now hurt its founder. Hence, global exchanges had to be reined in to preserve the economic basis of US international hegemony. Since its allies, and not the Soviet Union, precipitated its decline, the United States began to reassess its relations especially to Japan. These ideas, albeit never uncontested by liberal free-market orthodoxy, not only spread in the intellectual debates of the 1980s and early 1990s. They also gradually changed policy making and bureaucratic practices in the executive branch, injecting economic security considerations into the US relations with Japan.[42] Next to tariffs and other trade policy tools, export controls became one of the instruments used to recalibrate American openness toward Japan, as our case study on FDI regulation will show.

FDI, Computer Chips, and the National Security Toolbox of Knowledge Control

As discussed in chapter 1, well before the rise of economic security thinking, indeed beginning after World War II, US national security policy was (and still is) shaped by the conviction that technological superiority and lead time was the very bedrock of American global predominance. In the name of keeping the lead, the United States had since the 1940s developed a wide array of bureaucratic tools to control knowledge flows crossing national borders: classification, travel restrictions, export controls. The Exon-Florio Amendment of 1988 added money in the form of foreign direct investment (FDI) to this remarkably creative set of instruments to make something as elusive as knowledge controllable.

The control over transnational flows of money, understood as buying access to advanced knowledge, was thus a relatively late development in the US control system. It certainly has deeper historical roots, reaching back as far as to World War I and the early Cold War, but it was only in the globalizing 1980s that foreign direct investment became a serious concern of the national security state. The key institution for the implementation of FDI security screenings, the interagency Committee on Foreign

Investment in the United States (CFIUS), was also not entirely new. It was established with an Executive Order in 1975 to monitor the impact of OPEC "petrodollars" on US security, but it provided the authority only to review and not to block investments. Moreover, CFIUS hardly ever met—it was merely a "paper tiger."[43] That profoundly changed with the Exon-Florio Amendment.[44]

Exon-Florio amended the Defense Production Act of 1950 and was tucked into the Omnibus Trade and Competitiveness Act of 1988. This placement reflected two of the main concerns that had paved the way for the amendment: fears of an erosion of the US defense industry's ability to provide technology and goods central to US military power; and fears of the decline of US competitiveness. Thus, the amendment was already, on the face of it, very much an expression of "economic security" reasoning. The legal text, however, was short (taking up less than two pages) and hardly betrayed the complexity of the debate it resulted from. Basically it stated simply that "the President may make an investigation to determine the effects on national security of mergers, acquisitions and takeovers proposed . . . by or with foreign persons which could result in foreign control of persons engaged in interstate commerce in the United States." If such an investigation showed that a merger or acquisition "threatens to impair the national security," "the President may take such action for such time as the President considers appropriate to suspend or prohibit any acquisition, merger, or takeover" by or with a foreign person.

Clearly, the law gave the president wide-ranging competencies. The criteria he had to consider for blocking FDI were rather general and vague. Taking into account national security, "among other factors," he had to base his decision on the potential negative implications of foreign control on "the capability and capacity of domestic industries to meet national defense requirements, including the availability of human resources, products, technology, materials, and other supplies and services."[45] The reach of the US president's power—much of it in practice delegated to CFIUS—thus depended on his definition of national security, his threat perception, and his understanding of which parts of the national economy were central to the military. All these were politically complex, controversial, and messy questions. And even though the law mentioned "technology" only once, the nexus between high technology and scientific-technological knowledge, economic prowess, and military power was at the heart of the Exon-Florio Amendment, as the debates leading up to it showed.

Indeed, during the 1980s, FDI, specifically Japanese FDI, had increasingly been seen as a danger to the US technological lead in need of monitoring and restriction. Many instances of Japanese FDI raised eyebrows in the United States in this decade. But one of the most controversial cases, and central to the history of the Exon-Florio Amendment, was Fujitsu's attempt to buy the American semiconductor producer Fairchild. Semiconductor technology became highly politicized and something like the poster child of the "economic security" hawks.

At once a driver and an effect of globalization, FDI in the United States expanded at a pace not seen in the postwar era. After 1945, the outflow of FDI from the United States to foreign countries was generally much higher than the influx. As late as 1977, the overall stock of FDI in the United States was relatively small ($51.5 billion)—US companies had invested $146 billion abroad. But the year 1977 marked a historical turning point. From then on, FDI in the United States grew markedly faster than outgoing US investments, even though the latter always exceeded the inward flows. Fueled by the liberalization in countries like Britain and Japan of state controls over the movement of capital across borders,[46] in the following one and a half decades, the FDI stock in the United States expanded rapidly to $83 billion in the early 1980s and to $185 billion in 1985. In 1986, the year of the Fairchild controversy, the growth accelerated, the FDI stock swelling to $220 billion—to reach $403 billion in 1990 and $1.5 trillion in 2004.[47] By the end of the 1980s, this powerful surge had turned the United States into the "world's largest host of incoming foreign direct investment."[48]

Even though FDI in the United States was relatively small compared to the whole national economy (10.5 percent of the total net worth of all US nonfinancial corporations) and far below the levels in France (27 percent), West Germany (18 percent), and Britain (20 percent), the developments of the 1980s caused a debate about the benefits and dangers of foreign economic influence in the United States.[49] Indeed, in the 1980s, for "the first time since the nineteenth century, foreign-owned subsidiaries were becoming a significant presence on American soil."[50]

Much more of the FDI came from Western European countries, mainly the United Kingdom and the Netherlands, than from Japan.[51] Most critics had no doubt that most of the money flowing in had many beneficial effects on the US economy. Yet against the background of deteriorating US-Japanese relations, investments by companies like Fujitsu or Sony took center stage of the public discourse. The battle lines between opponents and supporters of Japanese FDI ran right through the Reagan administration, Congress, the business community, academia, and the national press.

Although it oversimplifies a complex debate, one can say that everywhere the question of Japan's investment pitted international-minded free traders against nationalist and protectionist "hawks," who called for great caution in the interest of national security, national interest, the national economy, and the national technological base. This clash roughly followed the fault line between liberalism and realism and shaped conflicting interpretations of the empirical facts of the Japanese multinationals' entering the US market. The liberal free traders argued that the international mobility of money was a key aspect of the global division of labor that, through the fostering of comparative advantages, would be advantageous to all trading partners. They pointed, for example, to the generation of new jobs in the United States, the positive effects of FDI on the deteriorating US balance of payments, and also the transfer of new technologies from abroad that made the US economy more competitive. In their understanding, the national origin of money did not (and should not) matter. In theory, but also in their actions, Japanese investors were not any different from, say, British ones. To claim otherwise only betrayed a hostile, even xenophobic double standard directed against Japanese FDI.[52]

The "hawks," by contrast, described Japanese investment as a Trojan horse that carried the aggressive trade competition into the United States, further hollowing out American competitiveness. Japanese companies, the story went, took control over a large segment of the US national economy and over a great number of American workers, thus exerting growing influence within the United States and curtailing US sovereignty and economic and political power. Whereas the liberals argued for an understanding of trade and FDI as an international win-win scenario, the hawks tended to describe them as a zero-sum game. Every Japanese success was a loss to the United States.

The hawks' arguments were at the center of a powerful wave of economic nationalism that drove much of the debate about Japanese FDI in the United States. Even though outnumbered by the liberals, the realist advocates of economic nationalism and protectionism—foremost among them the polemically so-called Japan-bashers like Clyde Prestowitz—wielded enormous influence on the public discourse, Congress, and the Reagan administration through the prolific publication of books critical of Japan.[53]

To the hawks FDI was a way to acquire American know-how by means of the checkbook, circumventing the existing mechanisms of knowledge regulation like export controls, classification, and, in the private sector, intellectual property laws. Against this backdrop, buying a company seemed to be a way of acquiring technology—the legal equivalent of economic espionage. The perceived challenge of FDI came to a head when Fujitsu announced in October 1986 that it wanted to buy the American semiconductor company Fairchild.

In the 1980s the semiconductor and the computer industries became one of the main battlefields in the conflict between Japan and the United States. All concerns over US competitiveness converged in this sector. US companies were constantly losing market shares to Japan, which was charged with unfair trade strategies like systematic dumping and market distortion in the form of state subsidies. An industry, which had been invented, and for decades dominated, by American entrepreneurs, constantly lost ground and seemed to follow the path into obliteration that other industries like consumer electronics had already taken. To the hawks, the seemingly unstoppable rise of the Japanese semiconductor industry was as much a real economic threat as a symbolic issue, a lightning rod that showed everything that was wrong with American technology and industrial policy and with the US-Japanese relationship.

As they took a stance against the Fairchild-Fujitsu deal, the national security hawks in the Reagan administration, as well as representatives of the economically embattled US semiconductor industry, did not mince words. They described Fujitsu's offer as an act of war. Stephen Bryen, the DoD deputy undersecretary for trade security, an especially vociferous hardliner, likened the Fujitsu offer to the "opening gun" of a battle, adding, "If one of the flagship companies of our semiconductor industry could fall into the hands of the Japanese, we could end up with no US semiconductor industry. We could lose the technology race by default." Similarly, industry representatives feared, in finest Cold War rhetoric, that selling Fairchild would have a "domino effect" or, with reference to another American war, would be "like selling Mount Vernon to the redcoats."[54] The strident language reflected Fairchild's symbolic significance as one of the founding fathers of Silicon Valley but also the protectionist lobbying skills of an industry that was indeed in dire straits and seemingly losing more and more ground to Japanese companies. Moreover, Fairchild was a major defense contractor. Many observers feared Fujitsu would gain access to classified technologies.[55]

The Defense Industrial Base

The Fairchild case coincided with two momentous turning points in Japanese-American trade relations. It was not only that FDI from Japan began to grow much faster than that from any other nation in 1986.⁵⁶ This was also the year in which Japan's global market share of integrated circuit exports was on a par with the United States' share for the first time. The curves of American decline and Japan's rise finally crossed. In this context, Fairchild's crisis conjured up all the negative implications the erosion of the "defense industrial base" had. Far from being an esoteric idea relevant only to the Pentagon's war planners, the notion of the "defense industrial base" (and its variations like "industrial base" and "technology base") played a key role in the competitiveness debates well into the 1990s—and is still influential today. In the 1980s it became one of the main strands of the "economic security" debate as it linked high technology, military production, national security, the national economy, and the increasing influence of global markets.⁵⁷ The concept can be boiled down to a set of four assumptions about how these elements interact.⁵⁸

First, the maintenance and improvement of modern military capabilities relied heavily on a constant infusion of high technology into weapon systems. Indeed, the concept of deterrence—at the very heart of the Cold War relations between the Soviet Union and the United States—was based on the constant mobilization of cutting-edge technology. It functioned as a force multiplier to offset the quantitative advantage of the much larger conventional forces of the Warsaw Pact. In the context of an incessant arms race, military power thus depended on staying technologically ahead of the enemy. As *Final Report of the Defense Science Board*, also known as "The Defense Industrial and Technology Base," summed it up: "Our national security is based on a strategy of deterrence. . . . The effectiveness of our deterrent depends upon our ability to maintain . . . technological superiority."⁵⁹

Second, research and development aiming at the most advanced military technology was the result of a complex public-private partnership. The American postwar system of innovation forged close contractual relations between the federal government, especially the military, universities, and the private business community. Hence, industrial innovation and military power were closely intertwined, so much so that in a large segment of the US economy they formed a powerful "military-industrial complex."⁶⁰

Third, the state's military capabilities depended on a healthy, strong, and reliable economic system that—despite the key role of the federal government—was firmly anchored in free-market and free-enterprise ideology and practices. US companies could be strong partners of the military only if they did well financially, generating revenues needed to invest in innovation and modern production facilities. In an increasingly internationalized economic and technological system, the health of high-tech companies depended on exports and international market shares.

Fourth, despite the importance of global markets, including international technology transfers, there was a need for a strong *national* economy. Since the international trade system might break down in the event of war, only national resources could be reliably and quickly mobilized to meet the needs of modern warfare. In other words, as economic globalization generated (technological) dependencies, it diminished the predictability military planners needed to rely on.

In short, the concept of the "defense industrial base" of the 1980s intertwined economic competitiveness and military power, and the main link between them was high technology. The Defense Science Board boiled their complex relationship down to one paragraph: "The National Technology Base is the essential foundation of our national industrial base. The competitiveness of our national industrial base depends on a continuous creation and infusion of technology just as our national security relies on technology to give our military forces the capability to defeat adversaries who can muster numerically superior forces."[61]

Of course, all this was not entirely new in the 1980s. This set of assumptions is at the very heart of the idea of "total war" and was therefore an integral part of economic and national security thinking at least since World War I. One could even argue that the intimate relationship between economics and the military is an essential characteristic of the Cold War. But the 1980s discussion about the "defense industrial base" recalibrated and rebalanced this relationship between national security and the national and international economy. This recalibration of national security and national economic policies was a reaction to several closely connected challenges posed by profound structural changes that had accelerated in the 1970s. These challenges are usually referred to as "globalization" and "technological change," both of which profoundly changed the semiconductor and computer sector.

The Crisis in the US Semiconductor and Computer Industries and Its Implications for Economic Security

Computers and their beating hearts—semiconductors, or "chips"—are children of war. In the 1940s and 1950s, they were predominantly developed for military purposes and with military money, and for a quarter of a century or so the military was arguably the most important customer of the computer industry.[62] Thus, the US federal government was the key player in this technological field, forging tight public-private partnerships with US companies like IBM. Clearly, this is also a story of US dominance. Without denying the contributions made in other countries, the rapid technological progress in computing was driven by US government funding and by US companies, which were the pacemakers and held the largest global market shares for decades.[63]

All this changed in the 1970s and 1980s. In the computer industry, the big push toward globalization was tied to technological diffusion: the spread of technological know-how, research and development, and production to ever more regions of the world. Japan and several Western European countries were not the only ones to catch up and reach production more or less on a par with US standards.[64] The larger trend involved the establishment of global assembly lines that internationalized computer production. American computers included more and more parts built in other countries, fostering complex networks of technological and economic interdependence. The relative weight of the United States in the computer world shrunk.

At the same time computer technology was thoroughly commercialized, diminishing the role of the US military. In the first postwar decades computer development was driven by the logic of "spin-off": the military pushed for cutting-edge technology that then migrated into civilian applications. Beginning in the 1970s, this relation was inverted. More often than not the most advanced technology was now the result of civilian research and development, and the military reaped the advantages of new products and processes, "spun-on" from the outside. Increasingly, the military bought technology off the shelf like any other customer. Computers were among the first dual-use technologies—meaning they were put to military as well as civilian uses—but starting in the 1970s the boundaries between these spheres became even more blurred than before (see chapters 4–6).

By the late 1980s, globalization, commercialization, and technological diffusion had not only civilianized computer technology; they had also turned the Pentagon into a customer of civilian *international* markets. Thus the US "defense industrial base" had become increasingly dependent on commercial technology that was developed, produced, and traded globally. Even though these structural shifts helped the Pentagon to cut costs and were also arguably beneficial for its keeping up with rapid technological change, they had worrisome implications for US national security.

These shifts were spelled out in a report of the Defense Science Board (DSB) that was put together for the Department of Defense almost at the exact same time as the public debates about Fairchild were heating up. It apparently provided and reinforced the arguments about imminent technological dependency the Department of Defense fielded against the Fujitsu-Fairchild acquisition.[65]

This document, entitled *Report of the Defense Science Board Task Force on Defense Semiconductor Dependency* and published in February 1987, was a sophisticated analysis of the role scientific-technological knowledge played in US technological leadership and competitiveness. The DSB stated that the erosion, or potentially even the loss, of the national knowledge base was the principal reason for US economic and military decline. At the same time, production and national control over knowledge were touted as key to future US power.

The report began with an alarm call. After repeating the creed that US military capabilities relied on technological superiority, the DSB stated, "The United States has historically been the technological leader in electronics. However, superiority in the application of innovation no longer exists and the relative stature of our technology base in this area is steadily deteriorating."[66] With a distinct sense of urgency, Charles A. Fowler, from the Office of the Secretary of Defense, pointed out in his cover letter that this was not just a military problem. The DSB report, he wrote in a dramatic gesture, "focuses on a critical national problem that at some time in the future may be looked upon in retrospect as a turning point in the history of our nation. The implications of the loss of semiconductor technology and manufacturing expertise, for our country in general and our national security in particular, are awesome indeed."[67]

Focusing on the most important and most sophisticated kind of chip, the dynamic random access memories (DRAMs), the DSB took stock of the American position vis-à-vis Japan. The picture was disconcerting indeed: In "slightly over a decade the US share of the most advanced generation of DRAM has fallen from near 100 percent to less than 5 percent." Japan had become the dominant producer because of unfair trade practices, the close cooperation of government and high-technology companies, superior capital market structures, and a larger "technical manpower base" (i.e., a higher share of engineers in the population). Moreover, Japan had outspent the United States in R&D. The trend was unmistakable: in one technological subfield after another, the United States' technological lead was slipping or had already been lost to Japan.[68]

The consequences were far-reaching. If the Department of Defense wanted to buy state-of-the art chips it had to turn to Japanese producers. Even though the DoD had in relative terms lost its role of a key customer of the semiconductor industry, in absolute terms, it still had a huge demand for computer chips. It bought about 3 percent of all semiconductors produced worldwide—in sales dollars its share was in the vicinity of 10 percent.[69] How much weapon systems of the latest generation depended on foreign technology was not exactly clear, but the DSB estimated that it was a "significant fraction . . . up to several tens of percent."[70] Especially worrisome was the situation in the field of the most advanced supercomputers. Dual-use high-performance computers played a prominent role for military command and intelligence functions, weapons design, and nuclear weapon testing and were therefore among the most tightly export-controlled technologies. In 1986, 100 percent of the memory capacity and 10 percent of the logic elements of US supercomputers were "derived from Japanese manufactured semiconductors."[71]

The loss of leadership also meant reduced American control over technology flows to the Soviet Union, with incalculable effects on the US military technological lead in the arms race.[72] But more importantly, there was the distinct danger that the United States and not the Soviet Union would be the target of technology denial. The DSB reasoned that "it would not be an illogical strategic business policy to delay release of the most advanced chips to competitors . . . , including the United States. Even if foreign manufactured chips are to be available to US manufacturers, it would appear likely that these chips will be a generation behind those" the Japanese would use in their products.[73] Indeed, the fear of technological denial—in peace as well as in war—was one of the most disturbing scenarios articulated in the DSB report.[74] Not only was the United States losing the lead—there was the distinct danger that it would be kept behind indefinitely.

To counter this threat and to "reverse the trend toward the export of semiconductor manufacturing and technology leadership," the DSB advocated the stabilization and expansion of the national technology base. The dangers of technological denial were "not a critical problem as long as the US has the knowledge and the resources to substitute domestic sources in a timely fashion should the supply of foreign products and technology be interrupted."[75] Knowledge retained in individuals working for US companies was seen as the decisive weapon to fight Japanese competition and American decline: "In order to retain infrastructure for ... industries such as those of computers and telecommunications, which supply DoD needs, action must be taken to maintain a strong base of expertise in the technologies of device and circuit design, fabrication, materials refinement and preparation, and production equipment."[76] The DSB understood "expertise" as a complex field of different kinds of knowledge developed and shared within national collectives of engineers and scientists. Central was applied know-how, used and acquired in production processes, not just theoretical knowledge. The shop floor in a semiconductor production facility was at least as important as a lecture hall or laboratory at a university. Face-to-face interaction was key: "In Japan, many engineering techniques are learned in the company, where engineers can acquire a deep, but narrow, expertise."[77] The DSB referred here to accumulated experience, which consisted not only of the entirety of acquired information, but also of "bodily" coded knowledge that could be gained only through individual practice. This experience could be communicated in written form only to a certain extent; sharing such knowledge required the physical presence and direct communication of people. Historians and sociologists of science call this "tacit knowledge."[78]

This meant that the "industrial base" was not simply about building technological products—it was also a reservoir of national knowledge. If this reservoir were depleted and lost—especially if it was related to the foundational technologies that other kinds of knowledge were based on and relied on—the United States would lose its technological superiority and its chance to catch up with Japan. Without skilled engineers there would be no semiconductor industry. But without a semiconductor industry there would also be no skilled workforce: "A competitive semiconductor industry is therefore essential in order to attract individuals necessary for maintaining a competitive technology base in the area. Further, the reservoir of human skills and expertise developed in the semiconductor industry is necessary not only for this industry, but also for new and perhaps not-yet-invented industries related to it. These skills cannot be retained and developed in academia alone."[79] The DSB advocated therefore the funding research and development in the field of industrial semiconductor *production*.

The DSB's report paved the way to the establishment of Sematech in 1987, a research consortium cofunded by the Department of Defense (through DARPA) and a group of fourteen US semiconductor manufacturers.[80] Sematech was a limited but crucial reversal of the traditional US government reluctance to intervene in the free market with measures of "industrial policy." It followed closely the well-established Japanese model of "industrial targeting" policies, which were based on government-coordinated R&D cooperation among private companies.

The path toward a partial abandonment of laissez-faire in the semiconductor field was also paved by the establishment of the National Center for Manufacturing Sciences in 1986. This public-private research consortium aimed at the revitalization of the US machine tool industry, which was also roiling under Japanese competitive pressures. Both consortia marked a "revolutionary step" in US technological policy as they institutionalized government support in dual-use technologies with the goal of enhancing US security by shoring up industrial competitiveness of private industries.[81] "The underlying premise of Sematech," explained Scott Kulicke, former president of the trade organization Semiconductor Equipment and Materials Institute who had served as the chairman of the DoC's Semiconductor Technical Advisory Committee,

> is that if the US loses the point of the pyramid that everything rests on—the ability to manufacture the most technologically advanced and commercially successful semiconductors—the US's historically preeminent position in the whole electronics industry is sure to be eroded or destroyed. . . . The belief in the importance of the point of the pyramid is precisely the rationale for Japan's multi-year, multi-billion dollar investment in semiconductors, a pattern that is now being copied in Korea and Europe. The historic predilection of the US government towards a free market and hands-off policy . . . is well known and may be theoretically defensible. Yet we are confronting an international competitive environment in which our trading partners have changed the rules.[82]

The Exon-Florio Amendment complemented defense industrial base policy in support of Sematech. Whereas Sematech was supposed to push the declining American industry into more effective innovation in competition with Japanese chip producers, CFIUS would shield US companies against outside intrusion and the loss of valuable national know-how.

The Exon-Florio Amendment in Action

Extensive criticism in America led Fujitsu to withdraw from the Fairchild deal in March 1987. But that did not stop the debate about the central problem: What could the United States do to avoid losing technological knowledge to competitors and enemies—which would be equivalent to losing both the Cold War and the economic battle against Japan? Even though the United States had time-tested export control tools, there were only limited legal instruments to stop foreign investors from acquiring knowledge by simply buying US companies. The fulminant growth of FDI in the 1980s and the increasing number of Japanese acquisitions in the US high-technology industries exposed a supposedly glaring gap in the national security toolbox of knowledge control in the anxious climate of 1986–87.

Exon-Florio filled this loophole. Indeed, it was quite literally a fill-in as it could be used only if existing legislation like the Export Administration Act, International Emergency Economic Powers Act, Defense Production Act, or antitrust law could not sufficiently address the risks of a foreign acquisition.[83] In the first four years after the enactment of Exon-Florio, the government investigated and reviewed more than seven hundred transactions. Members of CFIUS, which was chaired by the Department of the Treasury, were the Departments of Defense, State, and Commerce; the attorney general; the Office of Management and Budget; the US trade representative; and the chairman of the Council of Economic Advisers. If there was a need for special expertise (for example, in cases related to nuclear technology), other agencies like the Department of Energy would join CFIUS.[84]

The Committee on Foreign Investment in the United States was especially active in the high-tech fields. Of the seven hundred CFIUS filings in those first four years, 22 percent were related to the electronics industry; 16 percent to metals, energy, and mining; 7 percent to electric machinery; 6 percent to the aerospace industry; and 6 percent to chemicals. Given the history of the Exon-Florio Amendment, it is not surprising that the first case CFIUS screened was the acquisition of Monsanto's silicon division, which produced silicon wafers for semiconductors. In this case, the buyer was, however, not a Japanese, but a German company, Huels AG, and CFIUS decided not to intervene.[85] Nevertheless, Japan was the main geographic focus, with 192 of the filings, followed by 152 cases dealing with British companies, and seventy-four French, thirty-nine German, and twenty-four Canadian companies. Companies from other European countries were involved in 148 cases. The president himself blocked only one transaction. In February 1990, he forced the state-owned China National Aero-Technology Import and Export Corporation (CATIC) to divest the recently acquired MAMCO Manufacturing Incorporated, an aircraft parts producer from Seattle.

Export controls, especially over technical data, were central in the case against MAMCO. The company produced goods that were controlled by the US government, but since it produced exclusively for the US market it had never exported anything and therefore had never applied for a license. The export control implications of the acquisition by CATIC became apparent only during the CFIUS review. Moreover, since export controls rely on the compliance of the companies involved, there were inevitably security gaps. As GAO put it: "In cases where there may be reason to doubt a prospective purchaser's good faith compliance with the rules, the US export control process should not be expected to serve as adequate protection against technology loss." In the MAMCO case, CFIUS deemed it obviously too risky to trust the Chinese company to abide by the rules and reversed the transaction.[86]

In general, export controls and the protection of classified technological information played a prominent role in the CFIUS investigation and review process. Review filings had to include information if the company to be acquired manufactured "products or technical data subject to validated licenses or under General License . . . pursuant to" EAR and "defense articles and defense services" under ITAR. Also filings had to give notification of classified technologies the company produced and classified contracts it had with the US government. Moreover, CFIUS wanted to know about "any products and services (including research and development)" that had military applications or for which the acquired company was potentially the "sole-source supplier" of the Department of Defense.[87] CFIUS also asked if the acquiring company had ever broken US export control laws, in order to assess the risks of technology loss to third parties (i.e., especially the Soviet Union).

Accordingly, the "focal point" of the Department of Defense's participation was its leading export control office, the Defense Technology Security Administration (DTSA), specialized in the monitoring of transfers of militarily critical technologies.[88] Similarly, the Department of Commerce's Bureau of Export Administration was another key actor in CFIUS's deliberations. It not only assessed the technologies that would change ownership but also looked into the "export control violation history of foreign participants in a proposed transaction." Here, the main fear was that technology could fall into the hands of US foes. In any case, the Department of Commerce examined "the nature of the product or process of the US firm being acquired in order to determine how sophisticated and widespread the relevant technology is and how it might relate to national security." One of the DoC's technological focal points was semiconductor *manufacturing* equipment because in this sector FDI was on the rise. A key concern for the DoC, but also for the DoD, was to ensure "that the United States maintains an adequate defense industrial base." The calculus of its risk assessment included the questions of "dependency on foreign supply and possible supply interruptions, . . . the erosion of critical domestic production capabilities needed for national security purposes, and the denial to US industry of timely access to next-generation products."[89]

Thus, CFIUS's national security risk assessments were (and are up to the present) explicitly shaped by considerations of how to preserve US technological leadership and the defense industrial base (see our epilogue).[90] It followed in its assessment a broad understanding of national security that included a wide range of economic and technological considerations. By the end of the 1980s, then, the United States had established a far-reaching set of rules to defend America's "economic security"—not least against its own allies. The CFIUS technology control mechanism complemented the other regimes in the national security toolbox like export controls and classification.

The Tensions between Technological "Access" and "Loss": Pressures on the Export Control Systems

Clearly, the "economic security" debates about technology sharing and control in the 1980s changed the way the United States approached cooperation with its allies and the very principles of free and open markets. Basically these debates came down to two seemingly simple, closely related ideas: "technology loss" and "access to technology." Both ideas were infused with the fear that the United States was forfeiting its technological lead and thus its economic, political, and military power in the world.

The "loss" of technological knowledge was synonymous with the recipient of US technology learning too much, leapfrogging in technological development, and catching up with the American lead. Since the relative learning gains mattered, the United States controlled what could be shared across national borders. In a nutshell, this had been the logic of export controls since the 1940s. But now it was no longer exclusively targeting the Soviet Union and its allies. Western countries, first and foremost

Japan, became "countries of concern." Prior to the end of the Cold War, these concerns were, especially in the case of Japan, mixed with fears that the Cocom members' export controls were insufficient and would allow for technology shared through inter-West cooperation to leak to the Soviet Union. The 1987 Toshiba-Kongsberg scandal concerning the illegal export of Western submarine propeller milling technology by a Japanese-Norwegian consortium to the Soviet Union seemed to confirm how justified these US fears were.[91]

Questions of "access" to technology played, for example, a central role in debate about the Japanese-American codevelopment of the FSX jet fighter, which took place at the same time as the Fairchild case was animating the American public.[92] Similarly, the discussions about FDI time and again pointed out that Japanese companies took advantage of the openness of the US system while the Japanese government had installed serious impediments to American investors, leading to unequal access to technology. But as the discussions about the "defense industrial base" demonstrate, "access" was also intertwined with questions of dependency. If the United States did not invent and produce certain technologies anymore but imported them from abroad, critics feared, it faced the danger that it would be denied access to cutting-edge technology. As long as the United States had been the leading nation, access to technology was "automatic." But, as the Defense Science Board explained in a report of June 1990, during the 1980s it had become apparent that "changes in the world economy and in technology itself made assured access to technology very much an issue." It was a recurrent trope in the defense industrial base discourse that Japan could at any time renege on the Cold War alliance and deny technology to the United States. Ironically, what US observers feared the most was that Japan and other states could impose export controls against America: "DOD exports many subsystems as black boxes and forbids their transfer to third parties.... Foreign dependence complicates such actions; it allows others to 'say no' to us, and make it stick."[93]

The DSB alluded here to a widely discussed polemical book, *The Japan That Can Say No*, written by Shintaro Ishihara, a prominent member of the Japanese Diet's House of Representatives, and Akio Morita, Sony's cofounder and chairman, who soon would distance himself from the book. Published in Japan in 1989, this polemical essay circulated in political Washington first as a translation that the Department of Defense had commissioned and had made a political splash before a reworked official translation was published in 1991 under Ishihara's name.[94] Even though there was a debate in Washington and among the US public on how seriously the book had to be taken or how much it was an expression of the actual political climate in Japan, it struck a nerve.[95] Ishihara demanded that Japan assume a much more assertive position toward the United States in foreign policy as well as in the trade dispute. One of Ishihara's central arguments was that Japan should use its technological lead as a lever against the United States, not least in the semiconductor field. He claimed that the US chip industry was lagging behind Japan by more than five years. He had also read the DSB's report on semiconductor dependency. "Technology is so structured that once a gap opens, it keeps widening," Ishihara stated along DSB lines, adding: "The United States is desperate because electronics are the mainstay of national strength, including military power."[96] The biggest bombshell and "perhaps the most quoted passage in Ishihara's book"[97] was this paragraph: "In short, without using new-generation chips made in Japan, the US Department of Defense cannot guarantee the precision of its nuclear weapons. If Japan told Washington it would no longer sell computer chips to the United States, the Pentagon would be totally helpless. Furthermore, the global military balance could be completely upset if Japan decided to sell its computer chips to the Soviet Union instead of the United States."[98] Ishihara also boasted, "We control the high technology on which the military power of both countries rests. Unfortunately, Japan has not used the technology card skillfully. We have the power to say no to the United States, but we have not exercised that option. We are like a stud poker player with an ace in the hole who habitually folds his hand."[99]

The perception of the dangers of denial was pervasive in the US government, and a substantial number of US firms claimed that they had experienced difficulties in obtaining state-of-the-art semiconductor technology.[100] In any case, the fears of export controls being used against the United States in combination with a growing dependency on foreign suppliers put increasing pressure on the US export control system. Whereas the concerns over technology loss pulled in the direction of tightening and expanding export controls, the rationale of access pointed in the direction of loosening their grip. The alleged end of US technological supremacy, the global spread of technological capabilities, and the intensifying competitive pressures played into the hands of the advocates of a liberalization of international markets. In the debates about ways to save the defense industrial base, even security-centered actors like the DSB and the Office of Technology Assessment increasingly criticized the export control system as a serious impediment to the activities of US companies and thus to US competitiveness.[101] They echoed earlier calls of the National Academy of Sciences (NAS) in 1987 for a fundamental reform of export controls because it was "out of step with the rapidly changing environment in which they operate."[102] The NAS, which also subscribed to a "broadened definition of national security" that emphasized technology and global markets,[103] reported that in the mid-1980s, 40 percent of *all* US exports needed government approval and that 90 percent of all license applications pertained to shipments to free world countries. This added up to serious harm to the US economy (and the West). US exporters were often outcompeted by less stringently monitored Western companies especially because of unilateral American controls. The NAS estimated that the "short-run loss" due to these controls was "about 10 percent of the value of US exports."[104] A potential effect was "the erosion of competitive market advantages previously enjoyed by US industry and in some cases to the permanent loss of US markets." Confronted with a growing "perceived incompatibility between the execution of national security export controls and the realities of the global trading system," the NAS recommended a far-reaching relaxation and liberalization across the entire control system.[105]

How to resolve the tensions between markets and technological access on the one hand and national security and control against technological loss on the other would become the central question of the export control reform debates up to the present day. Despite the economic pressures, there was little doubt that there was a need for some controls to address security risks of knowledge sharing. The main bone of contention was how much and what exactly should be regulated. The Office of Technology Assessment and the DSB supported the approach that has come to be called "Higher walls around few(er) items."[106] As the DSB put it: "Refrain from excessive export controls, third-party resale restrictions, and classification regimes unless US technology is more than a full generation ahead of the best of the rest of the world."[107] Only the most advanced US technology and knowledge need to be protected.

The Threat from US Allies and the Byrd Amendment

The export control policy of the early 1990s was characterized by contradictory trends. The end of the Cold War pushed forward initiatives to reduce and relax export controls on the national as well as on the multilateral level. Despite lingering strong distrust of the Soviet Union, the US government gave in to persistent pressure from its Cocom partners and agreed in early 1990 to a significant reduction of the Cocom lists, thus freeing a lot of computer, machine tool, and telecommunication technology from oversight. Moreover, these reforms aimed at the creation of a license-free zone within Cocom to reduce the onerous effects of controls on West-West trade. However, the United States refused to refrain from enforcing reexport controls if technologies were transferred from a Cocom member state to countries of concern.[108] Nevertheless, the liberalization would proceed with great strides—and only shortly after, on March 31, 1994, Cocom ceased even to exist.

Simultaneously, however, there was a "tendency to liberalize trade with former adversaries and restrict trade with current allies."[109] In other words, the end of the Cold War did not mark the end of the economic security debate. Indeed, the debate raged on and paved the way to even tighter restrictions on the sharing of technology and foreign direct investment. At the very same time as the United States relaxed its export controls for high-technology trade with Cocom members, Congress pushed for stricter technology control through CFIUS for technology transfers to US allies.[110]

The trigger for this development was the attempt by the French state-owned company Thomson-CSF to buy the missile and aerospace division of the bankrupt US defense contractor LTV Corporation.[111] This marked "the first time a foreign firm has attempted to buy a complete weapons systems manufacturer" and was "the largest proposed acquisition to date by a foreign government." LTV's missile branch, based in Texas, was indeed the seventeenth-largest US defense contractor, and about 75 percent of its technologies and activities were highly classified.[112] After it became clear in April 1992 that Thomson-CSF had outbid the US defense contractors Martin Marietta and Lockheed, the by-now well-oiled machinery of the loose coalition of "hawks" among members of Congress, academia, and think tanks shifted into high gear and did its very best to torpedo the deal. The hawks' argument followed exactly the playbook developed in the economic security debate about Japanese FDI, adding virtually not a single new idea.

Linda Spencer, who worked for the Economic Strategy Institute, founded by Clyde Prestowitz and therefore unsurprisingly very much focused on the critical monitoring of Japanese investments, summarized before Congress many of the crucial arguments the hawks fielded against the French investor. Spencer wanted to "prevent the French from gaining cutting-edge technology, both classified and unclassified." French government control over LTV's "key expertise and technology" appeared to be a threat to US security for mainly four reasons. First, it would strengthen France as a commercial competitor who relied on unfair trade practices: "France has made it a matter of high national priority to match or surpass US technology and to achieve this goal the French government has a long track record of engaging in commercial espionage in order to gain access to cutting edge technologies." Second, the hawks stated that France would also proliferate missile technology to "countries hostile to the United States, including Libya, Iran, and Iraq." Spencer claimed that the US Customs Service had investigated Thomson for violating reexport regulations and diverting US lasers to Iraq. The third main argument was that FDI control was necessary to protect the American defense industrial base. The LTV acquisition was seen as a continuation of a process of hollowing out the US high-tech capabilities that Japanese investors had begun.[113] Norman Augustine, the CEO of Martin Marietta, played on this sentiment by calling the relationship between the French bank Credit Lyonnais, Thomson-CSF, and the French government "keiretsu-like," referring to the cooperative structures of Japanese companies and banks.[114] Fourth, as in the case of Japan, FDI critics saw the involvement of foreign governments in the ownership and the competitive activities of companies as a market distortion that put US firms at a serious disadvantage. With state-owned companies, national industrial policies, and subsidies, it was impossible to establish a level playing field.[115]

Much of the criticism was directed against the Department of Defense and the Bush administration. The hawks took umbrage with the DoD's approach to mitigate the risks of the acquisition through a "special security agreement" (SSA). As a standard procedure, the DoD used these agreements to limit foreign parties' influence on the business activities and access to classified and export controlled information and technologies of defense contractors that they held a share of.[116] The critics saw an SSA in the LTV case as insufficient to safeguard US technology. Grist to their mill was that the SSA was apparently controversial also within the DoD. In a report to Congress the Defense Intelligence Agency (DIA) had allegedly stated that there was "a 100-percent chance that US technology would be diverted to unauthorized parties if this sale goes through. The 100-percent score would mark the first time—after 199 previous reports analyzing previous takeovers by foreign parties of US firms—that the highest rating has been given by DIA."[117]

More importantly, the hawks criticized the way the Exon-Florio Amendment was implemented by the Bush administration. Thomson-CSF had notified CFIUS about its acquisition plans, but the critics did not expect that this would pose any hurdle for the transaction. For them, since the enactment of the amendment the executive had shown a distinct unwillingness to use it properly. Despite seven hundred CFIUS notifications, fourteen investigations, a string of high-profile cases preoccupying the political public, and the presidential action against MAMCO in February 1990, the Bush administration still allegedly adhered to the free-market ideology. "CFIUS could never surmount its political obstacles to fully implement the law, and has retained the 'paper tiger' image," George Washington University professor Susan Tolchin told Congress.[118] In her book *Selling Our Security* she had criticized Exon-Florio as "meaningless except in the symbolic sense," not least because "the president isn't interested" and the "congressional intent doesn't much matter."[119]

In fact Congress still cared very much. For instance, Senator Exon—less critical of the Bush administration than were Tolchin and others—declared, "The LTV/Thomson case is precisely the type of transaction which the Exon-Florio law was designed to handle." In his eyes, it was "perhaps the clearest test of the President's responsibilities." "If the President fails to execute his responsibilities in this now-notorious case, it is already clear that there will be a strong reaction from the Congress. The LTV case offers the President an opportunity to demonstrate that in fact he has a strategy for America's defense industrial base in the post cold war era." Congress acted quickly. In a gesture of distrust toward the administration, several bills were introduced in quick succession in Congress in an attempt to expand Exon-Florio's scope and to "force Presidential action under certain circumstances."[120]

The result of these congressional initiatives was the Byrd Amendment, which tightened up the Exon-Florio provisions. Introducing it on May 13, 1992, Senator Robert Byrd (D-WV) used phrases that had been used on multiple occasions against Japan: "I am introducing legislation today to help retard the hemorrhaging of America's critical industrial base and to stop a most unwise transfer of our advanced missile and associated technologies to Europe and most probably to a variety of Third World countries. There are no precedents for a sale of the magnitude that has been proposed in the case of the LTV Corp., and this is one time that we ought to firmly say 'no go.' "[121]

On July 5, 1992, Thomson-CSF withdrew its offer to buy LTV. Not only had the massive congressional pressure put the French company publicly on the spot. CFIUS had also changed its opinion. At the outset of the acquisition negotiations it had seemed certain that it would approve the deal under the condition of an SSA with the Department of Defense. But several months into the debate, CFIUS recommended blocking the deal. Thomson did not want to wait for the negative presidential decision and moved on.[122]

Despite this outcome, the Byrd Amendment was signed into law by President Bush on October 23, 1992. It expanded Exon-Florio in four ways. First, it put direct investment of foreign state-owned companies under close scrutiny. Formally, filings for an Exon-Florio review had been "voluntary," even if this did not exactly describe the practice that was shaped by government pressure and initiative. The Byrd Amendment, however, established mandatory CFIUS investigations "in any instance in which an entity controlled by or acting on behalf of a foreign government" proposed investment that would lead to control over a US company and could pose a threat to national security. Second, the amendment strengthened the export control dimension of Exon-Florio by adding criteria CFIUS had to weigh in its risk assessment. Now, direct references to the Export Administration Act's regulations for the proliferation of weapons of mass destruction, missiles, and nuclear materials and technologies were included, obviously reflecting worries about French export behavior. Moreover, the amendment obligated the president to report on "any case in which CFIUS engaged in an assessment of the risk of diversion of defense critical technology." Third, defense industrial base concerns were spelled out more clearly. CFIUS was called on to take into consideration "the potential effects of the proposed or pending transaction on United States international technological leadership in areas affecting United States national security." And finally, CFIUS's membership was broadened to include the president's national security adviser and the director of the Office of Science and Technology Policy.[123]

The Byrd Amendment would have little impact on CFIUS's practice.[124] And yet, the Exon-Florio process was certainly not a "paper tiger." By the early 1990s, CFIUS had made hundreds of reviews covering about 40 percent of all inward FDI and had become a very active part of the national security state and of the export control system—regardless of how unhappy the "hawks" were with its performance. The fact that CFIUS blocked only a small number of cases outright—by 1992, this had happened only once— misses the point. Not only were there additional withdrawals—Fujitsu and

Thomson-CSF are prominent cases that show how discouraging the very threat of a negative decision was. The negative publicity simply arising from the discussion of a company's name in connection to CFIUS was highly embarrassing. CFIUS also probably had a chilling effect, so that any number of takeovers may not even have been attempted to avoid government scrutiny.[125] CFIUS operates in secrecy, we do not know in detail how it worked (and works), and there is even a distinct lack of historical case studies using the open record. All the same, our analysis demonstrates that the impact of FDI screenings in the export control framework was substantial, pervasive, and systematic.

Into the Post–Cold War World

The Byrd Amendment demonstrated, once again, the power that the "economic security" paradigm had assumed since the early 1980s. It was not the end of the Cold War that changed the US understanding of what "national security" meant and encompassed. Rather the collapse of the Soviet Union and the following remaking of the international order were a catalyst that gave additional meaning and significance to profound economic, technological, and political changes that had asserted themselves ever more strongly on the United States and its position in the globalizing world. Japan's economic rise became for the United States the focal point of the challenges of globalization. In its open, aggressive, and emotionally charged confrontation with its closest ally in the Pacific, the United States began to redefine its Cold War concept of "national security." It shifted from traditional military and foreign policy concerns to concerns about economic competitiveness in global markets, especially in the realm of high technology. High technology like semiconductors, computers, aircraft, but also machine tools and consumer electronics, became the nexus that closely linked national economic performance to national political and military power.

Realist thinking propelled this conceptual shift. Realists had an enormous influence on the US debates about trade, military, industrial, and technology policies—and thus also partially redirected the US export control system. Realist ideas about trade and technological exchange as a zero-sum game, and the effects of interstate competition on relative power in an anarchic international environment, paved the way for policies and bureaucratic practices to curtail the sharing of scientific-technological knowledge not only with the Cold War enemy but also with America's allies in the free world. The logic of export controls, developed to fight the Soviet Union, was directed first against Japan and then also against the West European partners. The end of the Cold War and the loss of a shared enemy further blurred the slowly dissolving boundary between friend and foe.

This trend is one of the main reasons why the end of the Cold War did not lead to the demise of the US export control system. As urgently as nonproliferation concerns in the 1990s required the continuation of controls—as we will show in our next chapter—there was also a strong current of fear for the loss of US technological leadership and of an "erosion" of the technological knowledge base on which US economic and military power were built. Knowledge sharing was seen as inherently dangerous and risky—it had to be controlled if the United States wanted to preserve its hegemonic position in the international system.

Pushing against this policy, in the 1990s a new market opportunity also beckoned, an opportunity that, if exploited, could stimulate American companies in high-technology fields that had been battered in the 1980s and that were central to the emerging information economy: high-performance computers, but also telecommunications and space satellites. In the 1980s President Reagan characterized China as a "friendly, non-allied country that could work along with the United States to constrain Soviet military "adventurism."[126] Liberal internationalists seized the initiative when President Clinton entered office. Economic links were thickened as Beijing embarked on a program of economic modernization combining state-run heavy industries with a dynamic "socialist" market economy. As China's demand for advanced technology boomed, the Clinton administration encouraged a massive expansion of trade with the country. The relationship between regulatory instruments and economic security was inverted: protectionist measures that limited US access to global markets undermined the competitiveness of American firms in these sectors. In the name of economic security, export controls were increasingly removed on sensitive technologies to enable US companies to outbid rivals and secure a dominant position in the Chinese market, most notably in the high-performance computer sector. Infuriated, control hawks who had argued for tighter controls in the stand-off with Japan in the 1980s accused the president of putting business interests ahead of national security. An increasingly confrontational tussle arose between those who sought constructive engagement with Beijing, and their opponents, who saw China as a growing economic and military threat that had unscrupulously exploited its most-favored-nation trading status, that reneged on nonproliferation agreements, that violated human rights, and that threatened American allies militarily in the region (chapter 8). This engagement reached a climax at the end of the 1990s when the control hawks began to regain the upper hand (chapter 9). Their success, if only partial, was a foretaste of what was to come when President Donald Trump enrolled corporations and academia in a concerted technological trade war, backed by a suite of controls on knowledge circulation, with America's new rival for global dominance (chapter 10).

CHAPTER EIGHT

Paradigm Shifts in Export Control Policies by Reagan, Bush, and Clinton and the Evolving US-China Relations

The late 1980s and early 1990s were marked by several major geopolitical and technological transformations that cast a long shadow over the last decade of the millennium, and demanded a major reassessment of the goals and scope of export controls on dual-use technologies. The collapse of the Soviet Union and the Warsaw Pact forces removed the single most important argument for imposing export controls on high technology in the name of national security. At the same time, as we saw in the previous chapter, there was a new awareness that the United States' position as the leading global power was being undermined by aggressively competitive allies, especially Japan, who had cut deeply into the United States' share of global markets in key technologies of the information age. Military superiority was no longer adequate to ensure America's global supremacy. It also depended on having domestic industries in strategic sectors that could compete effectively in world markets. As two leading members of the Washington establishment, Admiral Bobby Inman (see chapter 5) and Daniel F. Burton, wrote in *Foreign Affairs* in April 1990, "national security can no longer be viewed in exclusively military terms: economic security and industrial competitiveness are also vital considerations," at least where technology was concerned.[1] The concept of economic security had moved to the political mainstream.

The reformulation of export control policies in an international system no longer defined by superpower rivalry was also shaped by political developments in the Middle East and China in the early 1990s. One was the aggressive behavior of Iraq's dictatorial leader, Saddam Hussein. The other was the rising economic importance of the People's Republic of China, stimulated by Deng Xiaoping's pragmatism and his opening to the West.

The Gulf War in 1991, in which a US-led coalition crushed Iraqi forces that had occupied Kuwait, was the first major international crisis since the end of the Cold War. It was a war that forced Western powers to acknowledge that "Iraq had successfully evaded the provision of virtually every extant nonproliferation regime. Iraq had succeeded in testing and weaponizing both chemical and biological weapons, improving the range of its Scud missiles, and undertaking a covert nuclear weapons development program."[2] This led the new administration of President George H. W. Bush to "refocus export controls away from a system based on anticommunist objectives to one based on non-proliferation."[3]

At the same time the adoption of a "socialist market economy" by the Beijing leadership in 1992 energized the expansion of a private sector that piggybacked on state-run firms to "grow out of the plan" and opened a colossal new market for American high-tech industry.[4] Building on the formal rapprochement between Beijing and Washington in 1979; on President Reagan's thickening military, political, and economic ties with China; and on the renewal of China's most-favored-nation trading status in 1991 by George H. W. Bush, the Clinton administration seized the opportunity to promote export enhancement in the name of "economic security." As his secretary of state, Warren Christopher, put it in 1993, "We must elevate America's economic security as a primary goal of foreign policy," adding that the new administration would stop treating economics as its "'poor cousin.'"[5] Tight controls on exports that contributed to the proliferation of weapons of mass destruction (WMDs) would be complemented by a major rollback of restrictions on trade with countries, including former adversaries, that abided by nonproliferation norms.

The tensions between these two goals were immediately apparent and boiled over during Clinton's second term. The key to economic security lay in the aggressive export by civilian firms of information technologies (items like telecommunications equipment, high-performance computers, and communications satellites) that produced high profit margins, funded advanced R&D, and equipped the US military with cutting-edge equipment. Yet precisely because these technologies were dual use, they exposed the government to charges that their export, and the export of the associated knowledge and know-how, strengthened the war-fighting capacity of America's enemies. As Strom Thurmond, the chairman of the Senate Armed Services Committee, put it in May 1995: "I don't want American forces to face their own technologies on a future battlefield because we have exported a large number of dual-use technologies to foreign countries."[6]

This chapter weaves together two narratives. Firstly, it describes major changes in domestic and export control policies that accompanied the paradigm shift from a national security regime based on Cold War logics to one that combined curbs on the proliferation of WMDs with the pursuit of economic security. The development and exploitation of increasingly sophisticated dual-use "information" technologies was the backbone of a so-called Revolution in Military Affairs. It also informed attempts to restructure the military procurement system in Clinton's first term, with the aim of producing a single national techno-industrial base that integrated civil industries into defense systems.

The second interwoven thread discusses the changing attitudes to collaboration in both Beijing and Washington. It analyzes the dilemmas facing successive US administrations as they tried to reconcile the opportunities for access to a huge market with their evolving export control policies and other constraints on trade with China imposed by "hawks" in Congress. The acquisition of dual-use technologies and the associated knowledge and know-how also drove the economic and military modernization of China from the 1980s onward, as it opened up its markets to Western capitalist firms. The important expansion of bilateral trade produced an increasingly vocal backlash from opponents against closer ties with China in civil society, in Congress, and in sections of the administration. They objected to Beijing's policies as regards Taiwan and Tibet; its suppression of human rights—which reached its most disturbing expression in the killing of hundreds of protestors in Tiananmen Square in 1989; and its weapons-proliferation behavior. These objections came to a head in Clinton's second term, as we shall see in the next chapter, when the president's enthusiasm for encouraging trade in dual-use technology with China was condemned as proliferating sensitive military technologies, above all in the space sector. Together these two chapters show that the seeds of the head-on confrontation between President Trump and China's President Xi Jinping (as discussed in our epilogue to this book) were sown in the late 1990s, when the Republican Party gained control of both houses of Congress for the first time since 1954.

The United States and the Modernization of China in the 1970s: A Brief Survey

The modernization of China spearheaded by Mao Zedong's revolutionary regime, which assumed power in October 1949, drew on two very different streams of expertise from abroad.[7] Firstly, with assistance from the Soviet Union and its Eastern European satellites, China embarked on what economic historian Barry Naughton has called "probably the largest coordinated transfer of technology across national borders ever known."[8] It was accompanied by the influx of about one thousand trained Chinese scientists and engineers returning to the mainland from abroad, many of whom played a major role in initiating a technologically advanced sector alongside the heavy industry projects transferred from the Communist bloc.

Ambitions for mobilizing science and technology for social development were thwarted by two ideologically driven assaults on elites: Mao's Great Leap Forward in the late 1950s, and the Cultural Revolution, which lasted from the late 1960s to the early 1970s. Naughton estimates that about eight hundred thousand intellectuals who had spoken out in the Let a Hundred Flowers Bloom campaign in the late 1950s were condemned, removed from their jobs, or sent to labor camps.[9] Thousands of Chinese scientists, notably senior men and women trained abroad, were intimated again during the Cultural Revolution. Zuoyue Wang estimates that hundreds were killed or committed suicide; many suffered persecution and public humiliation.[10]

As the worst excesses of the Cultural Revolution were brought under control, the Chinese leadership began once again to turn outward and seek capital equipment and know-how on a large scale. By the fall of 1974 "a wide range of sophisticated machine tools, of instrumentation, and of production and process equipment," including complete plants and plant complexes, "were being bought from more than a dozen countries," mostly in Eastern and Western Europe, and Japan.[11] The death of Mao Zedong, in 1976, was followed by the rise to power of the reformist-minded Deng Xiaoping. Deng was deeply concerned by the widening technological gap between China and the industrialized West. At the Fourth National People's Congress in 1975, Premier Zhou Enlai had called for "Four Modernizations"—in agriculture, industry, national defense, and science and technology. Deng's enthusiasm for the last pillar—science and technology—led to an outpouring of new ideas for a "technological revolution" to enable China to catch up with the West and paved the way for a new relationship with the capitalist world, and with the United States in particular.[12]

American foreign policy impeded US firms from having a significant stake in the Chinese market in the early 1970s. No American firm exhibited its advanced technology at the thirty-eight trade exhibitions held in China between 1971 and 1975, which were dominated by Japan, Western Europe, and Soviet satellites.[13] The OTA estimated that by 1979 the United States' share of imports by the PRC was only 7–8 percent.[14]

This absence was largely due to the severe restrictions on trade with China that were imposed by Washington when the PRC entered the Korean War in October 1950. The export controls were relaxed once the war was over, though the United States continued to (unilaterally) treat the PRC more severely than it did the Soviet Union, even in Cocom. This so-called China differential remained in place until the early 1970s, when President Nixon inaugurated a policy of "evenhandedness" in export restrictions vis-à-vis the two major Communist powers. His visit to Beijing in 1972 was of a piece with this major shift in foreign policy. The Shanghai Communiqué, which he signed with Chinese premier Zhou Enlai, committed both sides to facilitating people-to-people contacts and exchanges in a range of fields, including science, technology, culture, sports, and journalism.

Nixon's fall from grace put a brake on these initiatives until President Carter again reached out to China in the late 1970s. A high-level American delegation led by Frank Press, Carter's science and technology adviser, and including the NASA administrator and the director of the NSF, visited China in July 1978 to "offer to help China in developing its civil sector technology through commercial, university, and governmental cooperation."[15] These arrangements were formally consolidated in an agreement signed by Carter and Deng when the latter visited Washington in January 1979. It stipulated that, in developing collaborative programs with the PRC, "emphasis will be placed on topics which are less sensitive from the standpoint of technology transfer and foreign policy. In addition, relationships with long-term implications are encouraged; for example, in education, space cooperation, and energy development."[16] Space cooperation included an option to help China buy, and launch, a relatively low-capacity geosynchronous telecommunications satellite. Collaboration in high-energy physics and nuclear power was covered under the "energy" label. The trip was apparently a transformative experience for the

Chinese leader. As diplomatic historian Odd Arne Westad explains, visiting several American cities during his stay, Deng Xiaoping "was bowled over by the technology, the productivity, and the consumer choices he found. After returning home he told colleagues that he could not sleep for several nights, thinking about how China might achieve such abundance. One thing was clear to Deng: Working with the United States on foreign affairs opened gigantic opportunities for US technology transfers to China, both military and civilian."[17]

On March 1, 1979, the United States extended diplomatic recognition to the government of the People's Republic of China. The US Office of Technology Assessment (OTA) explained China's opening to the West by highlighting its "need to acquire new technology and new capabilities in its efforts to modernize and expand its economy."[18] Beijing also had a strategic interest in building an alliance against the Soviet Union. The Chinese leadership bitterly resented the sudden withdrawal of Soviet technical assistance in 1960. It was deeply disturbed by the violent Soviet repression of dissent in Prague in 1968. And in March 1969 ongoing Sino-Soviet border clashes brought the two countries to the brink of war when PLA troops ambushed Soviet border guards on Zhenbao Island near Manchuria. From Washington's point of view, the increasing Sino-Soviet split provided a new opportunity for American political leaders. As historian Robert Sutter writes, those who sought closer relationships with China after its three decades of isolation were driven by "the war in Vietnam, the growing challenge of an expanding Soviet Union, the seeming decline in US power and influence in East Asia and world affairs, and major internal disruptions and weaknesses."[19]

Opposing positions on Taiwan bedeviled rapprochement. Beijing insisted that Washington should respect their One China policy, which envisioned the eventual (re)integration of Taiwan into the mainland. Against them, a powerful "China lobby" in Congress demanded that the United States support Taiwanese autonomy politically and militarily. To facilitate rapprochement, both sides obfuscated their respective positions on Taiwan.[20]

Formal US recognition of the People's Republic of China paved the way for establishing closer trade ties. On July 7, 1979, the two countries signed an agreement granting China most-favored-nation (MFN) trading status. Export control reform facilitating military and technological collaboration was hastened by the Soviet invasion of Afghanistan in December 1979. Under the cloak of secrecy, and in collaboration with the CIA and the Pakistani government, Chinese and American light weapons were soon flowing to the Mujahedeen rebels to "turn Afghanistan into a quagmire for the Soviet Union," as Deng Xiaoping put it to Carter's defense secretary, Harold Brown, in January 1980.[21]

The strategic alliance built in the wake of the Soviet invasion of Afghanistan led to a number of visits in both directions by high-level military officials. William Perry, US undersecretary for defense research and engineering (1977–81), an official who had pioneered advanced weapons systems in both industry and government, led a delegation that visited China in September 1980 to explore its ability to absorb US military technology.[22] Perry informed the Chinese authorities that the US government had approved export licenses for four hundred items of dual-use technology and nonlethal military equipment with which to modernize their armed forces. He also emphasized that to modernize China's industry to meet the challenge of the information age, it was necessary for the country to fundamentally rethink the organization and goals of its innovation system, and to build a national, dual-use technological infrastructure.[23] This resonated with a vibrant debate in China on how to position itself to participate in the new technological revolution that was sweeping the advanced capitalist countries. In one of many similar interventions, in September 1984 Chinese premier Zhao Ziyang insisted that "the rapid transformation of information technology and the increasingly wide range of applications, such as making bio engineering practical, new materials, new energy sources, and marine engineering, will cause great breakthroughs in some areas and open up new applications, causing the global new technological revolution to reach a new stage. Our economic development strategy must develop policies to respond."[24] By redirecting their Mao-era military techno-industrial base toward economic modernization, the Chinese leadership hoped to leapfrog over the several levels of industrial development into the information age.

The Reagan Administration's Support for Chinese Modernization

China's economic modernization around these poles was facilitated by Reagan's liberalization of the export control regime, which led to a surge in trade with China. When Reagan was first elected, his main goals were to roll back Communist power and to consolidate relations with Taiwan. Push-back from senior officials, including his vice president, George H. W.

Bush, over how to deal with Beijing and Taipei were resolved in favor of those who argued that a strategic relation with the mainland to contain the Soviet Union should take precedence over fostering US-Taiwan relations (Chinese forces tied down forty-seven Soviet divisions on the Sino-Soviet Mongolian border). China was classified as a "friendly non-allied" country. The ban on arms exports was removed, and export controls were progressively relaxed to increase the "China differential" vis-à-vis the Soviet Union, but now in China's favor.

As a result, the value of US licenses approved for exports to the PRC soared from $374.3 million in 1980 to $5.5 billion in 1985 (though it declined to $3.37 billion in 1986). The electronics sector was particularly sought after. In 1986, approvals for exports of electronic computing equipment made up more than 80 percent of the total, according to OTA data.[25] To achieve these ends the United States secured a relaxation of export controls in Cocom, where its requests for "exceptions" to the multilateral export control regime almost doubled from 1,882 in 1983 to 3,653 in 1985. Of these, 80 percent and 96 percent, respectively, concerned American exports to China.[26] Meijer writes that in 1986 computers, aircraft, precision instruments, electronic circuit manufacturing, and telecommunications equipment made up about 80 percent by value of all individual validated export licenses for exports to China. By 1988 the United States allowed the export of computers to the PRC ten times more powerful than those being sold to the Soviet Union—a far cry from a proposal made by the Reagan administration in 1981 that the "China differential" should allow for sales of equipment and technology at technical levels that were just twice those authorized for export to the Soviet Union.[27]

In May 1983 China was moved into the most liberal of all export control categories, Country Group V, which included Western Europe, Japan, Australia, and New Zealand. A three-tiered licensing system was put in place in the name of national security. The Green Zone covered seven categories that posed a "minimal security risk," mostly electronics-related goods. Seventy-five percent of all exports to China in 1983 fell into this zone, with computers accounting for 41 percent of all license applications that were routinely approved by the Department of Commerce. The Yellow Zone dealt with sensitive technologies, requiring a case-by-case, interagency review. License applications would be approved unless there was a "clear threat" to US security interests. Finally, the Red Zone contained six mission areas that the Joint Chiefs of Staff (JCS) saw as posing a direct threat to US military interests. Here there was a "presumption of denial" for an export license. These "critical military capabilities" were nuclear weapons and their delivery systems, intelligence gathering, electronic warfare, antisubmarine warfare, power projection, and air superiority.[28] Beyond these the JCS saw little danger for a decade or more in assisting the technological modernization of the Chinese armed forces. By the end of the decade China was equipped with, inter alia, twenty-four Sikorsky S-70 and six Boeing Chinook helicopters; five General Electric gas turbine engines for two naval destroyers; artillery-locating radars; and avionics equipment to modernize fifty F-8 interceptor aircraft and antitank, antiaircraft, and antisubmarine missiles.[29]

One particularly controversial export agreement was that allowing for the launch of American-built communications satellites on Chinese rockets.[30] In 1985 China began to offer commercial launch services on the world market using its newly minted Long March 3 rocket. Three years later, in July 1988, Hughes Aircraft submitted license applications to the State Department to orbit three telecommunications satellites using the Long March. The Reagan administration agreed to the deal even though US Defense Secretary Caspar Weinberger was concerned about dual-use space technology migrating from launchers to missiles.[31] Weinberger also made it clear that US cooperation was conditional on China not proliferating strategic missile technology to other countries.

Beijing's missile proliferation practices in the Middle East were an increasing irritant in US-China relations. China sold Silkworm antiship missiles to Iran that were used to attack oil tankers in Kuwaiti waters in 1987. It also sold Dong Feng 3 intermediate-range ballistic missiles to Saudi Arabia in 1987–88. The Reagan administration decided to link the liberalization of export controls to China's cooperation with its nonproliferation policies. China quickly agreed to halt its proliferation practices in the region, and the US authorities just as quickly accepted their assurances to do so.

With these assurances secured, the US government agreed to using the Long March for launching US-built satellites after China had accepted three intergovernmental agreements. One dealt with Beijing's liability for damage caused by space launches under international law. The second attempted to reassure the US launch industry that it would not be seriously undercut by a Chinese competitor. It set price floors (no bid could be more than 15 percent lower than that of a competitor) and quota limits on the satellites that China could launch for international customers in the six-year period ending December 31, 1994.

The third agreement, adopted in March 1989, dealt with satellite technology safeguards. It included provisions for access by DoD personnel to the facilities housing the satellite and related equipment once it was in China, and "24-hour control and supervision by US personnel over the satellite and related equipment during transit to and within China, over preparations at the launch site, and during launch pad operations." The safeguards agreement also "detailed procedures to minimize foreign access to sensitive technical information and hardware, including significant satellite interface technology." Procedures would be defined to recover parts and debris in the event of an accident during or after the launch.[32] These controls over the transfer of technical data to the Chinese were sufficiently intrusive to satisfy the Pentagon that China's military capabilities would not be enhanced by using the Long March to launch US-built satellites. As for the proliferation dangers, these were handled by arguing that if China was encouraged to concentrate its resources in commercial space launch activities it would be easier for the US to put pressure on it not to proliferate: export licenses for satellites or related components would be conditional on China not sharing sensitive missile technology with third parties.

The increasing liberalization of export controls by the Reagan administration, including in dual-use domains known to have proliferation risks, attests to a shift in the forces driving US-China rapprochement. The original reason for normalizing the relationship with China—to foster a strategic alliance to counter Soviet military power—had gradually given way to a "thickening"[33] of bilateral economic relations, which were beginning to shape global US export policy as the Soviet Union tottered and then fell. Under pressure from US industry to secure a foothold in the Chinese market, the Department of Commerce and the State Department played down the risks to national security. The DoC and DoS also suggested that the benefits that China gained from liberalizing export controls provided leverage that could be used to contain its proliferation practices. These views were not universally supported in the Department of Defense. Stephen Bryen, who established a Defense Technology Security Administration (DTSA) inside the DoD in the late 1980s specifically to strengthen its oversight of export controls, was not alone in being hostile to the export of sensitive technology to the PRC. (We met Bryen in chapters 6 and 7: he was also vociferously opposed to Japanese firms gaining access to American semiconductor technology.) Though Bryen was overruled in the 1980s, his implacable hostility to trade liberalization with China gained considerably more support a decade later, as we shall see in the next chapter.

"Acquiring the Hen and Not Just the Egg": Key Characteristics of China's Import Control Strategy

As China began to assume the role of one the United States' major trading partners, American analysts increasingly turned their attention to the key considerations that shaped the Chinese Communist Party's import strategy for advanced technology. They emphasized that, in agreeing to the Four Modernizations, which privileged science and technology as instruments to modernize agriculture, industry, and national defense, the new Chinese leadership had to grapple with a fundamental dilemma. In turning outward they had to resolve the ongoing tension between the need to acquire technology, knowledge, and know-how from abroad that could lead to "foreign dependence," and the demand for self-reliance, which was rooted in a deep history of humiliating foreign conquest and territorial acquisition and amplified by the sudden withdrawal of Soviet support in 1960.

The ideological and economic dangers of foreign "influence" that had sparked the Great Leap Forward and the Cultural Revolution were still a force to be reckoned with by Deng Xiaoping and his pragmatic supporters. To isolate opponents they argued that opening to the outside world was a strategic measure for accelerating socialist modernization. As China's minister of foreign trade wrote in 1975, China was "putting into practice the principle of making foreign things serve China and combining learning with inventing in order to increase her ability to build socialism independently, and with her own initiative."[34] This emphasis on learning from abroad to enable China—eventually—to become independent in strategic sectors was repeated by Deng Xiaoping in a speech in March 1978: "Independence does not mean shutting the door on the world nor does self-reliance mean blind opposition to everything foreign. Science and technology are a kind of wealth created in common by all mankind. Any nation or country must learn from the strong points of other nations and countries, from their advanced science and technology."[35]

This reorientation away from a radical position that idealized self-sufficiency and economic autarky went along with making economic modernization a higher priority than national defense. In fact, Chinese defense expenditure actually shrunk in real terms in the late 1970s and 1980s. In the decade of the 1980s US arms deliveries reached a peak of only $106 million in 1989. The Chinese were not interested in purchasing complete weapons systems and placed a premium on buying US excess military equipment that was for sale or transfer at reduced prices.[36]

The Chinese had a two-pronged approach to economic modernization. They valued international education and scientific and technological exchange as a long-term strategy for modernizing their economy (seventeen thousand students and professors, mostly in science and engineering, were enrolled at US universities in 1985–86).[37] However, to achieve immediate, short-term gains, they concentrated—like the Soviets during détente–on buying know-how rather than finished products.[38] This was already evident in the early 1970s. Between 1971 and the end of 1974, for example, foreign firms held no fewer than thirty-two specialized exhibitions in major Chinese cities. The hosts ensured that these were educational rather than commercial events: a large amount of Western technical data was made available in glossy catalogues and in hundreds of technical seminars and in industrial films. These were supported by demonstrations and displays by highly qualified vendors hoping to get a foothold in the putatively vast Chinese market. Little actual business was done. At the end of the day foreign firms often sold their display models to save the costs of shipping them home; these were then "reverse-engineered" and copied.[39]

Self-reliance also meant purchasing manufacturing technology rather than finished goods, and preferring to enter into joint-venture and coproduction arrangements with foreign firms, backed by consulting and industrial training agreements. As an OTA report published in 1987 put it,

> Chinese policy has discouraged the acquisition of complete plants and equipment and has stressed the acquisition of know-how, "acquiring the hen and not just the egg," as the Chinese put it. Thus, modes of technology transfer that offer more intimate interactions with foreign technical personnel have come to be preferred. A wide variety of instruments of transfer, including licensing, joint ventures, cooperative ventures, wholly foreign-owned ventures, compensation trade, and the use of consultants and the procurement of technical services are being used. Much emphasis is being placed on foreign provision of training in contract negotiations of Sino-foreign technology transfer. As a result of this change, a much greater proportion of the technology imported since the end of the 1970s has been "unembodied" technology, or pure know-how.[40]

This analysis confirmed a report of the Defense Intelligence Agency in 1985 that concluded that "China tries to import only what it cannot produce for itself and to limit imports to advanced technology and equipment. In general the plan is to import technology that is as advanced as possible, yet still suitable to Chinese conditions. . . . The emphasis is on raising the technical level of existing enterprises rather than importing complete plants or equipment. Whenever possible China will attempt to acquire technology and know-how rather than finished products."[41]

In brief, China's import strategy subjugated the modes of acquisition of foreign technology, knowledge, and know-how to the pursuit of national autonomy, promoting what scholars call "techno-nationalism" whenever it could.[42] Of course China had every reason, as it sought to catch up with and overtake technologically advanced Western economies, to acquire "the hen and not just the egg," to import not just commodities, but "unembodied" technology or pure know-how. What created resentment was that China acquired its "hens" in disregard of established trading norms and practices (like intellectual property rights) and protected its growing market in a golden cage whose bars were multiple barriers to foreign access and acquisition. In an American-led global trading system that was underpinned by a principle of reciprocity, and a shared adherence to norms and values that were aimed to level the commercial playing field, China's techno-nationalism could not but breed resentment.

The Bush Administration: From Anti-Communism to Nonproliferation

When President George H. W. Bush entered the White House he inherited the generous trade relationships with China inaugurated by the Reagan administration. Pressure for further reforms was in the air. The dramatic dissolution of the Soviet "empire" between 1989 and 1991, and associated attempts to replace a Communist, state-driven economy with a capitalist, market system, removed the need for export controls to deny the Soviets advanced Western technology.[43] Soviet defense expenditure declined, stockpiles of conventional and nuclear weapons were reduced, and there was a phased withdrawal of forces from Afghanistan and from Eastern Europe. As a result, "at no time in the postwar era were Western governments seemingly *less* concerned by the prospect of Soviet military aggression in Europe, or adventurism globally than they were in 1990."[44]

In their London Declaration of July that year NATO governments asserted that they and the Warsaw Pact were no longer adversaries, eliminating a fundamental premise of the Cold War. The collapse of military competition along with a dysfunctional domestic economy went hand in glove with a delegitimization of Soviet Communist ideology that had served both as a glue holding together the Soviet Union and Eastern Europe and as a rationale for confrontation with the capitalist West.

As early as 1987 an important National Academy Study, *Balancing the National Interest* (the so-called Allen Report) called for limiting export controls to key technologies and for factoring foreign availability into licensing decisions (see chapter 7).[45] A second report, *Finding Common Ground*, though sensitive to the need to deny weapons to the Soviet military, explicitly called for "a new paradigm for the application of West-East export controls." In particular it urged the United States and the members of Cocom "to change the basis of their technology transfer and trade relationships with the Soviet Union and Eastern European countries from the 'denial regime' that has existed for more than 40 years to an 'approval regime' based on multilaterally agreed and verifiable end-use conditions."[46] The aim was both to stimulate US economic competitiveness in the global market place, and to assist in Soviet defense conversion, economic reform, and democratization.

The Bush administration resisted domestic pressures, as well as those from its European partners (notably West Germany) until early in June 1990. Then the House passed new export control legislation in a bill designed to "end restrictions on exports that were widely available or of little or no consequence to foreign military powers." It also limited the role of the Department of Defense in reviewing dual-use licenses.[47] The Cocom meeting shortly thereafter, and held just three months before German reunification, was judged by some observers to be "the most significant in the history of export controls."[48] Thirty out of Cocom's 116 industrial categories were deleted. Controls on the export of computers, machine tools, and telecommunications were liberalized. And it was decided to establish from scratch a new "core list" of the most sensitive items. The list was ratified in May 1991 and reduced the number of controlled items by a further 50 percent. These measures had dramatic effects. In 1989 the Department of Commerce approved about seventy-eight thousand individual validated licenses worth $116.4 billion.[49] By 1992–93 the number of license requests had dropped to about twenty-two thousand with a value of some $17 billion.[50]

Bush was slow to remove anti-Communist export controls. He was also slow to impose nonproliferation controls. The increasingly broad dissemination of WMDs emerged as a key foreign policy issue in the 1980s. The extent of the danger was already apparent from Saddam's Hussein's attacks against Iran and his own Kurdish population in the eight-year-long war that began in 1980. A series of inspections authorized by the United Nations following his defeat by an American-led coalition in 1991 confirmed that the regime had been able to acquire, through legal and illegal channels, a wide range of dual-use equipment from European and American firms to build and weaponize both chemical and biological weapons. Saddam Hussein had also improved the range of the Scud missiles that he had bought from the Soviet Union and was undertaking a covert nuclear weapons program using calutrons (an electromagnetic technique used at Oak Ridge in the 1940s, and subsequently abandoned in the United States) to enrich uranium.

To meet these threats governments formed a number of new multilateral regimes to complement controls against the spread of nuclear weapons organized in the 1970s under the auspices of the IAEA (the Nuclear Exporters Committee followed in 1974 by the Nuclear Suppliers Group). In the mid-1980s the Australia Group was formed in the framework of the OECD to curb the spread of chemical and biological weapons. In April 1987 the Missile Technology Control Regime (MTCR) was established with five members (the United States, France, Germany, Italy, and the UK), which increased to twelve by 1990.

The Bush administration was reluctant to use export controls for US foreign policy purposes until Saddam Hussein invaded Kuwait. The president had considerable difficulty justifying an American-led war against Iraq until public opinion and the congressional debate gelled around hostility to the proliferation of WMDs. Here was an issue of global importance that could inform a new "grand strategy" for the United States to replace the Cold War conflict against Communism. In January 1991 President Bush decreed that the Export Administration Act "also directs the establishment of enhanced proliferation controls, carefully targeted on exports, projects and countries of concern," preferably brokered multilaterally. A week later he asked Congress to authorize military action in the Persian Gulf, arguing that Saddam Hussein's "demonstrated willingness to use weapons of mass destruction pose [sic] a grave threat to world peace."[51]

Bush's commitment to curbing proliferation was expressed in his Enhanced Proliferation Control Initiative (EPCI), adopted in December 1990.

Its aim was to identify and control dual-use items useful for producing chemical and biological weapons, nuclear weapons, and their ballistic missile delivery systems. End use was thus the key criterion for deciding whether an export license was needed or not. The EPCI was accompanied by considerable nonproliferation legislative activities in Congress after the onset of the Gulf War, and it mandated multilateral action that led to tightening up proliferation export restrictions in the Nuclear Suppliers Group, the Australia Group, the MTCR, and Cocom.[52]

The relationship between Washington and Beijing was reassessed again in 1989, after government forces cracked down brutally on prodemocracy protestors occupying the huge Tiananmen Square in the heart of Beijing on June 4, killing hundreds of unarmed people and imprisoning and punishing countless others. The implosion of the Soviet system along with the ongoing reforms at home led the protesters to call for political changes to match the economic transformation of the previous decade. They were opposed head on by a number of hardliners in the Central Committee that had come to power in 1988 and who had embarked on an austerity program that "expanded centrally managed planning, fixed import prices and quotas covering about two-thirds of its trade, and unilaterally hiked prices on many items, including items of interest to the United States."[53] These men, while ostensibly in favor of reform, were in fact deeply hostile to some of its key elements and sought to take control of the economy. Their supporters also resented what they saw as the "contaminating" ideological effects of growing foreign influence in China.[54]

The opponents of excessive "foreign dependence," notably on the United States, were powerful enough to secure agreement to "39 Points" in November 1989, which called for an extensive retreat from the new economic model taking hold, above all in coastal cities and in the south, where entrepreneurial activity was transforming the local economies. They were, however, not powerful enough to implement their revisionist agenda, and by the end of 1990 "the pendulum had clearly swung in favor of a tentative re-endorsement of further reforms," which were consolidated in the Eighth Five-Year Plan (1991–95).[55] The austerity program imposed to curb market-oriented practices eventually amounted to little more than a temporary program of economic retrenchment. It only briefly reversed the surge in China's economy, which had grown at an average annual rate of about 10 percent from 1978 to 1988, commensurate with that of the four "Asian Tigers" (Singapore, South Korea, Hong Kong, and Taiwan).

The crisis in Sino-US relations triggered by the Tiananmen violence led Congress to take retaliatory measures against Beijing.[56] A first set of sanctions immediately adopted included the suspension of arms sales to Beijing and the postponement of all high-level military-to-military contacts. There would also be a presumption of denial on the sale of computers above forty-two million theoretical operations per second (MTOPS). A second set of sanctions postponed lending to China by international financial institutions. Then on November 21, 1989, Congress included a section in Public Law 101-246 that called, inter alia, for a halt to "exports of US satellites intended for launch by a Chinese launch vehicle, unless the president reports to the Congress that it is in US national security interest to terminate such a suspension."[57] Further sanctions on the use of Chinese launch systems would be imposed if the PRC (or any nation) violated the Missile Technology Control Regime (MTCR).[58]

Congress also moved to revoke China's most-favored-nation trading status.

The economic reforms over the 1980s had transformed China into an export-oriented economy that reversed the balance of trade between it and some of its key partners, notably the United States. By 1990 the United States ran a trade deficit with China of $10.4 billion. That year China also achieved its largest global trade surplus in the history of the PRC ($8.7 billion).[59] It was widely argued that this was the consequence of unfair trading practices that could no longer be overlooked.

Robert Suettinger, who served on the National Security Council in different capacities from 1989 to 1997, has said that the Tiananmen Square crackdown set the stage for "the increasingly rancorous and destructive battle between Congress and the executive branch over China policy—a battle that would last through the Bush presidency and beyond."[60] During the Bush administration, at least, that battle came down in favor of economic interests.

Consider first the sanctions on using China's launcher for satellites made in the United States, as authorized by the Reagan administration. The satellite manufacturers lobbied heavily to have these sanctions lifted for the contracts already in the pipeline. Bush tried to satisfy both Congress and the business community by first imposing them and then overriding them soon thereafter, using the discretion granted him in PL 101-246. Table 5 summarizes the situation.[61] On each occasion the president used his authority to lift the sanctions within about a year or less, and always in time to respect the planned launch window.

TABLE 5. **The Imposition and Waiver of Tiananmen and MTCR Sanctions under President Bush**

Date	Sanctions	Conditions (Abbreviated)	Waived	Motive for Waiver
11/21/89	Tiananmen	Can be waived if there is domestic political reform or if in the US national interest	12/19/89	Waived to allow launch of AUSSAT and Asiasat 1 satellites, worth $300m of business for Hughes
2/16/90	Tiananmen (additional sanctions)	Can be terminated in whole or in part if there is political reform, also in Tibet, and if in the US national interest	4/30/91	Waiver to launch AUSSAT (which had been delayed) and Freja satellites for Sweden
5/27/91	Category II MTCR	2-year sanctions on US exports to China Great Wall Industry, which built the Long March	3/23/92	Lifted after China gave written assurances to abide by the MTCR

Economic concerns also secured the maintenance of China's most-favored-nation status. This issue was discussed by the Subcommittee on Technology and National Security of the Congressional Joint Economic Committee, chaired by Jeff Bingaman (D-NM). It was particularly concerned by the US-China trade deficit and the unfair trading practices that, in the committee's view, had produced it.[62] The US trade representative to Japan and China, Joseph Massey, spoke of China's "predatory behavior" as regards intellectual property (IP), and its tolerance for the violation of the IP rights of US authors, composers, software designers, and others who created and owned intellectual property.[63] Richard Johnston, deputy assistant secretary for international economic policy, US Department of Commerce, recited a long list of unfair trading practices that tilted the playing field in China's favor. These included restricting import licenses for goods even when they were of a higher quality and lower price than domestic alternatives, and imposing tariffs of 120 to 170 percent on imports of products in which US firms were particularly competitive.[64] Restrictive practices were even built into China's five- and ten-year development plans announced in 1991. They stipulated that "Beijing will strengthen oversight of imports to curtail purchases of luxury goods, avoid imports of products that can be supplied domestically, and eliminate duplicate purchases," to quote the CIA.⁶⁵

All the same, none of the Bush administration officials who testified to Bingaman's subcommittee in June 1991 favored a confrontational approach to China. They did not want to invoke action against China's protectionist policies under section 301 of the 1974 US Trade Act.⁶⁶ They were also unanimously against removing its MFN status. Joseph Massey made the point eloquently to Bingaman's committee: "I think this is key Mr. Chairman, loss of MFN would probably not influence China to reform but would rather retard economic liberalization. The burden of denial of MFN would fall on the primary engine of economic reform in China—the economies of the southern and coastal provinces [and also] deal a severe blow to Hong Kong, [straining] the commercial and personal interchanges between individual American and Chinese businesses and people that can help liberalize trade practices in China further."⁶⁷

This reluctance was understandable. For all the disadvantages, American firms were performing extremely well abroad. Deputy Assistant Secretary of Commerce Melvin Searls pointed out to Congress that by 1988 US firms had "captured a substantial portion of the Chinese market in high-technology areas."⁶⁸ By 1991 over one thousand American companies had committed more than $4 billion in long-term investments in a large and diverse range of economic activities in the PRC.⁶⁹

In addition, no other Western trading partner was considering penalizing China for its violation of human rights. Japan was offering a loan package of $5.7 billion for development projects and was actively encouraging international banks to follow suit. CIA data showed that although in 1990 all of China's major trading partners had trade deficits with the PRC, not only was the United States' the greatest, but American exports ($4.8 billion) were lower than those of Japan ($6.1 billion) and the four major European Community countries, France, Germany, Italy, and the United Kingdom ($5.6 billion).⁷⁰ If the United States removed China's MFN status now, it would strike a blow to the political reforms it was supposed to promote, and its less scrupulous rivals could gain ever greater market shares, with serious repercussions on US economic security.

The administration also argued that it was gradually engaging China in the international nonproliferation system. In 1986 the PRC accepted IAEA safeguards over the export of nuclear materials to other countries. In 1988 it had promised to sell no more intermediate range missiles to the Middle East. In summer 1991 an official delegation to Beijing had secured a pledge to at least consider adhering to the Nuclear Nonproliferation Treaty and to the Missile Technology Control Regime (see table 5, last row). The State Department was optimistic that by encouraging trade with China it could persuade the leadership that it had to play by the rules of the international game if it wanted to pursue its export-driven economic reforms.[71] This was the wave that the new president rode when he entered office in January 1993.

The Clinton Administration: Reconciling Economic Security with National Security

With the election of Bill Clinton to the White House in November 1992, many industry leaders increased their lobbying efforts to relax and streamline export controls. During his election campaign Clinton had promised to tighten up export controls on proliferation, defining a policy that, he said, would "lay down a marker for the rest of the world." Meanwhile his campaign slogan—"It's the economy, stupid!"—indicated his determination to promote a new American economy led by exports in high technology. He combined these objectives in a major policy statement released as a "Fact Sheet on Nonproliferation and Export Control Policy" on September 27, 1993. His administration would "accord higher priority to non-proliferation" in the name of national security, making it an "integral element of our relations with other countries." But he also announced that "to strengthen US economic growth, democratization abroad and international stability, we actively seek expanded trade and technology exchange with nations, including former adversaries, that abide by global non-proliferation norms." Clinton thus followed Bush in seeking to build multinational organizations to curb proliferation, while also using enhanced trade as a foreign policy instrument "to integrate non-proliferation and economic goals" as his "Fact Sheet" put it.[72] To ensure his free hand in trade liberalization, he watched over the demise of Cocom and saw it replaced with the far less restrictive Wassenaar Arrangement, which included Russia and Eastern Europe and also controlled exports of conventional arms and related dual-use goods and technologies to "states of proliferation concern" rather than to any region or group of nations.[73] In practice the scope of proliferation controls were soon restricted to Iran, Iraq, North Korea, and Libya—or any other "rogue state."

Clinton's emphasis on economic growth, and his determination to remove barriers to expanded trade in the global market, was embedded in the broader strategic goal of strengthening the nation's economic security. To that end there was the need for the federal government to devise strategies to strengthen the United States' manufacturing base, to facilitate a strong American presence in global high-tech markets, to ensure that the United States was not dependent on foreign suppliers for key technologies, and to generate the new weapons systems needed to remain at least a generation ahead of military rivals.[74]

The pursuit of economic security was far more than a rationale for export enhancement. It dovetailed with the development of advanced war-fighting capabilities in the information-based "Revolution in Military Affairs" (RMA). And it inspired the aim of building a single national techno-industrial base that integrated the quest for competitive advantage in the market with the quest for military superiority in the battlefield.

The Revolution in Military Affairs

The so-called Revolution in Military Affairs (RMA) was field-tested in Operation Desert Storm in 1991.[75] It was devised to fight local wars using conventional (nonnuclear) weapons that were enabled by the synergistic combinations of microelectronics and advanced computing and networking technologies. The RMA used information dominance superiority or C^4I (command, computers, communications, control, and intelligence) as the crucial force multiplier to overwhelm the enemy. In the First Gulf War against Saddam Hussein's Iraqi forces, the US military combined stealth weapons (i.e., those having greatly reduced radar, infrared, acoustic, and visual signatures) and standoff precision-strike platforms with advanced intelligence, surveillance, and reconnaissance (ISR) systems. These fused multiple information inputs from the battlefield into a "real-time" awareness of the conditions on the ground.[76] It was an exercise in "non-linear warfare," now sometimes called "network-centric warfare,"[77] in which "the ability to exercise military control [was] shifting from forces with the best or the most individual weapon systems towards forces with better information and greater ability to quickly plan, coordinate and accurately attack."[78] It was used to devastating effect by the US-led coalition in Operation Desert Storm. The Allies lost about 150 people in the brief, six-week war in 1991. Iraqi military deaths are estimated by the US Defense Intelligence Agency to be of the order of one hundred thousand people. A similar or even higher number of civilians may have been killed.[79]

Operation Desert Storm was also the first comprehensive space war. Satellites were crucial for "enabling mobile communications in remote areas, as well as providing imagery, navigation, weather information, a missile warning capability, and a capability to 'reach back' to the United States for additional support."[80] Global Positioning Satellites (GPS) gave real-time navigational information to fighting units on the ground, at sea, or in the air. Communications satellites handled most of the voice and data transmission. Meteorological satellites provided real-time critical weather data. Reconnaissance satellites along with drones, aircraft, and other modes of intelligence located and tracked enemy forces. They provided "over-the-horizon" targeting for precision-guided weapons or stealth delivery systems that could be launched from tens or even hundreds of miles away (from whence the term "disengaged combat")[81].

The successful implementation of the RMA concept completely upended the technological infrastructure of war. As Michael Mazarr puts it, if previously the most advanced weapons were built exclusively for military use, here "the substructure of war will be information dominance, and its primary building blocks are computers, communication systems, satellites, and sensors."[82] If the RMA became current, conventional wars of the future would be fought with dual-use technologies developed to satisfy demand in an expanding civilian market, yet crucial to the waging of "network-centric" warfare. Competitive advantage would be achieved by the skillful integration and coordinated operation of civilian technologies, many of them readily available in the commercial marketplace.

Constructing a Single National Techno-industrial Base

The First Gulf War drove home the limits of the DoD's procurement system. The GPS receivers that US commanders used to communicate with soldiers on the ground were built according to the DoD's prevailing specifications (Milspecs). Each weighed seventeen pounds, cost on average $34,000, and took eighteen months to procure. When supplies ran out, the army was obliged to buy commercial receivers on the global market (with the help of Japan). Each weighed three pounds and cost $1,300—a few years later they could be bought for $800.[83] This experience encapsulated all the flaws of the prevailing procurement system and the "segregated" military-industrial base that supplied it. The DoD was burdened with an inflexible, time-consuming procedure that produced unnecessarily complex, costly products. In an emergency it forced the US military into dependence on foreign suppliers who could meet an unexpected surge in demand with a far cheaper and effective commercial equivalent. There was no better illustration that the technological autonomy required for military security depended critically on the "economic security" provided by an innovative, profitable commercial sector that competed successfully in export markets.

The roots of procurement reform can be traced back to the 1980s, when the DoD was increasingly obliged to rely on dual-use technologies following on cuts in defense spending and an increasing reluctance of many smaller firms to do business with a "sclerotic federal procurement system."[84] For the first three decades of the Cold War major US corporations and their subcontractors had produced the equipment required by the DoD to meet high performance specifications irrespective of cost. By contrast, the electronics industry that emerged in the 1980s flourished by producing high volumes of low-cost components on a rapidly advancing technological frontier that could be profitably sold in expanding commercial markets as well being integrated into weapons systems with little loss of performance (a chip costing $10 to meet military specifications could be replaced by a commercial equivalent costing $1).[85] Fifteen out of twenty-one items on the Department of Commerce's list of emerging technologies published in 1989 were also identified on the DoD's list of critical technologies released in 1991, and this provided a clear indication of the growing significance of dual-use commodities.[86] Securing technological leadership and market domination in research intensive industries like semiconductors, microelectronic circuits, software engineering, high-performance computing, and machine intelligence and robotics would generate profits in global markets that could be ploughed back into R&D. It would also ensure a domestic supply of key technologies for insertion into advanced weapons systems that could offset numerical inferiority with technological superiority.

The Department of Defense recognized that new technologies that were critical to military advantage were being driven by commercial, not military demand. As Clinton's then deputy secretary of defense William J. Perry told reporters in October 1993, "Basically, our strategy today in computers is to get on the shoulders of the computer industry and take advantage of the developments which are taking place."[87] Perry aimed to revolutionize the federal procurement system to meet this situation.[88] The Federal Acquisition Streamlining Act of 1994 effectively abolished the Pentagon's long-standing reliance on thirty-one thousand specifications and standards that had to be met by defense contractors. The armed services were told to use commercial performance and standards specifications instead of the existing Milspecs unless no practical alternative existed. Testifying to Congress in 1994, Anita Jones, the DoD's director of defense research and engineering, said that to erode a "segregated," "isolated," and "ghettoized" defense industrial base, defense contractors would have to learn to serve "multiple customers, not just one, to market products rather than respond to specifications, and to regard cost" as being "as important as performance."[89] The Pentagon was called on to adopt policies to support dual-use R&D (70 percent of new DoD investment in R&D would have dual-use applicability),[90] to integrate defense and commercial production alongside one another on the factory floor, and to design weapons systems from the outset to incorporate commercial rather than defense-unique materials. This dual-use technology strategy was intended to gravitate, wherever possible, "toward a single, cutting-edge national technology and industrial base that [would] serve military and industrial interests."[91] Generic dual-use technologies like software, computers, and semiconductors, as well communications and remote-sensing satellites, were the material infrastructure that wove trade liberalization and economic competitiveness into the pursuit of military superiority. From the 1990s onward the United States' military arsenal combined a mighty nuclear strike force with the capacity for rapid, lethal intervention in conventional warfare using the information-based Revolution in Military Affairs.

Reforming Export Controls on Sensitive Dual-Use Items

Clinton's vision for an American export-driven economy hastened organizational changes to dual-use licensing that had begun under his predecessor. In November 1990, with the opening up of the former Soviet bloc, and under pressure from industry, President Bush ordered that by June 1991 all items on the State Department's Munitions List that were also on Cocom's dual-use list should be transferred to the jurisdiction of the Department of Commerce unless they jeopardized significant US national security interests. The subsequent interagency "rationalization" exercise took longer than anticipated, lasting until April 1992.[92] The Department of Commerce insisted that all dual-use items should be under its jurisdiction. Against them, the Departments of Defense and State insisted that munitions controls were generally more stringent than the DoC's dual-use controls and resisted pressure to shift some sensitive items entirely under its control. In the event, about two dozen dual-use items were transferred from the Munitions List to the Commodity Control List, with the State Department retaining some control over particularly sensitive items. This did not settle the issue, however. Indeed, throughout the 1990s there were ongoing negotiations between the agencies over a number of sensitive dual-use technologies (hot-engine jet technology, stealth technology, high-performance computers, communications satellites) that led to investigations by the General Accounting Office on behalf of Congress into the functioning of the export control system.[93]

These ongoing turf struggles arose in part because the core licensing bodies had different and potentially conflicting objectives.[94] The Arms Export Control Act gave the State Department the authority to use export controls to further national security and foreign policy interests without regard to economic or commercial factors. In making licensing decisions it consulted with the Department of Defense and the intelligence community, but not the Department of Commerce. For its part, the Commerce Department, as the focal point for exports of civilian dual-use commodities, was given wide powers by the Clinton administration that deliberately tilted the balance in dual-use exports in favor of commercial interests. In 1994 the DoC advocated a revision to the Export Control Act that proposed that it "ensure that US economic interests play a key role in decisions on export controls and to take immediate action to increase the rigor of economic analysis and data available in the decisionmaking process" so helping to evaluate the financial losses to US firms by overrestrictive export controls.[95] By 1995 there were already strident complaints that "less than half of the applications that Commerce approved for missile technology exports for China were even seen by the Department of Defense"—and that was just one of many examples.[96]

In December 1995 President Clinton tried to calm matters down by opening up the DoC-led licensing process to all interested agencies. Executive Order 12981 still authorized the secretary of commerce to make the final determinations with regard to export licenses under his or her jurisdiction. However, it also decreed that "the Departments of State, Defense, and Energy, and the Arms Control and Disarmament Agency each shall have the authority to review any export license application submitted to the Department of Commerce," and it defined a procedure for settling disputes between the different stakeholders.[97]

The deregulation of dual-use exports in the post–Cold War period was extensive. The number of dual-use licenses applied for annually to the

Department of Commerce decreased precipitously from nearly one hundred thousand in 1989 to just 8,705 in 1996 and 11,462 in 1997.⁹⁸ The value of these exports was $4.9 billion in 1996, about 0.6 percent of the value of all goods and services exported that year ($846 billion).⁹⁹ This 90 percent decline in volume arose because far fewer export licenses were needed. In effect the system "bottomed out" close to the level required to restrict dual-use items that were "multilaterally controlled or items that [were] controlled to terrorist or other rogue states" that posed a proliferation danger (Iraq, Iran, Libya, and North Korea), that is, about nine thousand licenses annually.¹⁰⁰

The trade-offs between commercial benefits and national security risks were particularly contentious when exports to China were concerned. Two strategic arguments were explicitly given to justify the liberalization of trade in cutting-edge dual-use technologies with this partner. The first was economic security. Ian Baird, the deputy assistant secretary of commerce for export administration from 1992 to 2000, made the point clearly. As Baird explained, the Department of Commerce could count on the Department of Defense, which "understood that our defense industrial base was not a separate entity from our commercial industrial base, and that they were in fact completely interwoven, and that if you undermine your commercial strength and innovation you are having direct consequences on your defense industrial base. . . . Trade with China was an important aspect in maintaining that commercial strength."¹⁰¹ William Reinsch, undersecretary for export administration in the Department of Commerce, made the same point bluntly as regards satellite exports in congressional testimony in 1998. "As the line between military and civilian technology becomes increasingly blurred," Reinsch said, "a second-class commercial satellite industry means a second-class military satellite industry as well. The same companies make both products," he went on. "And they depend on exports for their health and for the revenues that allow them to develop the next generation of products."¹⁰²

The president himself also saw no intrinsic contradiction between export enhancement and national security. The General Accounting Office noted that in one of his reports Clinton had identified what he called "the *most serious national security issue*: the reliance of the US military on the high performance computer industry and the need to ensure that the industry is able to maintain worldwide market share to stay at the forefront of technological innovation."¹⁰³ For Clinton, economic security was a precondition for achieving national security goals.

The second key argument used to defend deregulation was that China was not a serious military threat. On the contrary, as Mitchel Wallerstein, who had been engaged in export control policy since the mid-1980s and who served as deputy assistant secretary for nonproliferation policy from 1993 to 1997, put it, "the Administration policy toward China is one of constructive engagement. . . . Our overall goal is to encourage China to become integrated into the world system and to meeting international norms of behavior, in non-proliferation and export controls, as well as other areas. We believe that expanding trade, business, academic, and government contacts with China is supportive of this goal."[104] Trade liberalization not only enhanced American economic security and military power. It was a lever that could be used to secure compliance with non-proliferation norms and practices—"rogue states" apart.

These arguments for promoting the export of dual-use high-tech commodities, and the related technical data and know-how, to China were vigorously contested by some in the State Department and the Department of Defense throughout the 1990s. They became particularly rancorous after 1995, when the Republican Party secured a majority in both the House and the Senate, and as a Democratic Party president became even more determined to promote trade with China. Political bargaining over where to strike the balance between commercial interests and national security pitted the proponents of trade liberalization against an increasingly influential and concerned contingent of "non-proliferation entrepreneurs"[105] or, more generally, "control hawks."[106] These were located in the administration, within Congress, and in the public sphere, where NGOs deeply concerned by the spread of WMDs had emerged in the 1980s.[107] They argued not only that China was a major proliferator of nuclear and missile technology. It was also, they insisted, a growing threat to US national security and to regional stability in East Asia. In effect, advocates of export enhancement (constituting what Meijer calls a "Run Faster Coalition")[108] were accused of aiding and abetting proliferation themselves by allowing sales of sensitive dual-use technology to China. In the next chapter we will study this conflict as it unfolded during President Clinton's second term of office and in the context of China's economic and military modernization.

CHAPTER NINE

The Conflict over Technology Sharing in Clinton's Second Term
The Cox Report and the Use of Chinese Launchers

The ambiguities surrounding the promotion of trade with China in the name of economic security came to the fore in President Clinton's second term. China's seemingly insatiable appetite for US advanced technology was a boon to firms that defined the cutting edge of the "information-based economy," and that could benefit from the relaxation of export controls to make major inroads into the Chinese market in pursuit of US economic security. At the same time, the end-use flexibility of their products undermined the argument that increased national security flowed from enhanced trade in dual-use commercial products, since they could also be deployed to strengthen the PRC's military capabilities. Trading with an emerging power whose values were orthogonal to those espoused in the West, that had no hesitation in exploiting the "free-market" practices of capitalism to extract and exploit advanced technology and knowledge to its advantage, and that was forging ahead economically was increasingly unsustainable. The president's decision early in 1998 to waive sanctions to allow an American satellite manufacturer, with close ties to the Democratic Party, to launch a huge satellite on a Chinese launcher was a turning point, exploited by his opponents to accuse him of putting business interests ahead of national security.

This chapter describes the debates surrounding the relaxation in the 1990s of export controls on telecommunications and high-performance computers (briefly), and on the launch of US manufactured satellites in

China, inherited from the Reagan and Bush years (in detail). The growing opposition to Clinton's policies was crystallized in a searing report (abbreviated the Cox Report, after the name of one of its chairmen), released in 1999. It detailed in highly provocative language the threat posed by China to US national security by its legal and illegal acquisition of advanced technology and know-how, including nuclear weapons and their delivery systems. This chapter analyzes in detail the criticism leveled by the Cox Report and a Senate Intelligence Committee on the risks of sensitive knowledge and know-how migrating from civilian space launchers to ballistic missiles. Even before the Cox Report's declassified version was released, a new law had been passed by Congress that placed an extremely high wall around the sharing of sensitive knowledge with space engineers and program managers in many foreign countries. It was the first time that Congress (as opposed to the executive branch) had defined detailed export control regulations for a specific technological market, specifying practices and procedures that had to be followed to control the transnational flow of knowledge with select powers, and that could be changed only by subsequent congressional intervention.

US-China Trade in Telecommunications Equipment and HPCs in the 1990s

Economic and military modernization went hand and hand in China in the 1990s. Already in 1982 Deng Xiaoping had announced a defense conversion program with his famous sixteen characters guidelines, including combining military and civilian activities, combining peacetime and wartime preparations, giving priority to military products, and letting the civilian sector support the military sector. This gained little traction in the 1980s, when, as we saw in chapter 8, the priority was placed on economic modernization, and the military was upgraded with conventional weaponry, much of it supplied by the United States to strengthen an ally in the anti-Soviet struggle. The situation changed in the early 1990s. During the Fourteenth Party Congress in 1992 the leadership formally adopted the concept of a socialist market economy that revolved around "the privatization of the state-owned sector and a decreased role of the state in economic management."[1] In parallel the First Gulf War alerted the Chinese military to the vulnerability of their forces to the Revolution in Military Affairs (RMA), which had crushed Saddam Hussein's army in 1991. According to Mark Stokes, the US assistant air attaché in Beijing from 1992 to 1995, "Since the conclusion of the Gulf War, a growing chorus of PLA officials have strongly advocated the aggressive pursuit of information-based warfare doctrine and systems." Stokes went on to quote a senior Chinese general at a high-level conference in December 1995 as saying that "as far as the PLA is concerned, a military revolution with information warfare as the core has reached the stage where efforts must be made to catch up and overtake its rivals."[2] The first steps were taken toward that end in 1991 when the Chinese authorities prioritized the development of an indigenous capability in microelectronics and telecommunications equipment. Later plans emphasized the need to build a high-capacity national information infrastructure using fiber optics, satellite communication systems, and systems integration and data fusion. This would need increasingly powerful computers that could handle massive amounts of data required for real-time information processing. The acquisition of foreign technology was essential if China wanted to modernize rapidly both its economic infrastructure and its defense capability.

During the 1990s China's defense budget doubled from RMB 63 billion (US$7.6 billion) in 1995 to RMB 121 billion (US$14.6 billion) in 2000, according to official statistics.[3] Military capabilities were strengthened along two axes. Firstly, China bought Russian and Israeli weapons to replace imports from the United States and the European Union, who had imposed arms embargoes on Beijing in retaliation for the violent suppression of prodemocracy protests in Tiananmen Square. Secondly, notwithstanding considerable internal resistance, there was increasing attention, from the late 1990s onward, on civil-military integration, "harnessing the technological and industrial capabilities of the civilian economy to advanced defense capabilities."[4] This was accompanied by the opening of the market, which led to a surge in foreign direct investment (FDI), which increased twelvefold from $3.5 billion in 1990 to $40.7 billion in 2000, mostly in industries that were relatively low tech.[5] Chinese global exports rose from $62.9 billion to $249.2 billion over the same decade, its imports from $53.9 billion to $225.1 billion.[6] Acquisition of dual-use information and communication technologies, including supercomputers, as well as of space technologies, including the provision of launch services, helped lay the foundation for integrating "the defense economy into the broader civilian economy to form a dual-use technological and industrial base that serve[d] both military and civilian needs."[7]

The Clinton administration's determination to liberalize export controls was a boon to China's 1990s modernization agenda. In the early 1990s computers and telecommunications equipment represented 88 percent of the value of all controlled US exports, many of them available from countries that were outside the prevailing control regimes.[8] Gregory Hughes, president of AT&T's Transmission Systems, testified to Congress in June 1993 that the firm would lose $110 million by the end of that year (and $50–$100 million annually thereafter) because of outdated export controls on transmission equipment, in which China and Russia were emerging as major markets. This was due to Cocom regulations imposed on exports of fiber optics transmission devices, software for digital telephone exchanges, and cellular telephone gear. Hughes called on President Clinton to eliminate all such controls for civilian uses in the two countries.[9] A few months later, in August 1993, a bill was laid before Congress to "liberalize controls on the export of telecommunications equipment and technology in order to promote democracy and free communication and enhance economic competitiveness."[10] The intelligence community added another argument: sales of telecommunications equipment to Russia and Eastern Europe, and to China (a market estimated to be worth some $50 billion), would assist in intelligence gathering. As one former CIA official put it in 2011, there was "an alignment of interests" between business, the Pentagon, and the intelligence community, who realized that if the United States traded with China "we would know the equipment which was being sold or utilized and would be able to sharpen our expertise in terms of being able to follow what was being said."[11] As a result, in 1994 the Commerce Department authorized the uncontrolled export of telecommunications equipment, such as fiber optic, radio relay, cellular communications equipment, and related advanced switching equipment to China.[12] These exports contributed to meeting the goals of the PRC's Eighth Five-Year Plan (1991–95), which called for the transformation of the country's telecommunications infrastructure from analog to digital coverage, making an investment of $29.4 billion in fixed-post and telecommunications systems.[13]

The Clinton administration adopted a particularly aggressive policy toward removing export controls on computers, which accounted for a good deal of US-China trade in the 1990s.[14] Export controls on computers were implemented by defining a threshold, measured by processing power in "million theoretical operations per second" (MTOPS), beyond which a device qualified as a high-performance computer (HPC) or supercomputer. With that benchmark established, "extraordinary licensing and safeguard conditions may be placed on the sale or transfer of any machine at or above that threshold."[15] In 1990 Bush's Department of Commerce set it at 100, 150, and 300 MTOPS depending on the destination country. The administration also reached an agreement with Japan. In the late 1980s, they established the Supercomputer Control Regime, to impose security safeguards on all computers of 195 MTOPS or more.[16]

Clinton radically revised these thresholds soon after entering office. He personally reassured Ed McCracken, the chairman and CEO of Silicon Graphics, that he was committed to developing an export control policy "that prevents dangerous technology falling into the wrongs hands without unfairly burdening American commerce." In September 1993, after reportedly having dinner in Silicon Valley with some two dozen computer industry executives, to their "shock and joy" the president lifted the threshold for supercomputers from 195 to 2,000 MTOPS and suggested removing all controls on computers with a power below 194 MTOPS unless they were being sold to the former Eastern bloc, China, or the "rogue states."[17] These were precisely the thresholds suggested by a representative of the computer industry to a House committee that held a "field meeting" in Santa Clara in August 1993.[18]

A second wave of liberalization followed in 1995 to accommodate advances in computing technology. The power of microprocessors (chips) that formed the basis of commercial computers increased exponentially from about 4.5 MTOPS in 1990 to about 4,000 MTOPS a decade later, thanks to improvements in photolithography, the heart of the chip-manufacturing process. In addition, the development of parallel processing, which broke tasks down into smaller parts that could be treated simultaneously by a number of interconnected processors, bypassed the need for large, standalone high-performance computers. These now had to compete with clusters of mass-market microprocessors or desktop computers, or with relatively low-power computers linked together by the Internet. Moore's law—the prediction made by George Moore of Intel in 1965 that computing power would double every eighteen to twenty-four months—seemed to be true.[19]

To meet the challenge, the administration broke the foreign market down into four tiers defined by the destination country of the export. The tier in which a country was placed was based on its adherence to proliferation control regimes, its relations with the United States, and the potential for transshipment or diversion to particular uses, notably the enhancement of WMDs.[20] China and Russia, along with about fifty other states, were deemed to pose proliferation risks and fell into Tier 3.[21] Here the limit above which a license was needed was set at 7,000 MTOPS for any recipient and any civilian use (as reported by the client) and 2,000 MTOPS for military and weapons end users and end uses.[22]

In January 1997 the Russian minister of atomic energy announced that his ministry had purchased five American supercomputers, four from Silicon Graphics and one from IBM, for two Russian nuclear weapons design labs, Chelyabinsk-70 and ARZAMAS-16. In addition, seventeen computers had been illegally sent to Russian weapons laboratories. Lifting the thresholds in 1996 had also enabled China to acquire at least forty-six American HPCs, one of which went to the Chinese Academy of Sciences (CAS). Testifying to Congress in 1997, Stephen Bryen, who had set up the Defense Technology Security Administration (DTSA) to monitor exports in the Reagan Administration (see chapter 8), said that "the sale of supercomputers to China should be regarded as a crazy policy." He pointed out that CAS was "deeply involved in nuclear programs." The machine that they had purchased from Silicon Graphics was "faster than two-thirds of the classified systems available to the Defense Department."[23] What is more, it was networked to workstations situated behind a firewall, using state-of-the-art digital telecommunications systems supplied by American companies. As the Chinese themselves put it, their new Silicon Graphics Power Challenge XL supercomputer provided the academy with "computational power previously unknown," which was available through networking to "all the major scientific and technological institutes across China."[24]

The nonproliferation entrepreneurs pressed their case at a meeting of the Senate Armed Services Committee in July 1998. Bryen described China's "current modernization program [as] potentially provocative and offensive in character."[25] He had been in the Taiwan straits in 1995–96 when the PLA had launched a number of missiles close to Taiwan on several occasions. Their aim was to intimidate Taipei, first, after a former Taiwanese president visited his alma mater, Cornell University, in 1995 to give a talk on democratization in his country, and then in anticipation of presidential elections on the island in 1996. He deplored the fact that by now perhaps as many as one hundred HPCs had been sold to China. They would give the country in-house military design capability for new weapons systems programs and help it to design smaller nuclear weapons for tactical systems and cruise missiles and to crack codes, particularly on commercially encrypted traffic. Gary Milhollin, the executive director of an NGO, the Wisconsin Project for Nuclear Arms Control, agreed. "By any objective standard," Milhollin said, "the supercomputer export control policy under the Clinton administration . . . has been a disaster. It has armed proliferant countries. It has decreased our national security by arming institutions that are making weapons of mass destruction around the world." And its response to export licensing disasters, like the sale of HPCs to Russian weapons labs, "is pretty much to sweep them under the rug."[26] In placing economic security ahead of all else, the Clinton administration was proliferating technologies that contributed to China's WMD posture.

The assault on the president and his administration for putting commercial interests ahead of national security reached a new level of intensity in the next two years. Reporting in September 1998, the General Accounting Office was dismayed that the number of HPCs exported to countries in Tiers 1, 2, and 3 had surged from 815 in FY1996 to 3,390 in FY1997. Only 42 of the latter required licenses (as compared to 395 licenses in FY1995).[27] A follow-up report in November 1999 noted that from November 18, 1997, to August 27, 1999, the United States had approved the export of 4,092 HPCs to Tier 3 countries, of which China had received 1,924, one of which had a power of 24,750 MTOPS.[28]

Clinton was unmoved. Indeed in July 1999 the president proposed raising HPC thresholds even higher for Tier 3 countries, from 7,000 to 12,500 MTOPS for civilian end users, with immediate effect, and from 2,000 to 6,500 MTOPS for military users, effective January 1, 2000.[29] No sooner was this deadline passed when, on February 1, 2000, President Clinton announced a further relaxation of export controls across the board, lifting the licensing threshold for Tier 3 countries like Russia and China from 6,500 to 12,500 MTOPS for military end users and from 12,500 to 20,000 MTOPS for civilian end users. The momentum was maintained partly because some of Clinton's most virulent critics, like Stephen Bryen, thought that it was too late anyway to reverse the liberal trade policies on computers. Other observers, like James Lewis at the Center for Strategic and International Studies, argued that it had become "ineffective and even counterproductive" to deny access to computing power in the name of national security after the 1980s. Lewis also pointed out that none of the multilateral nonproliferation regimes—the Nuclear Suppliers Group, the MTCR, or the Australia Group, all led by the United States—controlled computers, since only very low levels of computing power were needed for the design and manufacturing of WMDs anyway.[30]

The Clinton administration's successful deregulation of sales of HPCs and telecommunications equipment in the name of economic security hit a brick wall in the space sector. Here his willingness to allow US satellite manufacturers to use Chinese Long March rockets was both vociferously opposed and successfully reined in by his opponents, with major effects on US international space cooperation with friend and foe alike. This debate provides a particularly rich case study of the importance Congress attached to regulating the circulation of knowledge and intangible know-how to China, and of the draconian and intrusive measure that it adopted to achieve its aims. It also reminds us that strict controls over knowledge sharing with China were put in place almost two decades before President Trump made of them a key plank in his efforts to cripple the technological capacity of his major global rival, which we discuss in the epilogue to this book.

Communications Satellites and Long March Launchers

As we described in chapter 8, President Reagan authorized the launch of three satellites on China's Long March rocket, to compensate for the lack of domestic launch providers that occurred in the wake of the *Challenger* space shuttle explosion in January 1986. It is important to realize that the proliferation concerns raised by these transactions were not so much focused on the shipment of the satellites themselves, but primarily on the sharing of knowledge by the satellite manufacturer when the device was mated to the rocket, and if, or when, an accident might occur that was due to a malfunction of the launcher. Satellite providers, and the insurance companies that covered the launch of a satellite, suffered major financial losses when launch failures occurred. They, along with the launch providers, had a shared interest in enhancing the reliability and the performance of the rocket. In doing so, however, they could also potentially enhance the reliability and performance of ballistic missiles that shared many technological subsystems with space launchers. The risk of transfer across the civilian-military divide was heightened in China because the China Great Wall Industry Corporation, which provided space launch services for the PRC, also produced missile technology for the People's Liberation Army. In addition, it was a subsidiary of the China Aerospace Corporation, which had close links with the organization that produced China's ballistic missiles. In May 1998 the *Washington Times* reported that China had deployed eighteen ICBMs, some of which targeted the United States.[31]

The technological and institutional overlap between rockets and missiles led to concerns that dual-use knowledge would circulate freely within an engineering community that moved back and forth across the civil-military divide. To avoid that happening, even while benefiting from the economic advantages of launching in China, the administration put an increasingly comprehensive system of controls over knowledge circulation in place. It was intended to allow commercial interests to flourish while reducing the risk of the proliferation of sensitive knowledge to the barest minimum. It failed, however, to eliminate persistent doubts among some in Congress, in the administration, and in nonproliferation pressure groups that the Clinton administration's commercial satellite licensing system was biased in favor of business at the expense of national security—as with the export of high-performance computers.

Two questions took center-stage in the ensuing and increasingly acrimonious debates over export control policy: How effective was the regulatory system for exporting communications satellites that had been put in place in the 1990s? And what, if any, sensitive knowledge had been shared with Chinese engineers in successive launch campaigns—and three accident inquiries—that could enhance their missiles? These technical questions were embedded in a politically driven debate in Congress, and in the press, which included accusations that the Clinton administration was lax on security, and was granting favors to corporate executives with interests in China in return for their political support. To properly assess the relative weight of political and technical considerations in the policies eventually adopted for launching satellites in China, we need to enter in some detail into the messy complexity of the satellite export control system, and grasp the intricacies and ambiguities of how knowledge moves across a civil/military, launcher/missile interface.

In the late 1980s satellites and their components were treated as munitions and were licensed by the State Department through ITAR, in consultation with the Department of Defense. As pointed out in chapter 8, in 1990 President Bush instructed the relevant agencies to "rationalize" dual-use export controls, adapting them to a geopolitical context no longer dominated by superpower rivalry. In October 1992, jurisdiction over about half the existing types of satellites was transferred to the Department of Commerce (EAR/CCL)—a major move of control liberalization.[32] The others—communication satellites that incorporated one or more of nine dual-use components that were defined as militarily significant—remained in the realm of the much stricter ITAR and the US Munitions List. So too were the individual components themselves (say kick motors, if they could restart themselves) if exported individually (see table 6).[33]

TABLE 6. **Military Significant Technology Integrated in Commercial Communications Satellites, as Specified by the USML**

Characteristic or Component	Definition	Military Sensitivity of Characteristics Exceeding Certain Performance Parameters
Antijam capability	Antenna and/or antenna systems with the ability to respond to incoming interference by reducing antenna gain	Ensures that communications remain open during crises
Antenna	Allows a satellite to receive incoming signals	Antenna aimed at a spot roughly 200 nautical miles in diameter can become a sensitive radio listening device
Cross links	Provide the capability to transmit data from one satellite to another without going through a ground station	Permit the expansion of regional satellite communications coverage to global coverage; permits very secure communications
Baseband processing	Allows a satellite to switch from one frequency to another with an onboard processor	Onboard switching can provide resistance to jamming of signals
Encryption devices	Scramble signals and data transmitted to and from a satellite	Allow telemetry and control of a satellite to deny unauthorized access; can have significant intelligence features
Radiation-hardened devices	Provide protection from natural and man-made radiation environment in space	Permit the satellite to operate in nuclear war environments and to survive a nuclear explosion
Propulsion system	Allows rapid changes when the satellite is in orbit	Facilitates military maneuvers by allowing the satellites to accelerate fast to cover new areas
Pointing accuracy	Provides a low probability that a signal will be intercepted	High-performance pointing capabilities provide superior intelligence gathering capabilities
Kick motors	To deliver satellites to their proper orbital slots	If the motor can be restarted the satellite can execute military maneuvers by moving to cover new areas

The mechanisms to safeguard against knowledge loss that were negotiated by the Reagan administration with Beijing were tightened up in February 1993.[34] The Department of Defense was made responsible for overseeing the Space Launch Technology Safeguards Program. The DoD supplied monitors to attend "all technical meetings with US contractors and approve the release of technical data for all programs involving the launch of US satellites by China."[35]

The Clinton administration revisited export controls over communication satellites soon after the new president entered office. Over the next three years there was ceaseless pressure from his Trade Promotion Coordinating Committee (TPCC) to transfer all commercial communications satellites from the USML to the CCL. Clinton established the interagency TPCC by Executive Order in September 1993 to coordinate the government's "export promotion and export financing activities."[36] In 1995 the Department of Commerce's Office of Strategic Industries and Economic Security added their support for a more liberal policy. Even limited export controls against China had already had "disastrous" consequences for US satellite builders, who were trading with "the biggest emerging market in the world."[37] Not everyone went along with these pleas. A private defense firm that participated in the review opposed the relaxation of controls. Its representative was emphatic that "proliferation of the hardware and the basic technology" would be "the undeniable result" of any such trade, and could expose US forces to space-based weapons systems that they "will find difficult if not impossible to counter."[38] In October 1995 the secretary of state denied the Commerce Department's request to have jurisdiction over all communications satellites sold to civilian end users.[39]

The Department of Commerce refused to yield. It appealed the decision to the National Security Council and to the president. In November 1996 President Clinton accepted their demand, and *all* commercial communications satellites, including those with "military sensitive" technologies built into them (i.e., those identified in table 6), were transferred from the State Department to the Department of Commerce. (A license to export subsystems separately, rather than in a completely assembled satellite system, still had to be obtained from the State Department.) These steps toward trade liberalization were, however, limited by three measures to tighten up export controls.[40]

Firstly, the foreign availability clause, which authorized an export if similar technology was available on the world market, was no longer acknowledged as a reason to grant a license for a civilian communications satellite.

Secondly, new procedures for enhanced interagency consultation were implemented.[41] If before, the Commerce Department was not required to send license applications on to the DoD for review, now all applications were seen by five government agencies with a stake in the decision (the

Departments of Commerce, State, Defense, and Energy and the Arms Control and Disarmament Agency). Though none had veto authority, each had the right to appeal the decision via an escalating hierarchical chain, and ultimately to the president. By 1998 no agency had exercised this right. In the view of Mitchel Wallerstein, DoD deputy assistant secretary for counterproliferation policy, this new consultative procedure "put an end to the highly inefficient and counter-productive, internecine struggles within the executive branch of government."[42]

Thirdly, a more detailed and tighter safeguards procedure was put in place in 1996 to further reduce the risk of US satellite manufacturers providing sensitive hardware or knowledge to Chinese engineers that would enhance the performance of their space systems. It required the satellite contractor to draw up a Technology Transfer Control Plan (TTCP), to be approved by the DoD's Defense Technology Security Administration (DTSA). It also made obligatory the use of DoD-appointed monitors during every launch campaign and defined their tasks in some detail.

A launch campaign can take from one to three years. It consists of "a series of technical interchange meetings and other interactions between US satellite engineers and the launch service provider" that can last for several days and that take place both in the United States and at the provider's home base. Typically in the course of such meetings "the two sides exchanged detailed information on satellite and launch vehicle specifications, capabilities, technical requirements, satellite-launch vehicle integration and other matters critical to a successful launch to the designated orbit."[43]

Each TTCP included a "detailed transportation plan for shipping the satellite to ensure that only US personnel have access to the satellite at all times, and a detailed physical and operational security plan, including procedures for the supervised mating of the satellite to the launch vehicle." The Department of Commerce could license the release of technical data related only to "form, fit and function"—in oversimplified terms this referred to the satellite's "size and shape, how do you plug this thing into the launcher [sic]."[44] The sharing of all other technical data had to be licensed by the Department of State in consultation with the DoD, and in accordance with ITAR provisions. This included all data associated with the integration of the satellite payload with the launch vehicle, and any technical assistance provided by US companies to Chinese launch service providers, including help with any launch failure analysis.

There was also a change in policy regarding the presence of monitors. Monitors had been introduced under the Bush administration in 1992 but were necessary only if technology would be shared in the launch campaign that required a license from the State Department. In fact, no monitors were used for three launches in China under the Commerce Department's jurisdiction until 1996 on the grounds that all transactions had involved "purely commercial satellites."[45] The tighter system implemented in 1996 expanded the range of situations in which the presence of monitors was obligatory. Monitors were DoD employees recruited by the DTSA and were "mostly military personnel, sometimes civilian employees, who volunteer for this duty."[46] Their services were paid for by the satellite manufacturer, and they were present throughout the launch campaign to secure the hardware and to control the exchange of sensitive technical data, knowledge, and know-how shared in face-to-face exchanges. These monitors had to be present to ensure security at the launch site in China, and to participate in technical meetings between the US exporters' engineers and managers and the Chinese launch provider personnel.

Accepting and paying for these extremely intrusive modes of government surveillance were not the only hurdle to be overcome if a company wanted to launch US-origin satellites in China. The sanctions imposed after the bloody repression of prodemocracy protests in Tiananmen Square also stood in the way. These sanctions amounted to a weapons embargo and therefore applied to all items, including satellites covered by ITAR. Section 610 of Public Law 101-62 of November 21, 1989, allowed these "Tiananmen sanctions" to be waived if the Chinese authorities made progress on a program of political reform throughout the country or if the president judged that it was "in the national interest of the United States" to do so (see chapter 8). A different set of sanctions on the use of Chinese launch systems was imposed if Beijing violated the Missile Technology Control Regime.[47]

Table 7 summarizes all presidential actions regarding the imposition and then waiving of these sanctions until the end of 1998, including those taken by President Bush that we listed in chapter 8.[48] Between 1990 and 1999, Presidents Bush and Clinton together had approved twenty satellite projects for thirty-three launches in the PRC, declaring them to be in the "national interest." President Bush authorized fourteen of these launches (of which twelve took place). President Clinton authorized the launch of nineteen satellites, of which sixteen had been put into orbit by May 1999.[49] All decisions were reported to Congress, who never opposed them.

The tense equilibrium between the administration and its critics was shattered in 1998. It came to light that in analyzing the launch failure in 1996 of a Chinese Long March rocket carrying an Intelsat communications satellite built by an American company, Western experts had di-

THE CONFLICT OVER TECHNOLOGY SHARING 271

TABLE 7. The Imposition (Column 1) and Waiver (Column 4) of Tiananmen and MTCR Sanctions, 1989–98

Date and President	Sanctions	Conditions (Abbreviated)	Waived	Motive for Waiver
11/21/89 Bush	Tiananmen	Can be waived if there is domestic political reform or if in the US national interest	12/19/89 Bush	Waived to allow launch of AUSSAT and Asiasat 1 satellites, worth $300m of business for Hughes
2/16/90 Bush	Tiananmen (additional sanctions)	Can be terminated in whole or in part if there is political reform, also in Tibet, and if in the US national interest	4/30/91 Bush	Waiver to launch AUSSAT (which had been delayed) and Freja satellites
5/27/91 Bush	Category II MTCR	2-year sanctions on US exports to China Great Wall Industry	3/23/92 Bush	Lifted after China gave written assurances to abide by MTCR
8/24/93 Clinton	MTCR	2-year sanctions on exports to Chinese and Pakistani entities for missile proliferation	11/1/94 Clinton	Waived for Chinese entities, as being "essential" to US national security
N/A	Tiananmen	N/A	2/6/96 6/23/96 7/9/96 Clinton	Waiver to export Chinasat 7, 2 Chinasat, Mabuhay, APMT, Globalstar satellites
N/A	MTCR	N/A	11/19/96 11/23/96 Clinton	Waiver to export parts for Fengyun-1 and Sinosat
N/A	Tiananmen	N/A	2/18/98 Clinton	Waiver for Loral's Chinasat 8

vulged knowledge and know-how that could improve the PRC's ballistic missile program without first securing a license to provide technical assistance from the Department of State. The ensuing criticism of the administration and of the president by the General Accounting Office, in congressional debates, in two high-profile investigative committees, and in the press, severely damaged Clinton's authority and led to major reforms of the licensing process, spearheaded by Congress.

The Launch Failure of Loral's Intelsat 708 Satellite and the Collapse of Trust in the Regulatory System

On February 15, 1996, a Chinese Long March 3B rocket carrying an Intelsat 708 satellite manufactured by Loral Space Communications veered off course after liftoff before smashing into an inhabited village on a hillside adjacent to the Xichang launch center, killing over a hundred people and scattering debris far and wide.[50] The China Great Wall Industry Corporation (CGWIC) reported that its engineers attributed the accident to a loose electrical connection in the rocket's guidance system. At the behest of the insurance companies, Loral established an Independent Review Committee (IRC) of Western experts, including engineers from Hughes Electronics Corporation, to confirm the cause of the accident. The IRC held two meetings with CGWIC engineering personnel, one in Beijing, the other in Palo Alto. The minutes of these meetings, as well as a draft and final report, were all shared with the CGWIC. The IRC suggested a number of alternative causes for the rocket's malfunction. Their final report, which situated the cause of the accident in a different section of the guidance system, was faxed to China.

An official in the DTSA read about these activities in an industry publication and asked Loral's Washington office if they had received a license from the State Department to provide technical assistance to the CGWIC. They had not. Loral claimed that their report did not transfer significant or sensitive technology or information. The Pentagon's classified assessment in May 1997 concluded that, on the contrary, "Loral and Hughes committed a serious export control violation by virtue of having performed a defense service without a license."[51]

In April 1998 an article by Jeff Gerth and Raymond Bonner in the *New York Times* revealed that the Department of Justice was investigating whether Hughes and Loral had illegally given China "space expertise that advanced Beijing's ballistic missile program."[52] The article went on to say that another Tiananmen Square sanctions waiver that had been signed by the president "in the national interest" on February 18, 1998, had given "Loral permission to launch another satellite on a Chinese rocket and provided the Chinese with the same expertise that is at issue in the criminal case."[53] In a grueling battle that persisted into 1999, the administration insisted that, at least since 1996, the engagement of the State Department and the DoD in all licensing decisions, along with the implementation of safeguards, had stopped the flow of technology, technical data, and know-how across the civil-military divide. Their opponents challenged the possibility of a sharp civil-military divide and dismissed the government's efforts to hold the high ground by appealing to the rigor of its regulatory instruments.

The Clinton administration's handling of launch failures was mercilessly probed by its critics to expose the limitations of its regulatory practices. Flaws in the system had emerged twice before the debacle with

Loral in 1996. Firstly, there were the procedures surrounding accident inquiries into the launch of two satellites manufactured by the Hughes Space and Communications Company that were destroyed soon after liftoff in China: Optus-B2, launched on December 12, 1992; and Apstar-2, launched on January 25, 1995. Secondly there were serious doubts raised about the performance of DoD monitors in the launch campaigns.

Hughes engineers who discussed the causes of the launch failures of Optus-B2 with their Chinese counterparts attributed it to a weakness in the fairing that protects the satellite on top of the rocket. The company did not seek a license to proceed with the investigation, claiming that it had been told by a DTSA staffer, Lieutenant Al Coates, that a license was not necessary.[54] Years later Coates denied this was the case, and Hughes was accused of an "underlying pattern of misconduct" for not pursuing the need for a license for postaccident technical assistance to the Chinese because it was sure that it would never get one.[55]

The situation repeated itself with the Apstar-2 failure in 1995. Hughes engineers were convinced that the same problem with the fairing had again caused the Long March rocket to explode soon after liftoff with the satellite on board. Under pressure by insurance companies to get to the bottom of the problem, Hughes asked Gene Christiansen in the Department of Commerce, who had licensed the Apstar-2 export, to grant them a license to pass their final report to Beijing.[56] The report included recommendations for improving the fairings, and it should have been cleared by the State Department. Christiansen gave the company the go-ahead mistakenly assuming that the fairing was part of the satellite, not the rocket. Company officials did nothing to correct his error.

Serious doubts were also cast on the role of the monitors after the Loral crash in 1996. The launch agreement specified that Pentagon officials on-site were authorized to collect any debris from the accident. However, the Chinese authorities had kept the monitors locked in a room for five hours after the explosion, claiming that they needed to protect them from danger. Questions immediately arose over the whereabouts of the command box carrying radiation-hardened microelectronic circuits and encryption chips—items that were controlled under ITAR (see table 6). The National Security Agency, on the basis information provided by Loral, "determined that the cryptographic equipment that was aboard the Intelsat 708 satellite was also destroyed and not recovered."[57] Soon thereafter a Department of Defense Damage Assessment overturned Loral's version of events: it stated that "the command processor boxes as a whole were recovered, but not the circuit cards which contained the encryption."[58] The whole episode exposed the limits of using DoD monitors and safeguards to curb the loss of sensitive technology to the Chinese.

Much of the criticism of the Clinton administration claimed that it was putting business interests ahead of the president's supposed commitment to curb proliferation. For example, in August 1993 it was clearly established that China had shipped equipment related to its M-11 missile (not the missiles themselves) to Pakistan.[59] MTCR sanctions were imposed on Chinese entities for two years. The US aerospace industry protested at once. Norman R. Augustine, the chairman of Martin Marietta, wrote to Vice President Al Gore in September to say that sanctions "present US companies as an unreliable supplier."[60] Then in November 1993 the chairman and CEO of Hughes Aircraft Company publicly objected to imposing sanctions on sales of commercial communications satellites and asked the president to review the situation. Clinton obliged ten days later. After securing assurances from the Chinese authorities that they would not export MTCR-class missiles any longer, on January 6, 1994, the president announced that he was exempting commercial communications satellites from sanctions for missile proliferation imposed in August 1993.[61]

The policy change appalled the "'non-proliferation entrepreneurs." For Gary Milhollin, of the Wisconsin Project for Nuclear Arms Control, curbing proliferation was a moral issue. The United States' behavior toward China was signaling to other proliferators that they could share WMD technology with impunity. He was shocked by the State Department's reluctance to impose Tiananmen sanctions, claiming that it "simply is not interested in applying sanctions to China no matter what the facts are."[62] The Arms Control and Disarmament Agency (ACDA) shared Milhollin's view. In April 1998 it stated that "ACDA continues to believe that the United States should link its willingness to make Presidential national interest waivers for China's satellite launches to Beijing's taking concrete steps to resolve our concerns about its failure to implement the MTCR equipment and technology annex, and the ongoing missile-related exports to Pakistan and Iran. . . . History has proven that the only time we have gotten movement from the Chinese on missile non-proliferation has been in the face of a penalty being imposed. Carrots have gotten us nothing."[63] Against them, the Departments of State and Commerce insisted that sanctions were an instrument to be used with prudence in a relationship of "constructive engagement," which was gradually drawing China into compliance with Western nonproliferation norms.

Congressional Republicans weighed in. For Representative Floyd Spence (R-SC) the issue was clear-cut. "China possesses nuclear weapons and the missile on top of which those nuclear weapons are sitting are the very same type as those being marketed to the US satellite industry as providing cheaper access to space," he said. Every launch thus "directly or indirectly" enhanced China's space and military launch capability and posed a direct threat to national security.[64] A particularly bitter whistleblower in the DTSA, Peter Leitner, who had been recruited by Bryen in the Reagan years to oversee export controls, testified that "on several levels what passes for an export control system has been hijacked by longterm ideological opponents of the very concept of export controls . . . lull[ing] us all into a false sense of security while short-sighted business interests line their pockets at the expense of future generations of American soldiers and citizens alike."[65] Senator Strom Thurmond (R-SC) and Stephen Bryen saw the problem as structural rather than personal. They were not opposed to US satellite manufacturers using Chinese launchers. But they were convinced that the Department of Commerce's institutional mandate to promote trade was simply not compatible with the risks to national security, the protection of which should always take precedence over the pursuit of profit. These critics wanted the licensing authority returned to the State Department, and to be subject to a DoD veto, as in the 1980s.

Henry Sokolski, the executive director of the Nonproliferation Policy Education Center in Washington, DC, took a more nuanced position. Sokolski was one among many to point out that there was unambiguous evidence that controls on satellite launches in China were far less rigorous than they had been under Bush administration, where he had served as the deputy for nonproliferation in the Department of Defense until 1993. He was unusual, however, in insisting that there were no quick and easy solutions to the problems exposed.[66] In particular, attempts to improve the administrative process, including shifting licensing jurisdiction back to the State Department, and counting on industry itself to staunch knowledge loss, were beside the point.

Sokolski redirected the spotlight away from the regulatory process itself onto what it was supposed to control. Sokolski had no doubt that the Chinese space community had accumulated an enormous amount of American know-how, or rather "know why," that had dramatically improved the reliability and performance of their launchers and missiles. He explained what "know why" was:

> "Know-how" conveys a given technical procedure, such as satellite integration for a particular satellite and rocket launcher. "Know why" goes further to

explain in engineering and scientific terms why a given procedure is arranged the way it is (i.e. why certain steps must be followed in a given order by others and what fundamental problems or risks these procedures are designed to mitigate or resolve). Such "know why" would enable the Chinese on their own to engineer around such problems for other rocket or satellite systems. In short know how is relevant only to particular system, know why empowers the student to engineer around similar problems for a variety of systems.[67]

The evidence for this growing autonomy of Chinese engineers to manage a variety of launch campaigns successfully was plain to see, said Sokolski. He produced a chart showing that 78 percent of Chinese launches had failed to perform as hoped between 1970 and 1996. Thereafter, by contrast, every one of the next ten launches had been a success—and these used recently developed sophisticated launch systems and placed complex satellite payloads in orbit.[68]

What then was to be done? "Word out on the street," said Sokolski, was that the "fix" to the current crisis of confidence in the administration lay in changing jurisdiction over satellite launchings in China back to the State Department.[69] Sokolski disagreed. The issue was not "how much tougher State controls are than Commerce controls—which they are—but how profoundly deficient both are in monitoring unannounced meetings or data exchanges, emails, faxes, telephone calls or third-party nationals not covered by US laws—methods, all of which are easy conduits for the transfer of US satellite technology."[70]

Sokolski also emphasized that the safeguard system depended far too much on the good faith and word of industry itself. Even the best of controls, he said, "could not ensure contractor compliance in any but a few straightforward cases, that is, those cases in which it was absolutely clear what should and should not be transferred, and in which the contractor's cost for compliance was low."[71] Paul Freedenburg, a consultant for a law firm, previously an official in the Bureau of Export Controls, confirmed the point. In his view, the Technology Transfer Control Plan devised by industry placed "a huge burden on the US companies to create an administrative compliance structure" that was too difficult to implement on a day-to-day basis, and in each and every interaction with the Chinese.[72]

This granted, in Sokolski's view the only way to ensure that there were no losses of sensitive information to China was to bring all launches back to the United States. That decision was not to be taken in haste, however:

The House and Senate should insist that no further satellite transfers be allowed until its investigations and deliberations over this matter are complete. There simply is no substitute for determining what, if anything, of military significance was transferred to China, and whether or not such transfers could have been prevented by tighter controls. If nothing was transferred, then, all's well. If something was transferred, then, Congress must determine what it was, what else of value might be at risk, and whether or not these risks can be eliminated by tighter export controls or if barring future transfers is required.[73]

For Sokolski there was no substitute for establishing clearly what, if any, US sensitive knowledge and know-how had migrated into the Chinese missile program, for discussing if its transfer could be controlled, and for deciding whether all use of the Chinese launchers should be stopped.

Sokolski's advice was not taken. Congress opted for the administrative "solution." On October 17, 1998, President Clinton signed into law the Strom Thurmond National Defense Authorization Act for Fiscal Year 1999 (Public Law 105-261). It included a hastily added amendment that appeared as *Subtitle B—Satellite Export Controls*, sections 1511–16 of PL 105-261. Section 1513 announced a major policy shift. It stipulated that, as of March 15, 1999, "all satellites and related items" that were on the Commodity Control List of dual-use items under the jurisdiction of the Department of Commerce should be returned to the jurisdiction of the Department of State.[74] In a startling reversal of the liberalization decisions of 1992–96, all satellites were shifted back from EAR to ITAR.

There were no committee hearings on this amendment, although the Department of Commerce obviously strongly opposed it. The Department of State denied that it was in favor of the rider. Both houses voted on the act with little debate on the amendment.[75] President Clinton signed the act into law though he "strongly opposed" the transfer of authority and threatened to "to take action to minimize the potential damage to US interests."[76] The National Security Council, in consultation with the US satellite industry, suggested that Clinton issue an Executive Order granting the Department of Commerce a say in the State Department's decisions. They were roundly rebuked by the chairmen of six House and Senate committees, who warned the president against "direct contravention" of the legislation.[77] Henceforth the State Department, which placed greater weight on national security than economic opportunity, would determine whether any US-origin satellites could be launched from China, and it would ensure that those launches were tightly monitored to avoid the loss of sensitive knowledge.

Congress took an exceptional step in defining by law the detailed controls that were to be placed over foreign launches of US-built communication satellites. Normally the composition of the control lists and the practicalities of implementation of the export control acts were firmly in the hands of the executive branch. But in the case of satellites, Congress for the very first time in the history of export controls intervened to determine how one specific technology was to be regulated. If it did not take the more prudent path advocated by Sokolski and wait for the findings of various investigations then underway (see below) before changing the law, it was perhaps because a Republican-dominated Congress was able to capitalize on an organized and relentless critique of the White House's relationships with industry, with the Chinese community in the United States, and with the PRC itself. Senator John McCain (R-AZ) was one of a host of critics who called out "the extremely aggressive lobbying" by industry and who claimed that the Department of Commerce—"much more well known for acting on political considerations" than on behalf of national security—had made concessions in return for "campaign contributions made by individuals who may or may not be tied to the Chinese government."[78]

Two factors fueled this narrative: first, the passionate determination by a small coterie of Clinton critics, perhaps no more than forty Washington "insiders," who organized themselves into a "Blue Team" to bring down the president over his China policies; second, the White House's timing of important and controversial political decisions regarding space technology relationships with China that played right into their hands. Their campaign was further aided and abetted by a president who got caught up in a number of sex and property scandals that had no bearing on national security questions, but that generated extensive coverage on all aspects of his and his administration's behavior by a Washington press corps always on the lookout for a scandal.

The Blue Team's name was coined by the conservative congressional aide William Triplett, who took it from the color that China's military war-gamers assigned to the PRC's unnamed enemy.[79] They included "members of Congress, congressional staffers, think-tank fellows, conservative journalists, lobbyists for Taiwan, former intelligence officers, and a handful of academics," all of whom were opposed to the administration's policy of constructive engagement with China.[80] Triplett was a master of the well-timed leak and the planted question, and he did all he could to expose wrongdoing by China and to embarrass officials who were trying to improve relationships with Beijing. The Blue Team made optimal use of the conservative media but also benefited from extensive coverage in the *New York Times* and the *Washington Post*.

The seeds of the critique of the Clinton administration's space technology policies with China were sown during the president's first term when he waived Tiananmen sanctions on satellite exports and waived MTCR sanctions on China. A succession of events in 1996 tarnished his image further. Several Asians and Asian Americans were accused of illegally funneling money from foreign businesses and the Chinese government to the Democratic Party in an attempt to influence the presidential election. Clinton's close ties to the aerospace industry also came to haunt him. Why had the president waived Tiananmen Square sanctions in February 1998, allowing Loral to launch its next huge satellite in China, even though cautioned by the Department of Justice not to do so? Perhaps because, as *New York Times* journalists Jeff Gerth and Raymond Bonner wrote, Loral's "chairman and chief executive, Bernard L. Schwartz, was the largest personal contributor to the Democratic National Committee last year."[81] In a follow-up article on April 13, 1998, Gerth pointed out that Hughes's president, C. Michael Armstrong, had actively lobbied Clinton to waive sanctions on civilian communications satellites as soon as he entered office.[82] Both Armstrong and Schwartz had written to the president in 1995 urging him to transfer all satellites to the Commerce Department's jurisdiction. Both firms had also shown a clear tilt in political engagement, with contributions of $2.5 million to Democratic causes and candidates, compared to $1 million to Republicans.[83] And then there were the scandals surrounding earlier property dealings by the Clintons and concerning the president's sexual misdemeanors. In September 1998 the House Judiciary Committee announced that it intended to vote on a resolution to impeach Clinton.

This was the toxic context in which the Strom Thurmond Act was passed, and signed reluctantly into law by the president in October 1998. The majority of Congress, and Clinton himself, were not in favor of transferring jurisdiction over the licensing of launches for US-origin satellites back to the State Department. The State Department "neither welcomed nor supported this decision."[84] However, in Robert Lamb's view, "even if President Clinton had had the purest of motives and the greatest of proposals for export controls, his political opponents wanted to stop satellite exports to China, and the sex scandals and impeachment proceedings permitted him neither the time nor the political capital to oppose them."[85]

The Cox Report

In late 1998, the House appointed Representative Christopher Cox (R-CA) the chair of the Select Committee on US National Security and Military/Commercial Concerns with the People's Republic of China. It comprised five Republicans and four Democrats. Norm Dicks (D-WA) served as the ranking member for the minority party. The Cox Committee, as it came to be called, was charged with studying, inter alia, the threat to national security posed by exporting satellites for launch in China, as well as the PRC's efforts to obtain, via multiple channels, advanced missile and nuclear technologies. In October 1998 the committee's inquiry was extended to include alleged lapses of security and nuclear weapons espionage at America's four national weapons laboratories (Los Alamos, Lawrence Livermore, Oak Ridge, and Sandia). The classified version of the committee's findings was submitted to the White House on January 3, 1999. A declassified version was made available to the public on May 25, 1999. It became known as the Cox Report. A complementary investigation by the Senate's Select Committee on Intelligence produced a *Report on Impacts to US National Security of Advanced Satellite Technology Exports to the People's Republic of China (PRC)*, released just a few weeks before, on May 7, 1999.

In spring 1999, while the Clinton administration was assessing the classified report, two further security scandals were reported in the *New York Times*. On March 6 a sensational article by James Risen and Jeff Gerth was headlined "Chinese Stole Nuclear Secrets from Los Alamos, US Officials Say."[86] They claimed that, working with information provided by an unnamed suspect at the lab, the PRC had been able to make "a leap" in the miniaturization of its nuclear warheads. Two days later Wen Ho Lee, a sixty-year-old Taiwan-born American nuclear physicist working at the lab, was fired, his name was leaked to the press, and he became "publicly known as the scientist who betrayed his country and passed along nuclear secrets to the Chinese."[87] In December 1999 Lee was indicted by a federal grand jury on fifty-nine counts alleging that he had transferred sensitive data to unsecured computers and tapes at Los Alamos.[88]

On May 10, 1999, Gerth and Risen reported on another security breach committed by Peter Lee (no relation) that would be described in the soon-to-be released declassified version of the Cox Report.[89] He was charged with providing classified information to China on advanced technology for submarine detection. Lee also admitted to sharing classified information with the Chinese about nuclear testing using miniaturized fusion explosions. In the event, the scares surrounding Peter Lee and Wen Ho Lee evaporated. Peter Lee was sentenced to twelve months in a halfway house, a $20,000 fine, and three thousand hours of community service. Fifty-eight of the 59 counts against Wen Ho Lee were dropped, the judge publicly apologized for his treatment at the hands of the federal government, and Wen Ho Lee regained his freedom in September 2000.[90] Be that as it may, the cases raised enormous anxieties in the American public about the security of top-secret nuclear information.

Nuclear fear dominated the publication of the Cox Report. Cox claimed that "no other country has succeeded in stealing so much from the United States," with serious and ongoing damage to the country. House Majority Leader Dick Armey stated that "it's very scary, and basically what it says is the Chinese now have the capability of threatening us with our own nuclear technology."[91] Not everyone agreed. A group at Stanford University's Center for International Security and Cooperation was uncompromisingly critical of the Cox Report's "dramatic allegations," which had "drowned out the voice of those who seriously challenge the basis of much of the report's substance and conclusions, pointing to extensive problems with the factual content and unreasonableness or improbability of the dangers and risks assumed posed to US national security."[92] A multitude of other sources indicated that prudence was needed. John Spratt, who had served on the Cox Committee, publicly distanced himself from its claim that China had nuclear weapons "on a par with those in the US."[93] The president himself convened an interagency group to produce an Intelligence Community Damage Assessment that was confirmed by an expert panel chaired by Admiral David Jeremiah that Clinton set up in March to review its work.[94] The Jeremiah panel downplayed the role of espionage and concluded that China's main military strategy was defensive, not offensive. John Cirincione at the Non-proliferation Project at the Carnegie Endowment for International Peace agreed.[95]

As acrimonious as the atomic espionage controversies were, only about thirty-five out of almost nine hundred pages in the Cox Report dealt with nuclear weapons directly. The rest of the report was a blistering assault on the liberalization of export controls since the Bush era in regard to high-performance computers and machine tool and jet-engine technologies.

Large sections in its second volume dealt in detail with the launch failures of the Hughes and Loral satellites. It is to this crucial and understudied section of the Cox Report that we now turn. Our analysis of this material is interwoven with a discussion of the complementary report by the Senate's Select Committee on Intelligence, chaired by Richard Shelby (R-AL), which submitted its own findings on the impact of satellite exports on national security just a few weeks before Cox's declassified report was released in May 1999.[96]

What Sensitive Knowledge, If Any, Did Hughes and Loral Share with China?

We asked two questions at the start of this long discussion of the government's export control policies in the 1990s, to wit: How effective was the regulatory system for communications satellites that had been put in place over the 1990s? And what, if any, sensitive knowledge had been shared with Chinese engineers in successive launch campaigns (and three accident inquiries) that could enhance their missiles? The first has already been discussed. To answer the second, we will use the Cox Report, the Defense Department's and the State Department's Damage Assessments cited in that report, and the report by the Senate Intelligence Committee.[97] The analysis is broken down into four steps: What knowledge was shared? How was it shared? What difference did it make? And how could the risks to national security caused by knowledge transfer be reduced to the minimum?[98]

What Knowledge Was Shared?

It is striking to note how little emphasis was placed in these documents on the risks posed by the overlap of hardware between the civilian launcher and the military missile. It had been suggested that improvements to correct flaws in the structure of the rocket fairings that had caused the failures of Hughes's Optus-B2 and Apstar-2 launches could directly improve the shrouds protecting nuclear warheads on Chinese ballistic missiles. The Chinese vehemently denied these equivalences in a spirited response to the Cox Report.[99] The Senate Intelligence Committee also stated unambiguously that it had "found no evidence that technology transfers from American companies have been incorporated into the PRC's deployed ICBM force, which was largely developed and fielded before the first US satellites were approved for export to the PRC."[100] In fact a State Department document had already played down the significance of hardware transfer in the early 1990s. Instead it emphasized the risks involved in the transfer of technical data and the exposure of Chinese engineers to "procedural 'know-how' and systems testing/launch philosophy learned over decades of trial and error" while working for Hughes aerospace.[101] Sokolski also insisted that the loss of "know-how," which he called "intangible" knowledge, during a launch campaign with a civilian satellite was a major proliferation concern.[102] As he put it, "none of the hardware is, in fact, as likely to fall into Chinese hands or to be as militarily significant as is the intangible know-how many US firms want to transfer, in order to ensure the successful launch of the satellites."[103]

Coupling load analysis was a typical example. This technique was critical to ensure that a launcher passed through the sequence: ignition, stage separation, motor cut-off, and so on, without unduly shaking or shattering its sensitive payload. Sharing the techniques and procedures of coupling load analysis with the Chinese could help them *redesign* their rather rigid launchers to ensure that they safely orbited their most fragile and sensitive US satellites—and their own military warheads. By acquiring technical know-how Chinese engineers could *improve* the performance and reliability of their rockets.

The State Department's Damage Assessment also emphasized the importance of *learning* in the process of technology transfer. It identified multiple areas in which Hughes engineers had helped the Chinese to improve the Chinese spacelift program overall by remedying *deficiencies* in their technical grasp of rocketry. These included deficiencies "(in varying degrees) in the area of anomaly analysis, accident investigation techniques, telemetry (TLM) analysis, coupled loads analysis (CLA), hardware design and manufacture, testing, modeling, and simulation, and weather analysis."[104] The American company did not simply provide "descriptive accounts of failure analysis." "Throughout the course of the investigation, Hughes identified faults with Chinese practices and techniques" and "often corrected errors in incomplete or incorrect analysis or filled in gaps where the Chinese simply lacked the technical knowledge." Summarizing its findings, the State Department's assessment concluded that "essentially the APSTAR II failure investigation (and to some extent the investigation of the Long March 3B) served as a *tutorial* for the Chinese allowing them to improve on areas in which their spacelift program was weak."[105]

The Cox Report followed the same line of reasoning in its criticism of the charter provided to the Independent Review Committee established by Loral to investigate the catastrophic failure of the Long March 3B rocket. The IRC members were instructed "not only to go beyond reviewing the PRC failure analysis to making *an independent assessment of the most probable cause or causes of failure*" but also to "*review and make assessments and recommendations concerning the corrective measures to remove the causes of failure*."[106] The Cox Committee tabulated fifteen recommendations made by IRC in their report alongside corresponding corrective actions reported by CALT (China Academy of Launch Vehicle Technology) to show their similarity.

To defend themselves from charges that they had shared sensitive knowledge with their Chinese colleagues, the US providers claimed that they had done no more than share technical information in the public domain with them. This of course did not require a license from the State Department. The Cox Report disagreed. It emphasized that "there is as much experienced-based art as science in the successful application of the well-established numerical analysis and design methods available." Certainly the "basic theories and experimental methods for determining flight loads and environmental conditions are in the public domain." The elusive intangible knowledge that had been transmitted was not. "The successful application of these theories and methods in design often require know-how and engineering judgment derived from experience." "It was the benefit of this experience and know-how that Hughes engineers could have made available to their PRC counterparts."[107] By removing deficiencies and closing gaps in the PRC engineers' technical understanding, and by sharing know-how and engineering judgment, Hughes had helped improve the reliability of the Chinese launchers, and facilitated their access to space for both commercial and military purposes.[108]

How Was Intangible Knowledge Shared?

The documents being considered implicitly assumed that one of the most common means by which intangible knowledge or know-how and know-why was shared was in interpersonal discussions. The Cox Report stressed, for example, that the IRC and the Chinese teams had converged on an explanation for the Intelsat 708 launch failure after two major "face-to-face" meetings, the first in Palo Alto, the second in Beijing. Four Chinese engineers were present at a first three-day meeting, while twenty-two engineers and officers were at a second meeting. They were employed by several Chinese organizations that were involved in both civilian and military space activities, including the China Aerospace Corporation (CASC). CASC and its subordinate companies, research academies, and factories developed and produced strategic and tactical ballistic missiles, space launch vehicles, surface-to-air missiles, and cruise missiles, as well as military (e.g., reconnaissance and communications satellites) and civilian satellites. The meetings in Beijing were held in hotel rooms that were probably bugged by the Chinese intelligence services. The committee also held unmonitored technical interviews with over one hundred PRC engineers and technical personnel, and produced over two hundred pages of data and analyses.[109] The risk of sensitive information being shared with Chinese engineers who worked with both civilian and military space technologies was palpable.

The report identified a variety of mechanisms whereby knowledge was given and acquired in these engagements and revealed an acute understanding of the role of tacit, unformalized knowledge and experience in arriving at a solution to a problem: "The search for the true failure mode in an accident investigation is not a simple, straightforward procedure. In some respects, it is like finding the way through a maze. It is all too easy to start down the wrong path, and stay on it for too long. Insights, hunches and clues based on technical judgments and experience in prior failure mode analyses, simulations, and accident investigations can be helpful (particularly if they come from individuals or groups outside the organization)."[110]

It was not only technical data that was shared in face-to-face interactions then—that could be transmitted as easily in written format without interpersonal contact. Face-to-face engagements—a seminar-like process of learning—permitted the transmission of "know-how," of intangible, hard-to-specify professional judgment, hunches, and insights.

The IRC was convinced that the Chinese engineers had incorrectly diagnosed the cause of the Long March 3B near-horizontal liftoff because they had focused all their attention on the telemetry data for the first seven seconds after ignition instead of analyzing the data for the full twenty-two seconds of flight before the rocket and its payload smashed into the hillside. Accordingly, in discussions with them, the Western investigators had not simply asked the PRC engineers to reconsider their original diagnosis of the accident. They had insisted that their approach was incomplete and unconvincing. They had steered them away from their "protracted narrow focus on the wrong failure mode" of the inertial guidance system. Their "continuing skepticism" with the Chinese interpretation, and their "persistent calling attention" to the telemetry data for the full twenty-two seconds had eventually enabled their Chines colleagues to think out of the box and overcome their initial reluctance to accept the IRC's explanation of what went wrong.[111] The Western experts had not only identified the cause of the launch failure. They had transferred to the Chinese best practice procedures for launch-failure analysis that were readily applicable to both satellite launchers and ballistic missiles.

Did the Sharing of Know-How in Satellite Launch Campaigns Improve the PRC's Ballistic Missile Fleet?

The Senate Select Committee came to the conclusion that "technical analyses and methodologies provided by American satellite companies to the PRC during various satellite launch campaigns [had] resulted in the transfer to the PRC of technical know-how." But it added that although this transfer would enable the PRC to improve both its current and future space launch vehicles and its ICBMs, national security would be damaged only if this step was actually taken.[112] It was one thing to establish that Chinese engineers had acquired skill sets and know-how from Western experts that were portable across the civil-military divide. It was quite another to establish that it had actually been integrated into the missile program, so damaging US national security.

Indeed, "the technical issue of greatest concern was the exposure of the PRC to Western diagnostic processes, which could lead to improvements in reliability for all PRC missile and rocket programs."[113] It was in the nature of dual-use knowledge that it crossed the civil/military interface, but how did one assess the risk anyway? In its "findings and conclusions" the select committee admitted that the empirical evidence for "technology transfer" to the missile program was ambiguous or nonexistent. For one thing, even though there were significant exchanges with China during the recent launch failure analyses, "the integration of US technology and know-how into the PRC's ICBM force may not be apparent for several years if at all."[114] Even then, it went on, "indigenous improvements and improvements derived from non-US foreign sources will make it difficult to detect and measure to what extent technology transfers from American sources may have helped the PRC." In fact, the select committee boldly asserted that the extensive assistance that the Chinese had received from *foreign sources* "probably is more important for the PRC ballistic missile development program than the technical knowledge gained during the American satellite launch campaigns."[115]

It was clear then that Chinese engineers and managers had learned a range of skills from multiple sources, American and non-US, that could be used in both civil and military space programs. These might well be used to improve present as well as future launch capabilities. However, since the knowledge in question was intangible, embedded in heads and hands, and covered multiple features of a complex technological system, it would be well-nigh impossible to track and to assess its impact on national security. Not only was the technology that was transferred intangible. Its impact was difficult to detect and impossible to measure.

How Could the Risks to National Security Caused by Knowledge Transfer Be Reduced to the Minimum?

Testifying to a Senate subcommittee in March 2000, William Reinsch said that "with respect to China, if you are concerned about this, you really ought to make the decision up front as to whether you want to do business with them in satellites or not, because if you do business with them, there is going to be some leakage over time because of the nature of the transaction. And you either accept that and do your best to minimize it, or if you are troubled by it, you say no launches. In retrospect, I almost think that would have been the better way to go." Vulnerabilities were "inevitable in the nature of the transaction."[116]

The government was indeed faced with a dilemma. On the one hand it was evident that the Chinese were likely to integrate US engineering knowledge into their ballistic missile program eventually—the CIA liked to say that space launch vehicles were ballistic missiles in disguise. There was close integration between the organizations in China involved in launcher and missile development. There was motivation—after all, it was the stated intention of the PRC to upgrade and modernize its ballistic missile force. And there was evidence that US know-how had been incorporated into the space launch program, as Sokolski's chart showing an unbroken sequence of ten successful launches since 1996 suggested. If it could be done, it would be done.[117]

On the other hand, US satellite manufacturers were lobbying hard to use the Long March. Launch costs in China were considerably lower than those using Western launchers. Indeed the cost differential was often sometimes far greater than the 15 percent formally agreed on in the intergovernmental negotiations.[118] And the market was growing rapidly.

In 1997 a widely cited report by the Futron Corporation predicted an exponential increase in commercial satellite sales beginning in 1998, with a market value of over $120 billion by 2004.[119] American satellite manufacturers were emphatic that the future health of their industry—and so the economic security of the country—depended on their having a large slice of that market. How justified was it to impose export controls in the name of national security on them unilaterally, especially if William Reinsch's view was correct that "a second-class commercial satellite industry means a second-class military satellite industry as well"?[120]

It was agreed that the only way to limit the transfer of know-how in launch campaigns in China was to embed monitors in it from start to finish. Trained human beings who were involved in all aspects of the program could serve as a protective belt that not only ensured that hardware did not go missing, and that the satellites were protected from enquiring eyes. They could preempt the "active" transfer of sensitive information, know-how, and intangible knowledge in face-to-face interactions between Chinese and Western experts. The nature of the risk of "technology transfer" in a launch campaign was such that the permanent presence of a trained individual who knew the law, who understood the technology, and who was an expert in rocketry him- or herself was needed throughout the launch campaign. The monitor would be authorized to "censor" every meeting and every report that could help the Chinese eventually improve their ballistic missile program.

As matters stood, the existing safeguards program was hopelessly inadequate. Security at the launch site was irregular, and the monitoring of face-to-face interactions was spotty and erratic. A DTSA monitor at a launch in 1995 reported that "security of the entire building is poor with many unmonitored outside entrances. . . . The security of the perimeter around the entire complex was a joke."[121] The Cox Report devoted an entire chapter to this question, confirming the abysmal efforts being made to deny access to the satellite by unauthorized individuals. DoD monitors, paid for by industry, were not always used (as between 1993 and 1996), were inadequately trained, did not participate in all technical meetings, and often failed to file trip reports or to take notes at face-to-face meetings. Something clearly had to be done.[122]

The Cox Committee insisted that monitors should be properly trained, properly paid, educated in export control law and regulations, and aware of the limits of the knowledge that could be shared with foreign engineers, including that which should not be shared under any circumstances. They should also be obliged to file detailed reports at the end of each launch campaign describing all the activities they were responsible for. These reports had to be systematically archived and available for future reference. None of this could guarantee that no sensitive knowledge escaped, but it could reduce the risk of "technology transfer" to a minimum.[123]

The circulation of intangible knowledge between Western engineers and their Chinese counterparts in launch campaigns in the 1990s occurred via a multitude of channels using a range of procedures. What Fred Bucy called "active mechanisms" of technology transfer provided recipients with know-how that not only helped them understand how to improve the current performance of a technological system.[124] Once they had grasped the "detail of how to do things," they could intervene more effectively in the material world of engineering practice, gradually acquiring the know-why that became constitutive of their "experience" as rocket engineers.[125] When the know-how was "dual use" and portable across a permeable civil/military interface, like that which separated rockets from missiles, the distinction between providing technical assistance and offering a defense service was quickly blurred. So, too, was the distinction between what advice was covered by a license from the Department of Commerce, and what technical support needed an additional license from the Department of State. The federal grand jury assessing the circumstance surrounding the launch failure of Intelsat 708 decided that Hughes and Loral had interpreted that distinction to their advantage, but to the detriment of national security. The Justice Department penalized both companies for their misdemeanors.

On January 9, 2002, Loral announced that it had reached a $20 million settlement with the government. Of this, $6 million would be used to improve its export control compliance program. The balance was a civil fine of $14 million paid to the State Department "without admitting or denying the government's charges." Loral filed for bankruptcy the year after. In March 2003, Hughes Electronics Corporation and its parent company, Boeing Satellite Systems, announced that they would pay a civil penalty of $32 million to the State Department, admitting the seriousness of the charges brought against them, and that assistance to a foreign launch operator could have a negative impact on national security.[126]

Satellite Export Controls after the Strom Thurmond Act

The determination of Clinton's opponents to regain control over satellite exports triumphed with the passage in October 1998 of *Subtitle B—Satellite Export Controls*, sections 1511 to 1516 of the Strom Thurmond

National Defense Appropriations Act for FY1999. These few sections in a 360-page document had an immense impact on US relations in space, not only with China, but also with US allies. It also seriously disrupted international scientific collaboration in space science.[127]

As mentioned before, section 1513 of the Strom Thurmond Act stated the core policy reversal. It stipulated that "all satellites and related items" that were on the Commodity Control List should be transferred to the US Munitions List as of March 15, 1999. This was to respect the "sense of Congress," that US business interests were not to be "placed above" the United States' national security interests, particularly as regards the "exportation or transfer of advanced communications satellites and related technologies from United States sources to foreign recipients."[128] The Missile Technology Control Regime should be respected. There should be no blanket waiver of the post-Tiananmen sanctions on the export of satellites of American origin for launching on a rocket owned by the PRC.

Section 1514 discussed the safeguards needed to protect national security. All export licenses required a technology control plan approved by the secretary of defense and an encryption technology transfer plan approved by the director of the National Security Agency. The secretary of defense would monitor "all aspects of the launch in order to ensure that no unauthorized transfer of technology occurs, including technical assistance and technical data."[129] A license for a crash investigation was mandatory; the investigation would be monitored by officials of the Department of Defense. Technology transfer control plans had to be cleared by the Department of Defense, specifying in advance "all meetings or interactions with any person or entity providing launch services." The intelligence community had to verify the legitimacy of the stated end user or end users of any export license application. It also had to receive copies of all approved export licenses and technical assistance agreements. Congress had to be kept informed of various provisions. There was one exception made in section 1514 to these extensive safeguard provisions. They did not apply to a launches in, or by nationals of, a country that was a member of NATO, or a major non-NATO ally of the United States.

The monitoring program was expanded. It now covered launches by Russian as well as Chinese launch providers, and even some domestic launches from the United States. The "contents" of monitoring were spelt out in great detail. They included all "technical discussions and activities" from design to modification and repair, all activities related to the launch itself, and any activities related to the delay or failure of the launch, including postfailure launch investigation. Monitoring also became far more intrusive. Access by DoD monitors to a satellite of US origin was authorized "during the construction of the satellite, including the participation of monitors in telephone conferences, prior review of data to be exchanged and access to the manufacturers' databases."[130] James Bodner, the DoD's principal undersecretary of defense for policy, described the significant impact the new legislation was having. Previously missile engineers were sent off on temporary duty assignment to monitor meetings and launch sites but the implication had been that monitoring was not taken seriously. Now, by contrast, the Department of Defense had "a dedicated team of people. I think we have 33 today. We will be up above 40 next year. . . . When it comes to China and Russia," Bodner went on, "they attend every technical meeting where there might be transfer of tech data. That is certainly one of the most vulnerable points. It is not just at the launch site or in the case of failure. It is in the design of the system in the first place because some of the most critical losses are the tech data that might be lost." As Bodner put it, "they monitor these things from the cradle to the grave."[131]

Section 1515 addressed the controversial question of granting presidential waivers to the Tiananmen sanctions in the name of national security. A report to Congress justifying any waiver had to identify the "militarily sensitive characteristics integrated within or associated with" the satellite to be launched by the PRC. These characteristics were defined in section 1515 (b) as just those nine that had been deemed obsolete in 1996. No fewer than nine other subsections called on the president to provide a "detailed justification" of the economic benefits of the export. These ranged from the very detailed to the very generic. At one end of the spectrum we have section 1515 (b) (5): "The impact of the proposed export on employment in the United States, including the number of new jobs created in the United States, on a State-by-State basis, as a direct result of the proposed export." At the other end of the spectrum we have section 1515 (b) (11): "The increase that will result from the proposed export in the overall market share of the United States for goods and services in comparison to Japan, France, Germany, the United Kingdom, and Russia." Other stipulations were even broader, asking the president to assess the impact of the proposed export on transforming the Chinese market and liberalizing its trade practices toward the United States.

Once the new system was in place the State Department took initiatives of its own. The Strom Thurmond Act did not modify the ITAR regulations, which made no special exception for NATO allies. The State Department thus reserved, and exercised, the right to apply the sweeping controls in the act to everyday transactions with allies.[132] The department went even further and expanded the reach of its own regulations by declaring "that not only were communications satellites now munitions, but their components were now munitions as well." State also decided that "all satellite technology, including fundamental research," that had been exempted by the Corson Report's Fundamental Research Exclusion in 1985 also required a munitions license for export. "Deemed export" directives were sent to NASA and to universities requiring licenses to collaborate with foreign researchers on fundamental research.[133]

The effects of extending the reach of regulations on knowledge sharing from satellite launches with China to international collaboration in space science were described by Charles Elachi, the director of NASA's Jet Propulsion Laboratory from 2001 to 2016. As he put it in an interview with Krige in 2009, applying the ITAR to space cooperation with the European Space Agency (ESA) changed the whole nature of face-to-face exchanges:

> The bigger impact, in my point of view, was more on the interaction between people, more than actually getting a piece of hardware, because now if we want to talk with the ESA, we have to be careful what we talk about and so on. It's not an issue of do we send a transistor from the US to Europe, even if that's a factor. But it's really the interaction, and, I'm guessing that's where maybe people like us are unhappy, and that's where I am unhappy about also this thing [ITAR], because the strength was in building trust and a good relationship and exchange of ideas that kind of put a limitation on doing that.[134]

The expansion of the reach of export controls on satellites from trade with China to scientific collaboration with Europe is indicative of the immense interpretive flexibility of the formal, legal provisions granted to the executive. As we stressed before, applying export regulations over technology and knowledge circulation is not straightforward. It requires charting one's way through a constantly changing list of technical restrictions, it calls for at least a working familiarity with the technical aspects of the knowledge that is being shared, and it involves extensive consultation with firms, trained scientists and engineers, and multiple arms of the administration. The export control regime is not static. It is a behemoth at the heart of the national security state, whose reach and potential impediments to free trade and to unhindered international collaboration in

THE CONFLICT OVER TECHNOLOGY SHARING   293

science and technology are in a constant state of flux. It is also, for that very reason, a powerful and flexible political tool for securing American global leadership. This is perhaps not always clear in a historical narrative that follows its general evolution across a long time scale, as we have done in this volume. Our last chapter, an epilogue that deals with the place of knowledge regulation in Sino-American relationships during President Trump's administration between 2017 and 2020, will demonstrate what we mean. Seen in the light of the history of export controls, these three years involved an enormous acceleration, expansion, and concentration of export control mechanisms in the hands of the national security state, driven by the determination to refashion the global circulation of people, ideas, and things to ensure the future of a Pax Americana.[135]

We began this volume with a brief analysis of the punitive measures taken by the Department of Commerce against an American firm, Hydrocarbon Research Incorporated (HRI), and its CEO, Percival Keith. Keith signed a $17 million contract with the Rumanian government to build a number of petroleum refining plants in the country in 1962 without acquiring an export license to share unpublished technical data with them. Much of this chapter, the last in our deep history of the construction of export controls to regulate the circulation of knowledge by the US national security state, analyzes in detail the government's condemnation of two major American companies who launched their telecommunication satellites in China in the 1990s without acquiring an export license to share sensitive technical data, knowledge, and know-how with Chinese aerospace engineers and project managers. What can we learn about the history of export controls that we have described at such length in this book by comparing these two cases separated by over three decades?

There are four striking features of the fashioning and refashioning of export controls over the mere thirty-five years that separate the two cases, which we have brought to light in this volume.

Firstly, the reach of export controls to embrace sensitive technical data, and tacit knowledge and skills shared in face-to-face exchanges, has become a taken-for-granted feature of international trade with many countries. In 1962, Percival Keith could not imagine that export controls on unpublished technical data applied beyond the limited domain of fundamental, secret know-how held by individuals and in the company. The Department of Commerce seized on the occasion to publicly interpret the meaning of the term in the *Federal Register* and to make an example of HRI. On that occasion the DoC decreed that it covered "any information or documents which a competent, experienced engineer thought he could furnish or prepare with his own mind using well-recognized engineering principles, texts, technical articles, and patents."[136] This concern with the transfer of tacit knowledge and skills to economic and military rivals was foregrounded in the 1970s by corporate leaders like Fred Bucy. He argued that "the detail of how to do things" be formally enshrined in export controls of militarily critical and dual-use technologies, and that they encompass both allies and enemies. Hughes and Loral, unlike HRI, were well aware of these regulations. They did not act in ignorance. Their failing was to take advantage of the determination of the Department of Commerce to advocate for business to quietly get around such regulations if they could. Export controls have been used by the national security state since the 1940s to secure the United States' scientific and technological lead in peacetime and to consolidate the reach of its global power. They are part and parcel of the everyday activities and responsibilities of American firms that commercialize cutting-edge technology and knowledge and that market it abroad (and more recently of American research universities that encourage international collaboration). The absence of any consideration of export controls in multiple fields of academic research in this phase in US history is a symptom of the failure to engage with the relationship between national security and knowledge production and circulation in the corporate world, and between the national security state and sectors of American industry that exploit advanced science and technology.

Secondly, the rationale for state intervention in the market using export controls has not only persisted; it has become amplified over time. The HRI case arose in the context of trade with the Communist bloc in the first decade of the Cold War. Restrictions were imposed to isolate economically and technologically a foreign power that was seen as an existential threat to the republic and its values. The appeal to national security justified their imposition on the free play of market forces, and as a necessary constraint on individual entrepreneurship. Presidential support for Hughes's and Loral's active engagement in trade with China was part and parcel of a deliberate government strategy that fused an expanded conception of national security, now embracing economic security, with a determination to strengthen America economically and militarily in a post–Cold War world. Export controls on the circulation of sensitive knowledge have been and remain not only instruments of trade policy.

They are integral to (changing definitions of) national security and are embedded in the toolbox of foreign policy and grand strategy by which the United States positions itself as the leader of the free world.

Thirdly, the HRI case engaged one government agency, the Department of Commerce, and, apart from serving as a warning to US industry and its foreign partners, had little impact beyond specialist law journals. Hughes's and Loral's trade relationships with China mobilized the Department of Commerce, but also the Departments of State and Defense, the intelligence community, and Congress. By the 1990s export controls on sensitive knowledge and know-how had become bureaucratized across multiple agencies of the national security state and increasingly dissolved the boundaries between the executive and legislative arms of government.

Finally, policies for regulating knowledge flows have become increasingly politicized over the decades. The HRI case was a "routine" administrative matter in which the government rapped a small company over the knuckles for its trade practices with a Soviet satellite. The dispute between the Department of Justice and Hughes and Loral was a high-profile, widely publicized event that engaged multiple actors in the state and in civil society, a hot-button topic in the media, sensationalized in the Cox Report to Congress, and entangled with a bruising conflict between Republicans and Democrats that reached its climax with the decision to impeach President Clinton in 1998.

Export controls in peacetime were children of the Cold War national security state. There was talk of abolishing them altogether when the Soviet regime imploded in the early 1990s. They not only survived "the end of history." They were refashioned to meet an expansive conception of national security (economic security) in a neoliberal, global age, when the dynamism of American corporate and academic R&D propelled the United States to a position of immense power. The globalization of the world economy also exposed the United States to the dangers of sharing cutting-edge technology and knowledge with a new emerging power, whose leaders were convinced that scientific and technological innovation held the key to realizing "the Chinese dream of the great rejuvenation of the Chinese nation."[137] Notwithstanding the time-sensitivity of much of the material in our epilogue (chapter 10), it was impossible for us to end this study without engaging, if only in a limited way, with the hugely expanded role of regulations in the circulation of knowledge and know-how by President Trump and his administration in their dealings with Beijing between 2017 and 2020.

PART FOUR

CHAPTER TEN

Epilogue

*Export Controls, US Academia,
and the Chinese-American Clash during
the Trump Administration*

In 2019, the Chinese company Huawei, the world's largest provider of telecommunication equipment and the second-largest smartphone producer,[1] was a target of the US export control system. On May 16, the Bureau of Industry and Security, the export control agency within the Department of Commerce, put Huawei and sixty-eight of its affiliates in twenty-six countries on the so-called Entity List, effectively banning American businesses from exporting dual-use technologies to Huawei.[2] The effects of this bureaucratic decision were felt almost immediately. Within three days, on May 19, Google halted all transactions involving the transfer of proprietary hardware, software, and technical services except for those that were publicly available via open-source licensing. Even though the full implications of this decision were difficult to gauge, there was no doubt that it would directly affect millions of Huawei cell phone users outside China as their devices ran on Google's mobile operating system Android and used Google applications.[3] The mobile phone parts producer Lumentum Holdings, which derived 18 percent of its revenues from doing business with Huawei, followed suit and stopped its shipments to the Chinese company.[4] Huawei was highly dependent on American chips and other components.[5] Indeed, several large makers of computer chips—Intel, Broadcomm, Qualcomm, and Xilinx—froze their exports to Huawei "until further notice." Thirty of Huawei's ninety-two major foreign suppliers were located in the United States. These reactions were not entirely unexpected by Huawei. Six months ahead of it being blacklisted, the company had begun to stockpile semiconductors and other components for up to twelve months of production.[6]

The ripple effects of the Entity List decision immediately crossed the Atlantic and reached Europe, and not only because Europe was Huawei's biggest foreign market: it sold 26.3 million cell phones there in 2018 alone.[7] Equally important was that, owing to the exterritorial, global reach of US reexport controls—which impacted all transactions that included a certain amount of American parts or technology—even companies like the German semiconductor producer Infineon were also forced to take action. Although Infineon stressed that most of its products did not incorporate American technology, the prospects of it being affected by some reexport controls raised concerns.[8] Infineon and other European and American chip companies took hits in the stock market.[9] Very quickly, Huawei's position on the European market showed signs of erosion. One month to the day after the Department of Commerce's blacklist decision, Huawei's international sales had dropped by 40 percent.[10]

These dramatic developments in just a few days and weeks in 2019 demonstrate the enormous disruptive power that the US export control system could wield over global business networks, especially in the telecommunication industry, and its international supply chains, production system, and practices of trading and modes of technology sharing. Just one bureaucratic decision by the Department of Commerce threw an entire industry into disarray and appeared to question the very logic of globalized capitalism.

The Huawei case became, during the Trump years, *the* public symbol of the US-China clash, and it rapidly changed the political economy of global knowledge sharing. This visibility must be situated in the incredibly deep and broad context of the deterioration of Chinese-American relations during the Trump presidency and the changes induced in the basic concepts and the scope of export control legislation and practice. Indeed elsewhere in this volume we have shown how, at least since the 1980s, US debates about national economic policy became increasingly "securitized" while national security was more and more "economized." We have also shown that export controls and concerns about the international sharing of US scientific-technological knowledge were at the heart of this conceptual shift toward "economic security"—a term that consciously conflates and merges national security and economics. Seen in this light, the Huawei case is just another instructive example of how the lines between national security, foreign policy, trade policy, and technology have become blurred.

Officially the Department of Commerce "blacklisted" Huawei in 2019 because it found "reasonable cause to believe that Huawei . . . has been involved in activities determined to be contrary to national security or foreign policy interests of the United States." In concrete terms, the department accused Huawei of circumventing the sanctions the United States had imposed against Iran after leaving the international Joint Comprehensive Plan of Action (also known as the Iran Nuclear Deal) in May 2018.[11]

Huawei was much more than simply a sanctions breaker and a clear-cut national security problem. The events in May 2019 were the high-water mark of a much longer history of American suspicions that Huawei was not an independent company, but in fact an agent of the Chinese government, whose technology served as a Trojan horse to infiltrate the American government, citizenry, and corporate world. Washington and the company had a string of conflicts that seriously curtailed Huawei's ability to establish a foothold in the US market. As early as February 2008, the US government had used its Committee on Foreign Investment in the United States (CFIUS) to block Huawei's attempt to acquire a 16.5 percent stake in its US rival 3Com.[12] In October 2012, the Permanent Select Committee on Intelligence of the House of Representatives published an investigative report on Huawei and ZTE (another prominent Chinese company in the telecommunication technology industry). Because they could not "be trusted to be free of foreign state influence and thus pose a security threat to the United States and our systems . . . the risks associated with Huawei's and ZTE's provision of equipment to US critical infrastructure would undermine core US national-security interests." The report demanded that CFIUS "block all acquisitions, takeovers, or mergers involving Huawei and ZTE" and insisted that the US government, including contractors, should not buy or use any equipment from them.[13]

We see then that economic issues like US foreign investment policy played a prominent role when the House report talked about national security concerns, like the protection of "critical infrastructure" against foreign manipulation and cyber espionage. This economic undercurrent was even stronger when the report dealt with Huawei's putative unfair trade practices and market advantages due to support from the Chinese government. It paid special attention to Chinese infringements of American companies' intellectual property rights, not least through economic espionage in cyberspace.[14] Summing up these concerns, the report quoted an assessment by the national counterintelligence executive, which warned that "foreign attempts to collect US technological and economic information . . . represent a growing and persistent threat to US economic security."[15]

Security and economics have gelled ever since 2012 along with Huawei's rise as the technological leader in the field of 5G, the key technology used to speed up telecommunications that is also used in high-profile projects like autonomous vehicles. The US government increasingly interpreted Huawei's market success and technological lead as a threat to American national security and to economic and technological preeminence in the global economy. Even though, or exactly because, it lacked a clear definition, "economic security" was, once again, increasingly used as a term for describing the interplay between the global economy, the international political economy of technology transfers, and national security. Part of the political appeal of the concept was that it reduced the messiness and complexity of this interplay to a simple, but nevertheless ambiguous formula with considerable political punch.

A closer look at this conceptual change of national security since the 1980s (see chapters 7–9) makes it clear that the Trump administration's policies and initiatives were not an exception or aberration from the "normal" course of US policy but only variations on widely accepted (but never uncontested) ideas and patterns. What made the Trump administration stand out was mainly its style, volume, and rhetorical grandstanding, not the novelty of its concepts and approaches. The conflict with Huawei did not come out of the blue and was not even a surprise—as we mentioned above, the firm itself had anticipated restrictions on its trade with the United States. Nevertheless it is remarkable to what extent the Trump administration, starting in 2017, used export control policy as a key weapon of US trade, national security, and technology policy in the increasingly tense American-Chinese confrontation that dominated the international news during his term in office. Even though Huawei was the most visible target, export controls had a much further reach and influenced, at times subtly, sometimes assertively, a wide array of issues, spanning from the competitiveness of US companies in comparison to Chinese firms, to concerns about China's military buildup and its power projection in the Pacific, to discussions about the costs and benefits of academic exchanges between American and Chinese academic institutions. Huawei is thus part of a much bigger picture, some of which we will capture in this epilogue.

The chapter consists of two sections. The first discusses how the concept of "economic security" paved the way for profound changes in the statutory basis of the American export control system. Two major laws, the Export Control Reform Act and the Foreign Investment Risk Review Modernization Act, marked the first major overhaul of export control legislation since 1979 and 2007 respectively. They redirected export controls, including foreign direct investment controls, along economic security lines and focused specifically on a tighter regulation of emerging technologies like artificial intelligence.

The second part will discuss how the current export control debates targeted, once again, the American academic community. We argue that even though the public debate was very much focused on the threat China allegedly posed to US industry, universities and research institutions were arguably one of the main battlefields of Trump's trade war.

Enacting "Economic Security": The Export Control Reform Acts of 2018

Less than a year before the Department of Commerce added Huawei to its Entity List, the US export control system was subject to a major revamp. On August 13, 2018, President Trump signed into law the Export Control Reform Act of 2018 (ECRA) and, closely related to it, the Foreign Investment Risk Review Modernization Act of 2018 (FIRRMA),[16] both rolled into the National Defense Authorization Act for Fiscal Year 2018. Though at first sight these two pieces of legislation were much less spectacular and controversial than the developments in the Huawei case, they were no less important to the overall trajectory of the US national security and economic policy. ECRA marked the first statutory reform of export control legislation related to dual-use technologies since 1979. In fact, since the Export Administration Act had never been made permanent, a lack of congressional consensus over the direction of export control policy after the end of the Cold War led to a failure to reauthorize the Export Administration Act in 1994. Presidential Executive Orders, based on the International Emergency Economic Powers Act of 1977 (IEEPA), had kept export controls alive ever since, and until 2018. During these more than two decades, several attempts to reform and update export control legislation failed to result in statutory changes.[17] Thus, it was remarkable that bipartisan agreement in Congress not only reauthorized the law but also reformed it—notwithstanding the backdrop of the deeply partisan climate of the Trump years that threatened to paralyze the government.

Even more remarkable was that Congress combined ECRA with yet another law, FIRRMA. This law was the first reform of the national security review mechanism for foreign direct investment (FDI) in eleven years. The statutory frame for the Committee for Foreign Investment in the United States (CFIUS) had, since the enactment of the Exon-Florio Amendment in 1988, adapted to new security challenges in the wake of 9/11 by passing the Foreign Investment and National Security Act of 2007.

Neither ECRA nor FIRRMA marked revolutionary changes to the existing export control system. They mainly codified and formalized current practices. But they also expanded the regulatory reach of export controls and further developed older concepts and tailored them to the new political context of heightened US-Chinese tensions.[18] These conceptual changes marked a contrast to the export reform efforts of the Obama administration, which had focused on institutional and procedural reform in order to make the system more transparent and less intrusive on everyday business practices.[19] Moreover, not only were both acts written at the same time; they were conceived as twin legislation with complementary functions. Both acts equally focused on emerging technologies and framed their control as the core problem of "economic security."

The first legal change of 2018 was that ECRA made dual-use export controls permanent for the first time.[20] After World War II, export controls were initially seen as emergency legislation to ease the transition from a wartime to a peacetime economy. This established a precedent that was never revised during four decades of the Cold War. While the legal foundation of CFIUS had been permanent since 1988, the different versions of the Export Control Act had always had sunset provisions. ECRA changed this, finally acknowledging the obvious fact that export controls had become a central element of US national security policy. In the age of China's rise, export controls were here to stay.

The Export Control Reform Act did not mention China once. But FIRRMA showed that the People's Republic was the main target. It demanded that the secretary of commerce report every two years on FDI transactions made by Chinese entities in the United States. Congress was especially interested in learning more about possible patterns of Chinese investments and wanted to know if such patterns "align[ed] with the objectives" of the "Made in China 2025" program, a far-reaching industrial policy initiative of the Chinese government to push the economic development of ten key technologies (more on this later).[21] Concerns about the investments made by Chinese state-owned enterprises (SOEs) and companies with close ties to the Chinese government like Huawei and ZTE also informed the decision to deviate from the principle that a review by CFIUS should be—at least in theory—a voluntary process requested by the companies involved in a merger or acquisition. FIRRMA made the review mandatory for transactions when "a foreign government has, directly or indirectly, a substantial interest" in the acquiring firm.[22]

Probably the most important legal change of 2018 was the greater emphasis ECRA and FIRRMA placed on controlling "critical technologies." The ECRA bill called for the establishment of a "regular, ongoing interagency process to identify emerging critical technologies that are not identified in any" export control list, "but that nonetheless could be essential for maintaining or increasing the technological advantage of the United States over countries that pose a significant threat to the national security of the United States." This interagency process was to be built around the well-established export control expertise of the Departments of Commerce, State, Defense, and Energy (which was involved in the control of nuclear technologies). But the ECRA bill also called for the mobilization of the expertise and resources of all other relevant government agencies, US industry, and academic institutions to "identify and describe" the (emerging) critical technologies.[23]

The Foreign Investment Risk Review Modernization Act directly referred to this section of ECRA as well as to ITAR and EAR to define the US "critical technologies" that it aimed to protect against acquisition by economic actors who posed a national security threat.[24] The new foreign direct investment review law broadened CFIUS's review rights by allowing it to look into *any* investment of a foreign person if it pertained to any American business that "produces, designs, tests, manufactures, fabricates, or develops one or more critical technologies" or that produced or designed critical infrastructure. Here, of special concern were those transactions that facilitated foreign "access to any nonpublic technical information in the possession of the United States."[25] Similar to what the export control regulations stipulated, this kind of information was defined as providing "knowledge, know-how, or understanding, not in the public domain, of the design, location, or operation of critical infrastructure; or . . . is necessary to design, fabricate, develop, test, produce, or manufacture critical technologies, including processes, techniques, or methods."[26]

Neither of the two laws specified what kind of technologies they envisioned as especially sensitive and important for maintaining the United States' "technological advantage," or lead time. But as soon as the Department of Commerce tackled the translation of ECRA into new export regulations, it was apparent that it wanted to cast a wide net. In November 2018, it published an advance notice of proposed rule making (ANPRM) in the *Federal Register*, which not only asked for "public comment on criteria for identifying emerging technologies that are essential to US national security" but also provided a list of technologies for which it wanted to determine whether they fell into the new category of "emerging critical technologies." This list encompassed no fewer than fourteen technologies, which were then further broken down into forty-six subcategories and examples. The main categories included biotechnology; artificial intelligence (with eleven subcategories like computer vision, speech and audio processing, and AI cloud technologies); position, navigation, and timing technology; microprocessors; advanced computing; data analytics technology (e.g., automated analysis algorithms); quantum information and sensing technology (like quantum computing and encryption); logistics technology (for example, mobile electric power and distribution-based logistics systems); additive manufacturing (e.g., 3-D printing); robotics (such as microdrone and microrobotic systems, swarming technologies, and also "Smart Dust"[27]); brain-computer interfaces; hypersonics (flight control algorithms, propulsion technologies, etc.); advanced materials (including functional textiles and biomaterials); and advanced surveillance technologies (such as voiceprint and faceprint technologies).[28] Clearly, all these categories named technologies with a potential military or intelligence application. But obviously the Department of Commerce also scouted out large swaths of the civilian US high-technology landscape for technologies to be included in the realm governed by export controls. Even though the list was probably deliberately overbroad so as to prod the high-tech industry to provide public comments, it showed the scope of the new thinking about emerging technology. And without any doubt it had a chilling, even alarming effect on the entire US high-technology sector.[29]

The public debate in the United States about the features and implications of the perceived technological race with the "strategic adversary" China touched several items on the Department of Commerce's catalogue. Apart from semiconductors and 5G—which overlapped with several of the categories listed—no technology received more attention than artificial intelligence (AI). Shortly after the department's specification of emerging technologies, President Trump issued an "Executive Order on Maintaining American Leadership in Artificial Intelligence" in February 2019. The order announced a federal initiative to push for more research and development to maintain the United States' position as the AI "world leader." For the White House, AI was a key technology that promised "to drive growth of the United States economy, enhance our economic and national security, and improve our quality of life." This call for more R&D had a defense flip side: export controls. The Executive Order declared that while the United States "must promote an international environment that supports American AI research and innovation and open markets," it had also to protect "our technological advantage" and "our critical AI technologies from strategic competitors and adversarial nations."[30] The next day, the Department of Defense chimed in with its "Artificial Intelligence Strategy: Harnessing AI to Advance Our Security and Prosperity." It stated that "AI is poised to transform every industry, and it is expected to impact every corner of the Department." It then warned that "other nations, particularly China and Russia, are making significant investments in AI for military purposes. . . . These investments threaten to erode our technological and operational advantages."[31]

"Economic Security" as the Conceptual Framework of Export Control Policy

The Export Control Reform Act's reshaping of the concept of national security was no less important than the establishment of a permanent statutory basis for export controls and the heightened emphasis on expanding export controls to catch new cutting-edge technologies. The law emphasized the economic dimension of US security interests, going beyond conventional thinking about the economic underpinnings of modern military planning and procurement.

At first ECRA's statement of policy referred to the traditional mainstays of export control thinking, listing the fight against weapons proliferation and terrorism as the main goals of export controls, next to the furthering of US foreign policy, concerns about the unwanted strengthening of foreign military programs through trade, and the protection of critical infrastructure. But, considerably extending the scope of the export control system, ECRA then argued that international markets and US companies had to be seen as an integral part of national security and export control considerations: "The national security of the United States requires that the United States maintain its leadership in the science, technology, engineering, and manufacturing sectors, including foundational technology that is essential to innovation. Such leadership requires that United States persons are competitive in global markets." This line of thinking was—as we have shown in earlier chapters—not new. But ECRA codified the idea of "economic security" and thus elevated it to the status of an official doctrine and leading principle of US export control policy. The conceptual shift had potentially enormous implications, as ECRA incorporated the entire private economy of the United States and all elements of the national innovation system into the export control world. Export control goals such as preserving "the qualitative military superiority of the United States" and strengthening the US "industrial base" had in this context potentially far-reaching implications that blurred the line between "civilian" and "military."[32] Indeed, ECRA seemed to expand the challenges of controlling "dual-use" technologies to the national economy as a whole.

This was a major departure from the Export Administration Act of 1979. This older act attempted to demarcate, not blur the lines, calling for export controls only on goods and technologies "which would make a significant contribution to the military potential" of countries threatening US national security. Economic considerations that potentially included the civilian sector had, since 1949, been limited to the use of export controls to regulate short supply challenges—for example, in regard to raw materials. Indeed, the 1979 act placed export controls in a free-trade framework and attempted to limit, not to expand their scope. In 1979, there was no word of using export controls as a tool to strengthen US companies' competitiveness in global markets or the defense industrial base; American technological or economic leadership was not even mentioned.[33]

The Foreign Investment Risk Review Modernization Act was less explicit than ECRA in its use of economic security concepts. It stressed the enormous benefits the United States derived from its openness to foreign direct investment. But at the same time the act followed ECRA's way of thinking when it described these benefits as "including the promotion of economic growth, productivity, competitiveness, and job creation, *thereby* enhancing national security." The intersection of national security with economic prowess was the reason for FIRRMA's reform of FDI policy, justified because the "national security landscape" had "shifted in recent years" and had changed the "nature of the investments that pose the greatest potential risk to national security." Interestingly, FIRRMA warned against expanding CFIUS's purview to "issues of national interest absent a national security nexus." Yet at the same time, the law, with a clear nod to China, targeted countries "of special concern" that have "demonstrated or declared a strategic goal of acquiring a type of critical technology or critical infrastructure that would affect United States leadership related to national security." Moreover, FIRRMA was concerned about the future of the defense industrial base, or "the control of United States industries and commercial activity by foreign persons as it affects the capability and capacity of the United States to meet the requirements of national security, including the availability of human resources, products, technology materials, and other supplies and services." It was difficult to discern where the realm of the defense industrial base began and how it could be distinguished from a purely civilian sphere, since the term "availability of human resources" was to "include potential losses . . . resulting from reductions in the employment of United States persons whose knowledge or skills are critical to national security, including the continued production in the United States of items that are likely to be acquired by the Department of Defense or other Federal departments or agencies for the advancement" of national security.[34]

The bipartisan enactment of ECRA and FIRRMA rode on a wave of official statements, as well as punditry and media coverage, that established economic security as the dominant, though sometimes contested, intellectual concept to interpret the political meaning of international trade relations especially in regard to the political economy of technology sharing.[35] One of the most prominent renditions of this conceptual shift was the Trump administration's *National Security Strategy* of December 2017. One year after the presidential election, it attempted to give US national security contour and substance for the next three years. Even though the strategy dealt, of course, with traditional national security issues like terrorism, the proliferation of weapons of mass destruction, and the projection of US interests through diplomacy and military power, the document's main focus was on economic and technological challenges in the international realm.

One of the main goals of national security policy, the strategy proclaimed, was to "promote American prosperity"—the phrase served as the title of the second of five main chapters. This chapter's motto, "Economic security is national security," was ascribed to Donald Trump, giving it the full authority of the US presidency. The strategy stated, "A strong economy protects the American people, supports our way of life, and sustains American power." At first sight the economic model the text touted looked like the well-worn free-market ideology the US government had subscribed to after World War II: "For 70 years, the United States has embraced a strategy premised on the belief that leadership of a stable international economic system rooted in American principles of reciprocity, free markets, and free trade served our economic interests." But this model was in a serious crisis. The White House perceived the structural changes of globalization as a threat to the US economy. "Over decades, American factories, companies, and jobs have moved overseas. After the 2008 financial crisis, doubt replaced confidence. . . . The recovery produced anemic growth in real earning for American workers. The US trade deficit grew as a result of several factors, including unfair trading practices." The most disconcerting factors weakening the system, however, came from abroad: "Today American prosperity and security are challenged by an economic competition playing out in a broader strategic context." The United States had, under the premise that economic liberalization would facilitate political change, expanded the free-trade system "to countries that did not share our values."[36] By 2017, the Trump administration claimed, this premise had "turned out to be false."[37]

The states in question were Russia and, more importantly, China, which the White House designated as "revisionist powers"[38] whose main goal was to "challenge American power, influence and interests, attempting to erode American security and prosperity." Instead of becoming more like the capitalist West, these illiberal states simply exploited the benefits of free trade, adhering only "selectively" to its rules and agreements. China's and Russia's long-term objective was, in the strategy's view, to change the US-led international order, "to make economies less free and less fair, to grow their militaries, and . . . to repress their societies and expand their influence."[39] The White House perceived their behavior in the global market place as "economic aggression" that targeted directly the United States and its international predominance.[40]

The term "economic aggression" was soon to be pushed and popularized by a report from the White House Office of Trade and Manufacturing Policy, published in June 2018. The office was established in April 2017 in order to "defend and serve American workers and domestic manufacturers while advising the President on policies to increase economic growth, decrease the trade deficit, and strengthen the United States manufacturing and defense industrial bases."[41] It was headed by economist Peter

Navarro, who had been one of the harshest critics of China's economic practices and the terms of US-Chinese trade relations for more than a decade.[42] The 2018 report defined "economic aggression" loosely as "aggressive acts, policies, and practices that fall outside the global norms and rules"[43]—which had historically been defined by institutions shaped and dominated by the United States. The scope of practices deemed "aggressive" was very wide: protectionist measures to shield domestic markets, industrial policies to strengthen Chinese exports (including the support for "national champions"), efforts to secure access to or control over natural resources, and even the fact that China dominated certain manufacturing industries like the production of refrigerators, TV sets, and air conditioners. The main focus of the report was, however, China's strategies to acquire foreign and especially US technologies, listing more than fifty "aggressive" Chinese practices, spanning from economic espionage (often, but not only, in cyberspace), evasion of export controls, and foreign direct investment, to technology transfer through "coercing" foreign companies into joint ventures with Chinese firms, reverse engineering, product counterfeiting, and piracy. Navarro's office had no doubt that China's long-term goal was to secure global dominance: politically, economically, militarily, and technologically—and that all these dimensions of national power were closely linked. Conversely, the report implied that US international power rested on the same dimensions and that there was a need to push back against Chinese inroads in all of them. Proprietary technology was at the heart of this challenge because it straddled and connected all components of power. The report ended with a dire warning and call to arms: "Given the size of China's economy and the extent of its market-distorting policies, and China's stated intent to dominate the industries of the future, China's acts, policies, and practices of economic aggression now targeting the technologies and the IP of the world threaten not only the US economy but also the global innovation system as a whole."[44]

The Export Control Reform Act and FIRRMA have to be placed in this larger context of the Trump administration's framing of US-Chinese economic relations as a head-on clash of two adversarial powers. The administration's views were deeply rooted in the assumptions of the realist theory of international relations. *National Security Strategy* announced the return of "great power competition" and global geopolitical struggle.[45] The White House claimed it was an expression of "principled realism . . . because it acknowledges the central role of power in international politics, affirms that sovereign states are the best hope for a peaceful world, and clearly defines our national interests."[46]

In this context, the protection of the US technological lead—*National Security Strategy* particularly mentioned fields like AI, cyber, and space technologies, embedded in the "defense industrial base"—was simply a function of state power. Moreover, the building blocks of the national innovation system—federal R&D funding, the private economy, and the academic sector—were all conceptualized as instruments the American state could and should use against its main competitors, China and Russia. ECRA's and FIRRMA's goal to shore up the protection of American technology was therefore part and parcel of a national security policy that "integrate[d] all elements of America's national power—political, economic, and military" against geopolitical enemies.[47]

The Trump administration's economic security approach made it difficult to differentiate between "aggressive" and innocuous regular market behavior. Another example of this conflation of categories was the report of the National Counterintelligence and Security Center entitled "Foreign Economic Espionage in Cyberspace," published in July 2018. This report also saw China (along with Russia and Iran) as the main culprits whose economic espionage posed "a significant threat to America's prosperity, security, and competitive advantage." According to the US counterintelligence community, China engaged in a "complex, multipronged technology development strategy" that mixed "licit and illicit methods to achieve its goals." Also, the lines between the state and private sector were blurred: "Chinese companies and individuals often acquire US technology for commercial and scientific purposes. At the same time, the Chinese government seeks to enhance its collection of US technology by enlisting the support of a broad range of actors spread throughout its government and industrial base." As for China's strategic goals, the report identified "comprehensive national power," an "innovation driven economic growth model," and "military modernization." It then identified ten channels of technology acquisition, covering a vast spectrum from "non-traditional collectors" (who were in the parlance of the Trump administration businessmen, but also students and scholars coming to the United States), joint ventures, academic and industrial research cooperation, FDI, talent recruitment campaigns, and covert and overt information collection by Chinese government intelligence services. Since the report treated all these forms of contacts as dimensions of "economic espionage," it blurred the boundaries between trading and stealing, legal and illegal, legitimate and nefarious American-Chinese interactions.[48]

In its most extreme rendition, this conflation of categories turned potentially every interaction between China and the United States into the equivalent of an act of war. Indeed, *National Security Strategy* demanded that the United States buckle up for a new kind of competition with states like China and Russia. They "recognize that the United States often views the world in binary terms, with states being either 'at peace' or 'at war,' when it is actually an area of continuous competition. Our adversaries will not fight on our terms. We will raise our competitive game to meet that challenge, to protect American interests, and to advance our values."[49] This new war was fought not only with all means modern *states* had at their disposal, not least through the mobilization of national economies. The Trump administration envisioned such a war as being also a clash of antagonistic *societies*. In widely quoted testimony before the Senate Select Committee on Intelligence, FBI director Christopher Wray offered some insights into this line of thinking. When Senator Marco Rubio, an anti-Chinese hardliner, asked him about the "counterintelligence risks" posed by "Chinese students, particularly those in advanced programs in the sciences and mathematics, Wray said that "almost every field office that the FBI has around the country" had noticed China's "use of non-traditional collectors" of intelligence, "especially in the academic setting, whether it's professors, students, scientists." "They're exploiting the very open research and development environment that we have, which we all revere, but they're taking advantage of it," Wray said. Having blamed American academia for being naïve about this threat, Wray called for a completely new framing of Chinese activities in the United States and an all-encompassing offensive against Chinese espionage. As he put it, "one of the things we try to do is to view the China threat not just as a whole of government threat, but a whole of society threat on their end. I think it's going to take a whole of society response by us. So it's not just the intelligence community, but it's raising awareness within our academic sector, within our private sector, as part of the defense."[50] Indeed, as we will show in the second part of this chapter, universities and academic research institutions became increasingly embroiled in US-Chinese competition for technological leadership and global power during the Trump administration.

The Research Community, Intellectual Property Theft, and Economic Security

Beginning in 2018 there was a steady stream of accusations that Chinese nationals, in particular, were stealing intellectual property from American universities. The assault was spearheaded by members of the national security community like FBI director Wray; it had bipartisan support in Congress, where Senator Marco Rubio (R-FL) was particularly vocal; and it enjoyed widespread publicity in the daily press and specialized journals (like *Inside Higher Ed* and *Science*).

The Trump administration took at face value the Chinese government's stated ambition to become the globally dominant manufacturing power by 2050. It insisted that, to do this, Beijing was exploiting every legal and illegal path to acquire US science and technology to reduce the time and the cost of innovation, so securing an unfair competitive advantage over its archrival. Academic research laboratories at the cutting edge of R&D in emerging technologies were prime sites for an organized program of economic espionage and IP theft. The values of scientific internationalism and openness to collaboration rendered them easy prey for foreign nationals, notably from China, who "walked out the door" of American laboratories with IP that they commercialized back home.

This campaign was noteworthy for repeated accusations of what was loosely called "theft" (or of what "boils down to theft") and for expanding the concept of "intellectual property" to cover far more than knowledge that had immediate financial value and that could be patented. The government's argument gained further traction as one case after another revealed malpractices by researchers working with China, both relatively minor (like failing to list all sources of foreign funding on a grant application) and downright reprehensible (like sharing a confidential grant application submitted for peer review with a colleague abroad).

In this context, research universities were once again called on to give more weight to national security at the expense of scientific openness, at least in their dealings with China. These developments continued the trends we described in chapters 5 and 6. Beginning in the Reagan era, US administrations were time and again deeply concerned that research universities were transferring militarily relevant technology, knowledge, and know-how to foreign nationals. President Obama personally mentioned his concerns at the end of a state visit by Chinese president Xi Jinping to the United States in September 2015, announcing that they had mutually agreed to protect innovation and intellectual property. His conciliatory attitude to Beijing subsequently gave way to what has been called President Trump's "increasingly zero-sum, unilateralist, protectionist, and nativist 'America first' approach to the relationship."[51]

Sino-US Scientific Collaboration: The View from the Bench

The Trump administration was very aware of the benefits that the US innovation system gained from international collaboration. The FBI's assistant director of counterintelligence, E. W. (Bill) Priestap, told the Senate Judiciary Committee in June 2018 that "the vast majority of the 1.4 million international scholars on US campuses pose no threat to their host institutions, fellow classmates, or research fields. On the contrary, these international visitors represent indubitable contributions to their campuses achievements, diversity of ideas, sought expertise, and opportunities for cross cultural exchange."[52] Priestap pointed out that in 2017 immigrants contributed $36.9 billion to the US economy and supported 450,000 jobs. They had founded almost a quarter of all new US businesses, nearly a third of US venture-backed companies, and a half of Silicon Valley's high-tech start-ups. During congressional hearings in April 2018 it was also mentioned that, by usually paying full tuition fees, international students helped an underfunded higher education system to balance its books.[53]

These functional arguments for international cooperation ignored academic considerations that fostered Sino-American scientific collaboration in knowledge production. It was these bonds with China that were of the greatest concern to the administration. What was the "scale and scope" of the ties between the Chinese and the American innovation systems by 2020? Was there essentially a one-way flow of knowledge from America to China, or was the relationship reciprocal and of mutual benefit? How important was the role of the Chinese authorities in setting the agenda for collaborative activity with the United States? Did they manipulate it and Chinese participants to serve the Communist Party's long-term economic and military objectives to the detriment of US national security?

In 2015, during the Obama presidency, the FBI distributed an educational circular remarking that very few foreign scholars in the United States were "actively working at the behest of another government or competing organization."[54] The tone changed with his successor, who took a decisively less conciliatory approach to China. The Trump administration's intelligence community claimed that by intent, or by accident, many US-based researchers were part of—or pawns in—a deliberate effort to acquire intellectual property from American universities, often while working in projects funded with American taxpayer's dollars. The dominant narrative held that "for the Chinese government international scientific collaboration is not about advancing science, it is to advance China's national security interests."[55] Chinese talent recruitment programs, designed by Beijing to reverse the "brain-drain" of gifted researchers who had moved abroad, were decried as highly funded, state-driven, intellectually targeted initiatives that lured outstanding researchers from the United States with financial incentives and lavish research resources to steal intellectual property and violate export controls. No attempt was made to inquire what benefits the US research community may have gained from working with China.

The National Science Board's 2020 report on science and technology indicators (hereafter NSB 2020), released on January 15, 2020, throws light on this latter issue.[56] It painted a picture of an American innovation system that was being outspent and outperformed by its Chinese competitor, but one that was also its preferred international partner. NSB 2020 noted that while gross domestic expenditures on R&D in the United States increased from $PPP 280 billion (purchasing power parity dollars) in the year 2000 to $PPP 549 billion in 2017, China had experienced an exponential rise over the same period from a mere $PPP 40 billion in 2000 to $PPP 496 billion in 2017. Its "manpower" pool had grown apace such that in 2015 China awarded thirty-two thousand doctorates in natural science and engineering, bypassing the United States' thirty thousand. China's science and engineering publication output had also risen tenfold since the year 2000, so that its output in terms of absolute quantity exceeded that of the United States when the report was written.[57] This had been accompanied by a growth in quality. Citation rates of papers published with authors resident in China increased almost threefold from 2000 to 2013.[58] US-Chinese coauthorship of articles in science and engineering journals had also increased sharply. In the year 2018, 39 percent of papers with an America-based author were coauthored with someone in another country. Of these, 26 percent were with authors from China, more than with any other nation. Chinese authors were also being increasingly cited as first authors in collaborative work, an indication that they took the lead on conducting the research and writing the paper. Two scholars with extensive experience in analyzing publication and citation data concluded in 2019 that China was "a major player in US-China research collaboration, via growth, via funding, and via intellectual leadership."[59]

Many factors accounted for the enhanced quality of the Chinese research capacity that in turn made it a desirable partner for US-based researchers. Two were particularly important. The first was the imposition by the state of merit-based criteria to maintain the standard of research that was performed. The Chinese research environment became immensely competitive, putting a premium on publication-based success measures. As Richard Suttmeier explains, "The awarding of advanced degrees, promotions and salary increases, and success in competition for research funding are all being linked to publication in SCI [Science Citation Index] journals."[60] The visibility achieved, especially by publishing in high-impact journals, and in English, facilitated engagement in a transnational research community.[61]

The second main factor was the crucial role played by American researchers of Chinese origin in enhancing excellence by building collaborative bonds with colleagues on the Chinese mainland. Copublication of articles by teams comprising overseas Chinese in the United States and researchers on the Chinese mainland was encouraged by what Zuoyue Wang calls an "ethnically based transnational research community." This was inspired in part among overseas Chinese by "cultural nationalism in the sense of identification with the developmental aspirations of their country of origin."[62] These affective bonds fostered the self-organization of research agendas that were of mutual scientific interest.[63]

The Trump administration presented US-Chinese scientific cooperation as a one-way transfer of knowledge and technology from the metropolis to the periphery. The picture that emerges here is of a dumbbell. The desire to do good science bridged two huge, dynamic, and increasingly interdependent communities, who researched and published together in the spirit of scientific internationalism. Cultural nationalism provided an additional motivation for overseas Chinese researchers to build ties with the mainland. The Chinese government's talent recruitment programs, like similar initiatives in the Western world, mobilized the intellectual capital that overseas Chinese and foreign scholars had accumulated abroad to inject new ideas and expertise into the indigenous innovation system. US researchers gained access to a well-funded and ambitious community of researchers who were rewarded for publishing high-impact papers while engaged in a techno-scientific project of modernization on a gigantic scale.

An account of Sino-American collaboration from the bench, an account that highlights self-organization by a transnational network dedicated to doing good science, is compatible with researchers in both the United States and China being engaged, deliberately or not, in state-driven agendas in pursuit of national security. In fact, even the most prudent of observers concluded that the integrity of the US research system had been corrupted by "foreign influences through rewards, deception, coercion, and theft . . . to some degree," and that, notwithstanding the paucity of official unclassified information regarding China, "there are enough verified instances to warrant concern and action."[64] It is to the sources of these concerns and the actions taken to mitigate them that we now turn.

Chinese Students and Visitors in the United States: Scholars or Spies?

The publication of the White House's *National Security Strategy* in December 2017 was quickly followed by a number of highly visible government initiatives that turned the spotlight on the dangers to economic security posed by collaborative research with Chinese partners. In addition to buttressing CFIUS and export controls, the administration pledged to enhance "counterintelligence and law enforcement activities to curtail intellectual property theft." It would "explore new legal and regulatory mechanisms to prevent and prosecute violations." And it would review visa policies "to reduce economic theft by non-traditional intelligence collectors," possibly restricting access by "foreign STEM students from designated countries to ensure that intellectual property is not transferred to our competitors."[65]

Congress also moved into high gear, staging a string of hearings that debated the risks of academic collaboration with China. In July 2018, for instance, the House Permanent Select Committee on Intelligence discussed "China's threat to US research/innovation leadership."[66] The FBI's Bill Priestap testified before the Senate Judiciary Committee and one of its subcommittees in June and again in December 2018 to discuss how to protect America's free and open academic environment from China's "non-traditional espionage."[67] Echoing his boss Christopher Wray, he enumerated the multiple channels exploited by foreign powers (notably China) to secure cutting-edge knowledge that would contribute to the country's emergence as an economic superpower by 2049. Exploiting the openness of the US research system by "theft, plagiarism, and the commercialization of early stage collaborative research," "foreign state adversaries" had saved their country time, money, and resources to achieve generational advances in technology. Of particular concern, said Priestap, was the "use of foreign [notably Chinese] academics by their home countries' intelligence services in furtherance of this exploitation."[68]

In November 2018 the Permanent Subcommittee on Investigations of the Senate Committee on Homeland Security and Governmental Affairs released a staff report on the threat posed by China's talent recruitment plans to the US research enterprise.[69] It described how American taxpayer-funded research had contributed to China's global rise over the last decades, and it chided the federal government's grant-making agencies (and the FBI) for not doing enough to stop Beijing from recruiting US-based researchers who transferred knowledge and intellectual capital to China in exchange for financial and other benefits.

Priestap's and Wray's statements were not isolated interventions, but part of a much larger current in American public discourse that held that many of the 350,000 Chinese students enrolled at US universities were, willingly or not, agents of the Chinese government. President Trump is reported as having said at an informal dinner with the CEOs of thirteen major American companies that "almost every student that comes over to this country is a spy" (given the context of his rambling remarks, everyone knew that he meant those from China).[70] Investigative journalist Daniel Golden quotes the FBI's former head of counterintelligence in Tampa, Florida, as saying that the implicit message from the Chinese authorities to Chinese researchers in the United States is "Don't come home empty-handed."[71] Mark Warner (D-VA), the vice chairman of the Senate Intelligence Committee, has gone further, suggesting that remaining abroad was an increasingly difficult option for Chinese students. He claimed that Chinese spy services were increasingly "weaponizing" students, by "literally threatening their families if they don't come home after they have done that advanced research degree in the United States or elsewhere in the West, and come home with intellectual property."[72]

The charter for the hearing *Scholars or Spies? Foreign Plots Targeting America's Research and Development*, held in April 2018, was "to explore foreign nations' exploitation of US academic institutions for the purpose of accessing and engaging in the exfiltration of valuable science and technology (S&T) research and development (R&D)."[73] The presiding chairman, Ralph Abraham (R-LA), explained at once that exfiltration was a "new word being used to describe the surreptitious removal of data, as well as R&D."[74] This, he said, was facilitated by the "lax security posture of our academic institutions," which enabled "individuals, including professors, students, researchers and visitors—some with strong ties to a foreign nation," to target "the innovation and intellectual property from our country's greatest minds and institutions" and to share it with "foreign governments, universities, or companies." This "nefarious activity" was all the more reprehensible when the "target" was funded by a federal grant program. In that case US taxpayers were "unwittingly funding the technological advancements and innovative breakthroughs that allow foreign nations to improperly gain a competitive economic advantage." While China was not the only culprit, it was the most aggressive. "China steals our fundamental research and quickly capitalizes by commercializing our technology," Representative Abrahams said.[75]

We want to make two points about this hearing. Firstly, participants were at pains to stress the important contribution made by international collaboration to the dynamism of the United States' innovative research system. The second point to note is that the examples given by the witnesses in support of their accusation that China was stealing IP *from American universities* were anything but convincing. Michael Wessel of the US-China Economic and Security Review Commission, chose a few specific cases that had become public over the past three years "regarding efforts to target US academic institutions for intelligence collection." Apart from his misleading interpretation of the case of Ruopeng Liu, which we will discuss later in more detail, Wessel referred to a 2015 case when "Chinese professors" were among six defendants charged by the Department of Justice "with economic espionage and theft of trade secrets . . . for the benefit of universities and companies controlled by the PRC government."[76] Contrary to the impression that this gives, the six were not professors at American universities, they were not accused of stealing protected knowledge from American universities, and they were (probably) not directed by the PRC government to do so. They were professors at Tianjin University in China who had graduated with PhDs at an American university, and who were "alleged to have stolen valuable technology from Avago Technologies and Skyworks Solutions," to quote the DoJ's press release of May 19, 2015.[77] Avago and Skyworks are two American companies where two of the six professors worked after graduating from a university in southern California in about 2005. According to the indictment, beginning in 2006 they explored ways to commercialize the technology of the kind developed by the two US firms in a joint venture between Tianjin University and the Chinese authorities. Certainly, in September 2020, one of them was sentenced to eighteen months in a minimum security prison and ordered to pay back $477,000 to the firms that he stole IP from.[78] What matters for this argument, though, is that the American academic system is involved only to the extent that two of the six earned a PhD in the country before working in the commercial sector. And let us not forget that they are accused of industrial espionage at two small US firms, not theft of IP from a US university.[79]

While the carelessness of the quoted congressional testimony cannot be condoned, it is important to probe deeper into some of the reasons for the frustration that has engendered it. We will do so after analyzing the government's position on China's talent recruitment programs, which seek to reverse a "brain drain" of gifted scholars to the Western capitalist world.

Knowledge Flows from the United States to China: The Talent Recruitment Programs

In January 2020 Charles Lieber, the chair of Harvard University's Department of Chemistry and Chemical Biology, was charged by the US Department of Justice with "one count of making a materially false, fictitious and fraudulent statement" regarding his involvement in China's Thousand Talents Plan (TTP).[80] Wire fraud is, apparently, "a backstop charge that is used [by the Department of Justice] when economic espionage is difficult to prove."[81] The maximum sentence for such a charge is five years in prison, three years of supervised release, and a $250,000 fine.[82] Lieber's prestige at Harvard (he was one of some two dozen distinguished University Professors), his academic reputation (he was once tipped as a possible Nobel Prize winner for his pathbreaking research in nanoscience and technology), and the scale of support for his research in Boston (he had received over $15 million in research funds from the National Institutes of Health [NIH] and the Department of Defense since 2008) ensured that the charges against Lieber made headline news.[83]

So too did the depth of his cooperation with the Wuhan University of Technology (WUT) in China. Lieber was recruited as a Thousand Talents foreign expert. He was given $1.5 million to establish a mirror laboratory at WUT, where he was expected to spend up to nine months of his time each year, "declaring international cooperation projects, cultivating young teachers and PhD students, organizing international conference[s], applying for patents and publishing articles in the name of WUT."[84] The Wuhan laboratory was to focus on "advanced research and development of nano wire-based lithium-ion batteries with high performance for electric vehicles."[85] Lieber was also expected to host Chinese researchers in his Boston lab for two months a year, according to the FBI. The Chinese university paid him a salary of up to $50,000 a month plus about $158,000 over three years for living expenses. It must be stressed that Lieber was not accused of espionage but of lying to the federal authorities. In fact he was fully within his rights when he signed on to the Thousand Talents Plan.

The Thousand Talents Plan (TTP) is one of about a dozen programs that have been put in place by the Chinese authorities since the mid-1990s to encourage citizens who received advanced qualifications abroad to return to the mainland.[86] It was inaugurated by the Chinese central government in 2008 to attract (back) people who were trained and living abroad using a range of generous incentives.[87] According to the official website, by the end May 2014, more than 4,180 overseas experts had been recruited by the TTP.[88] The number had risen to "more than 7000 people overall" by 2018.[89]

Those who devised the plan for the government showed a deep understanding of the range of skills required to build and sustain a dynamic research system. They were not simply interested in people with outstanding intellectual track records. As the program's "History and Background" statement puts it, when gifted recruits "go (back) to China" they are expected to play "a positive role in the scientific innovation, technological breakthrough, discipline construction, talent training and hi-tech industry development, [becoming] an important force in the construction of the innovative country."[90]

The FBI pointed out in 2015 that "associating with these talent programs is legal and breaks no laws."[91] The risks run by researchers who participate in them lie in the terms of the contracts signed with China that violate American standards of research integrity. A report by a US Senate Permanent Subcommittee on Investigations cited extracts from these contracts to show the importance attached by the Chinese authorities to acquiring and controlling ownership of intellectual property generated by TTP contractors. For example, one contract stated that if Chinese scientists contributed to any joint discoveries, the partner institutions, one in the United States, the other in China, would "jointly own, protect, and manage the commercialization of these jointly made discoveries."[92] In fact, according to Lawrence Tabak, the principal deputy director of the NIH, the "key qualification" for being recruited on a talent program is, quite simply, "access to intellectual property."[93] The "Made in China 2025" plan defined the areas of expertise the recruitment programs should seek out and the related IP they hoped to generate domestically.

Made in China 2025

Made in China 2025 was launched in 2015 with the goal of upgrading China's manufacturing industry. It was drafted by the Ministry of Industry and Information Technology over two and a half years, with input from 150 experts from the Chinese Academy of Engineering. "Its guiding principles," to quote Scott Kennedy of the Center for Strategic and International Studies (CSIS), "are to have manufacturing be innovation-driven, emphasize quality over quantity, achieve green development, optimize the structure of Chinese industry and nurture human talent."[94] It singled out ten key sectors in which the government sought to secure a dominant share of the global market, with firms setting their own technology standards and engaging more directly in international standard setting.[95] These sectors were central to the so-called fourth industrial revolution, integrating big data, cloud computing, the internet of things, and other emerging technologies into global manufacturing supply chains. Some of their foci—artificial intelligence, robotics, autonomous vehicles, augmented and virtual reality, financial technology, gene editing—"will be foundational so that many applications or end-use technologies will be built upon them."[96] They were also dual use, with sales in the global market fueling innovation in the domestic civilian and military sectors.

According to the China State Council, Made in China 2025 aspires to Chinese industry becoming increasingly independent of global supply chains, "aiming to raise the domestic content of core components and materials to 40 percent by 2020, and 70 percent by 2025."[97] The Thousand Talents Program was, one assumes, mobilized to recruit experts from all over the world to achieve these ambitious goals, and to foster emerging technologies in the new manufacturing innovation centers foreseen in the plan: fifteen by 2020, and forty by 2025.

Made in China 2025 was one facet of a comprehensive "National Innovation-Driven Development Strategy" announced by the Central Committee of the Communist Party of China and the State Council in May 2016. It attributed the "backwardness and beatings of China in modern times" to its lack of participation "with previous scientific and technological revolutions." Its "weak science and technology" had undermined its "national strength." China would never again sit on the sidelines, allowing another revolution to pass it by. "Innovation drive is the destiny of the country," an essential force "to realize the Chinese dream of the great rejuvenation of the Chinese nation."[98]

China's holistic approach to becoming a manufacturing superpower posed a direct threat to US economic security. The Chinese leadership had the visionary goal, the political will, the industrial infrastructure, and the highly trained cadre of scientists and engineers needed to reach its objectives. It also had the financial means derived from its huge and—until recently—increasingly favorable balance of trade. In 1990 the difference between exports to and imports from China and the US was $10.4 billion in China's favor. It had increased to $84 billion in 2000, notwithstanding a major export drive to the PRC spearheaded by the Clinton administration. By 2010 it had soared to $203 billion in China's favor, while in 2018 the trade deficit had reached the staggering figure of $419 billion (with imports to the United State from China reaching $539 billion).[99]

In August 2017 the Office of the US Trade Representative launched an official investigation under section 301 of the Trade Act of 1974 to see if China's policies on IP, innovation, and technology were unfair and harmed US stakeholders.[100] In his submission to its hearings, James Lewis of CSIS explained why plans like Made in China 2025 were so harmful to American economic security: "What is new is that unfair trade, security and industrial policies, tolerable in a smaller developing economy, are now combined with China's immense, government-directed investment and regulatory policies to put foreign firms at a disadvantage. . . . China now has the wealth, commercial sophistication and technical expertise to make its pursuit of technological leadership work."[101] China's determination to transform itself from a big manufacturing country to a powerful one in global markets, along with its quest for independence from foreign sources of core components and materials, threatened to completely reshuffle the distribution of power in the global economy. As one study by the US Chamber of Commerce concluded in 2017, Made in China 2025 "aims to leverage the power of the state to alter competitive dynamics in global markets in industries core to economic competitiveness."[102] This threat posed to US economic security was amplified by Xi Jinping's promotion of a policy of "military-civil fusion" similar to that undertaken by the Clinton administration (see chapter 8).[103]

There was a notable paucity of empirical evidence that would enable the public to assess many of the concerns raised by Congress and the intelligence community regarding the risks involved in university collaboration with China. Excessive zeal and carelessness by the FBI increased suspicion of the government. The case of Xiaoxing Xi, in which the interim chair of the Physics Department at Temple University was arrested at home at gunpoint and mistakenly accused of sharing proprietary knowledge on a so-called pocket heater that makes thin superconducting films, acquired high public visibility. And as trust in the FBI receded, rumor and hearsay proliferated unchecked in the public space when reliable knowledge was not available.[104] A new "Red Scare" was said to be "reshaping Washington," eliding the distinction between the Chinese people and the Chinese government and the Communist Party.[105] Leading members of the research community felt obliged to confirm their commitment to America's core values. In May, 2019 Yale's president Peter Salovey circulated a public letter affirming "Yale's steadfast commitment to our international students and scholars," regretting their "sense of unease" at the increased scrutiny of academic exchanges pursuant on escalating tension in US-China relations.[106] A month later MIT president Rafael Reif stated that "faculty members, post-docs, research staff, and students tell me that, in their dealings with federal agencies, they now feel unfairly scrutinized, stigmatized, and on edge—because of their Chinese ethnicity alone."[107] The chairman of the Committee of 100, an American organization of prominent Chinese Americans, as well as Representative Judy Chu (D-CA) deplored government policies that, she said, targeted "an entire ethnic group of people for suspicion that they're spies for China"[108]

The administration's emphasis on the enormous risks of international academic collaboration with China, even while recognizing its importance, were indicative of the ambiguity and confusion surrounding where, and how to balance scientific openness with the needs of national economic and military security. The insistence that the Chinese authorities were making a concerted, top-down effort to systematically exploit IP produced in American research labs was a warning to academia and to federal grant-giving agencies that policies were urgently needed to shift the balance in favor of tighter restrictions in the name of economic security. The problem was finding the most appropriate instruments to define the line beyond which transgressions of Western norms of research integrity had to be punished.

Fundamental Research and IP Loss: Criminalize or Mitigate Risk?

If we take at face value FBI director Christopher Wray's statement that "China seems determined to steal its way up the economic ladder at our expense,"[109] we would expect to find that a large number of academic researchers had been brought before the courts and sentenced for criminal behavior. In fact, a survey of the information in the public domain of academic researchers who were charged with criminal behavior between 2013 and September 2020 produced only about fifteen cases, many of which ended with plea deals and an admission of guilt to a far less important offense than the individual was originally charged with. This does not mean that there is no substance in the FBI's accusations. The government has successfully prosecuted many Chinese nationals for industrial espionage. The small number of cases of this kind that we find in academia simply means that the norms that defined the practice of fundamental research in American universities today, with their stress on freely sharing knowledge that was destined for publication, were vitiated by multiple loopholes that subverted the possibility of successfully prosecuting violators in court.

The infamous example of Ruopeng Liu makes the point.[110] In 2006 Ruopeng Liu began work on his PhD in Dr. David Smith's laboratory at Duke University. He knew that Smith was a world authority on "metamaterials." These are "weird" materials that do not exist in nature, as Smith put it, and that can be used to render objects invisible to microwave signals. Smith's fundamental research was funded by the DoD and the intelligence community. Liu was a brilliant, ambitious, and entrepreneurial student. He arranged for reciprocal visits between Duke and his Chinese mentor's research group, during one of which, visitors from China photographed Smith's experimental equipment without his permission and reproduced an exact replica of it back home. After Liu had left Duke University with a PhD in 2009 and had returned to China to start his own business, Smith discovered that, unbeknownst to him, Liu had been working toward commercializing his research abroad, an unethical practice that would have cost Liu his PhD at Duke.

Did Liu steal IP from Duke University? An FBI official has claimed that Smith's technology was on the Chinese government's "shopping list." Liu dismissed such charges as "ridiculous" and "far away from the truth."[111] He claims to be innocent of any malpractice. David Smith himself says that Liu, who was a prolific patenter, notified the authorities at Duke about potential inventions, and "we were able to preserve the IP."[112] Duke University's vice president for public affairs and government relations has said that "there is no indication that any intellectual property owned or controlled by Duke University was stolen."[113] The FBI opened a case against Ruopeng Liu in 2010 to investigate whether he had stolen IP. It was closed a few years later for lack of evidence.

By 2017 Ruopeng Liu was a billionaire, feted at home as "China's Elon Musk." He enjoyed close relationships with the government, including Xi Jinping himself. David Smith is remembered as a brilliant academic who was, in his own words "incredibly naïve" about the need to protect IP in the "natural chaos that occurs in US academic environments" so enabling Liu to "walk out of the door" of his laboratory with Pentagon-funded research. As for the loss of intellectual property, Smith said, it is "something we're all grappling with—where to draw the line."[114]

Smith's difficulty in knowing "where to draw the line" admits to the ambiguity and confusion created in the fundamental research space in a community in which "the majority of collaborations are just informal, people getting together at meetings and brainstorming," as Smith put it.[115] A cultural revolution was required in laboratories like his, in which, "unlike in the corporate world, university researchers are rarely required to sign nondisclosure agreements or terms of collaboration, which many professors view as volatile of the spirit of academic openness," particularly in dealing with foreign students and visitors.[116] This made it very difficult to criminalize IP theft in academia. As the FBI's Bill Priestap explained, the "contractual paucity" that pervades the tradition of international academic collaboration "makes proving foreign intellectual property theft challenging, since *US economic espionage law requires the victim of the theft to demonstrate that he took reasonable precautions to protect the secret stolen.*"[117] In addition, as one FBI special agent put it, "When you're writing affidavits explaining what the intellectual property is and how they're stealing it—while protecting each company's property rights—that was all scientifically very challenging to articulate."[118] Building a case for the prosecution is further complicated by the logistics of an enquiry that stretches back into China, which has a different concept of what counts as IP, and where officials are reluctant to cooperate anyway. Rather than seeking to criminalize researchers with accusations of economic espionage, the FBI thus emphasized the need for US businesses and universities to "protect their information and mitigate lost or stolen information."[119] Priestap pointed out that this meant that colleges and universities needed to implement "clearer—and in some cases more restrictive—guidelines regarding funding use, lab access, collaboration policy, foreign government partnership, nondisclosure agreements and patent applications."[120] Following this injunction, many if not most research universities revisited their policies and protocols governing federal research grants and took measures to protect intellectual property in response to the threat posed by China.[121]

Measures to Mitigate Risk in the US Innovation System Up to September 2020

Three major stakeholders have taken important steps to mitigate the risk of sensitive knowledge leaking out of US research sites into China (as well as Russia, North Korea, and Iran). Firstly, there are the federal grant-awarding agencies—among whom the Department of Defense (DoD), the Department of Energy (DoE), the National Institutes of Health (NIH), and the National Science Foundation (NSF) are particularly prominent. Secondly, there is the State Department, which has defined new, restrictive visa policies, notably for STEM students, as called for in the White House's *National Security Strategy* of 2017. And then there is Congress, which has debated and adopted new measures to protect the national innovation system. At the time of writing (fall 2020) it seemed there was no overarching, coordinated approach being taken to the problem of economic espionage. Each actor had adopted measures appropriate to its domain of responsibility. This was particularly striking as regards the federal funding agencies, where one might have expected a more coordinated response as grantees often received research support from more than one of them, sometimes simultaneously.

The NIH explicitly engaged an approach that sought to "mitigate and prevent rather than criminalize."[122] This agency invests some $39 billion annually in medical research and distributes about 80 percent of that money through some fifty thousand grants to more than three hundred thousand grantees or principal investigators in universities, medical schools, and universities both in the United States and around the world.[123] In August 2018, NIH director Francis Collins was asked about the risks posed to research integrity by China's talent recruitment programs at a hearing of the US Senate Committee on Health, Education, Labor, and Pensions. Collins testified that "the robustness of the biomedical research enterprise is under constant threat" and that the magnitude of the risk was increasing. He added that he had just sent letters to over ten thousand senior representatives of NIH grantee institutions asking them to review their records for evidence of malfeasance, some involving foreign governments.[124] Collins also mentioned that he was working along with other government agencies and the broader biomedical research community to identify steps that could be taken to help mitigate "unacceptable breaches of trust and confidentiality that undermine the integrity of US biomedical research."[125]

Collins's letter of August 20, 2018, stated that the NIH was aware that "some foreign entities have mounted systematic programs to influence NIH researchers and peer reviewers" and to exploit the "trust, fairness and excellence of NIH-supported research activities."[126] He identified three specific areas of concern: the diversion of IP in grant applications or of findings produced by NIH-supported biomedical research to other entities, including other countries; the sharing of confidential information on grant applications by NIH peer reviewers with others, including foreign entities; and the failure to disclose substantial research support from other entities, including foreign governments, while working at NIH-funded institutions. Collins warned the recipients of the letter that they might hear again from the NIH regarding the research activities of their personnel and urged them to "reach out to an FBI field office to schedule a briefing on this matter." To tighten up its inhouse procedures the NIH Advisory Committee to the Director set up a working group on Foreign Influence on Research Integrity, which submitted recommendations in December 2018 on how best to deal with the breaches identified in Collins's letter.[127]

The NIH director's initiative was not prompted by "some big explosive episode," he told reporters from *ScienceMag*, but by a "gathering sense that it's time to take action."[128] In fact in mid-2016 the NIH was told by the FBI that a probe had uncovered a researcher at the MD Anderson Cancer Center who was sharing NIH grant proposals that were supposed to be treated as confidential with several other people. Looking more closely at the files of some of their funded researchers at American universities NIH grant managers noticed that some of them "were publishing papers that listed a foreign institution—often in China—as the primary affiliation and cited foreign funding sources in the fine print." There were also clear conflicts of commitment: one researcher given eight months of NIH funding at an American university also had nine months of external time commitments in the same year.[129]

The government has extensive powers to investigate suspected violations of research integrity. The actions at MD Anderson, which is part of the University of Texas research system, were triggered by a chain of events that began in the summer of 2017, when the FBI notified the cancer research center that it was investigating the "possible theft of MD Anderson research and proprietary information."[130] A federal grand jury then issued a subpoena authorizing the intelligence community's access to five years of emails by some of the center's employees. This was followed by a voluntary agreement reached with the center's new president in November 2017. He authorized the FBI to search the network accounts of twenty-three employees "for any purpose ... at any time, for any length of time and at any location." This was followed by FBI interviews with four Chinese Americans who worked at MD Anderson, asking whether they or others had professional links to China's Thousand Talents Program. One of them claimed afterward that, in fact, the FBI was less concerned with espionage or IP loss as such than with loyalty: "They wanted to know, in effect, are you now, or have you ever been, more committed to curing cancer in China or in the US?" MD Anderson subsequently authorized the FBI to share any relevant information from its employees' accounts with the NIH or any other federal agencies. These investigations by the FBI, and the subpoenas in particular, seem to have provided the evidence that the NIH used in its letters to MD Anderson identifying five researchers who were suspected of violating the NIH's rules. The compliance office moved to terminate the contracts with three of the five, two of whom resigned before the process was completed.

NIH officials used a very comprehensive concept of theft to characterize the forms of abuse they unearthed among their grant recipients. Jodi Black, the deputy director of its Office of Extramural Research, speaking to the Annual Conference of the Association of Public and Land-Grant Universities in San Diego in November 2019, insisted that violations of research integrity "boil down to theft," and she listed multiple possible forms this theft could take. "Employee theft" occurs when a grant recipient spends "excessive time away" or "works for a competing employer" abroad on "company time." Sharing confidential grant proposals submitted for peer review with colleagues abroad is deemed to be "theft of nascent ideas" and can also be regarded as "theft from the public," whose tax dollars financed the grant. Undisclosed financial conflicts of interest lead to a loss of royalties and are described as "economic development theft."[131]

This language—stealing "company time," "economic development theft"—indicates the extent to which NIH sees the biomedical research it funds as deeply embedded in the market. In fact, NIH's Lawrence Tabak describes fundamental research as "pre-patented material" and says that it "is the antecedent to creating intellectual property," which, "'in essence,' is 'stolen' when someone takes his knowledge back to China."[132] Jodi Black remarked that nearly all the violations of research integrity that had been investigated by the NIH involved "pre-clinical research" (i.e., they were on the path to drug development), and so they also "boiled down to theft." For senior administrators in the NIH, then, all research normally covered by the Fundamental Research Exclusion (NSDD 189) is apparently so pregnant with commercial possibilities that it has to be treated as if it were patentable and so "effectively" stolen when foreign researchers acquire it.

Black told her audience in San Diego that by November 2019 "at least 120 scientists, not all ethnically Chinese," in "70+ institutions, many fields of biomedicine, all over . . . the US" had been involved in one or more of the violations that "boiled down to theft."[133] That is a tiny fraction of the three hundred thousand people funded annually by the NIH. Although the number may be small, the NIH was in no position to downplay its significance. In fact the agency was under relentless pressure from Senator Chuck Grassley (R-IA), who chaired the Senate Committee on the Judiciary from 2014 to 2018, and who became the chairman of the Senate Finance Committee in 2019. Grassley was not satisfied with the NIH's vetting procedures for grant recipients. He urged federal agencies to do more to thwart foreign governments' "real, aggressive and ongoing attempts" to steal the fruits of US-funded research.[134]

The NSF's mission to advance the progress of science situated it in a field that was less entangled with hot-button domains like biomedical research. It commissioned a study from the JASON group to assess the benefits and the risks of the openness associated with fundamental research in the face of the threat from China. JASON was originally established in the 1960s to encourage younger scientists to engage in military-related research. NSF director France Córdova described them as "an independent group of high-level academics that interfaces with the intelligence community."[135] Their unclassified report was released in December 2019.[136] It emphasized the immense contributions made by foreign nationals to the US innovation system at least since the 1930s—adding that the United States had "benefitted enormously from brain drain right up to the present day."[137] It pointed out that China was not unusual in having talent recruitment programs: both developed and developing nations sought to repatriate citizens and to attract foreign talent in what had become a global competition for "intellectual capital." And while they regretted that it was difficult to assess the "scale and scope" of foreign influence on the US innovation system, they confirmed that there were genuine reasons for concern regarding the actions of the Chinese government and its institutions when it came to respecting US practices of research integrity—"research transparency, lack of reciprocity in collaborations and consortia, and reporting of commitments and potential conflicts of interest, related to these actions."[138]

Prudence characterized the JASON Report's recommendations. On the one hand, the report proposed that "the scope of expectations under the umbrella of research integrity should be expanded to include full disclosure of commitments and actual or potential conflicts of interest." This aligned it with the NIH's determination to ensure that grant applicants listed all their foreign ties in detail when they sought funding. Coupled with this, the JASON group recommended that the "NSF should support reaffirmation of the principles of NSDD-189, which make clear that fundamental research should remain unrestricted to the fullest extent possible." The JASON Report also reminded its readers that NSDD 189 had been reaffirmed by Condoleezza Rice in 2001 and by Ashton Carter in 2010. The report saw no reason to break with that tradition and its mechanism to restrict the circulation of fundamental research in the name of national security, namely "the general principle of creating high walls i.e. classification, around narrowly defined areas."[139] In sum the FRE was sacrosanct, and the NSF should do all it could to exempt fundamental research from government control.

The NSF backed away from some of the JASON Report's recommendations. Its published response gave only qualified support for the Fundamental Research Exclusion. It insisted that "in some cases, there may be a need to protect certain data and information for national, military, or economic security purposes," adding that the NSF would work with other government agencies to define the boundary between open research, and research that had to be protected for security reasons. It also established the new post of chief of research security strategy and policy.[140]

At the time of writing it was hard to say if the FRE would survive in its present form, as the JASON Report hoped. Neither the NIH nor the NSF was advocating for it. The Trump White House's *National Security Strategy for Critical and Emerging Technologies*, released in October 2020, used language that undermined its philosophy altogether. As the strategy statement put it, one priority for protecting the United States' technological advantage was to "protect the integrity of the R&D enterprise by fostering research security in academic institutions, laboratories, and industry while balancing the valuable contributions of foreign researchers."[141] The specificity of basic and applied research in universities was not only ignored. The priority now was to protect the "security" of this research space, not to foster its openness. Access to it by foreign nationals was not guaranteed, but it would be weighed, "balanced" against the associated security risks. There were echoes here of the recommendations made in the staff report to the Senate Committee on Investigations in November 2018. It called on the administration to "consider updating NSDD-189 and implement additional, limited restrictions on US government funded research." It also suggested that federal agencies ask whether "openly sharing some types of fundamental research is in the nation's interest."[142]

There were also attempts made to mitigate the risks associated with the circulation of "knowledgeable bodies." The NSF, as well as the Department of Energy, took steps to restrict the travel of their employees and contractors to sensitive foreign entities, so denying those countries face-to-face access to American experts. The DoE is responsible for the country's seventeen national laboratories, which host thirty thousand visitors annually, about ten thousand of whom are from China. In June 2019 it prohibited its employees and contractors from joining talent recruitment programs run by China, Iran, North Korea, and Russia. Those who wished to participate in such a program had first to discuss its implications with the appropriate DoE officials. Those already in a program had to disclose their participation within thirty days or risk being penalized and could even be dismissed.[143] In July 2019 the NSF also barred its personnel from participating in such programs and called on grant applicants whose home institutions had campuses abroad to justify why it was preferable for them to do their research there rather than at the American campus.[144] Some actors have called for a more punitive approach. Michael Wessel, whom we met earlier, proposed in the *Scholars or Spies?* hearings that "participants in China's 1,000 Talent Program should be prohibited from receiving future federal support in terms of grants, loans or other assistance."[145]

Measures to control the transnational flow of US expertise to China via the talent programs were complemented by new State Department visa policies that restricted the access of foreign nationals to research in the United States. It kept the promise made in the White House's *National Security Strategy* of December 2017 to review visa policies "to reduce economic theft by non-traditional intelligence collectors," possibly restricting access by "foreign STEM students from designated countries to ensure that intellectual property is not transferred to our competitors."[146] As of June 2018 the State Department limited visas for graduate students from China who wished to study in aviation, high-tech manufacturing, and robotics to one year, renewable, rather than the previous five years.[147] All these fields were identified in the Made in China 2025 plan. Twelve months later the

State Department went further. Visa application forms in use after May 31, 2019, required every prospective traveler and immigrant to provide up to five years of information on travel histories, telephone numbers, email addresses, and social media identifiers, including pseudonyms, on about twenty platforms, twelve of which were based in the United States and included Facebook, LinkedIn, Twitter, and You Tube.[148]

Finally, universities across the country also played their role in mitigating the risk of losing sensitive knowledge to China.[149] Maria Zuber, planetary scientist and vice president for research at MIT, took the lead.[150] She established a review process for (a) projects funded by people or entities in China (including Hong Kong), Russia, and Saudi Arabia, (b) projects involving MIT faculty, staff, or students who conduct work in these countries, and (c) collaborative projects with people or entities in these countries. All such projects were subject to an internal "elevated risk review" as regards intellectual property; export controls; data security and access; economic competitiveness; national security; and political, civil, and human rights; as well as consistency with MIT's core values. The PI was required to produce a satisfactory "risk management plan" addressing identified concerns, failing which the collaborative project with the targeted countries could not proceed. MIT has also decided not to accept any further funding from the Chinese telecommunications companies Huawei or ZTE, or their subsidiaries, because of their violations of sanctions restrictions.

Two features stand out from this survey of the variety of initiatives that were launched to control the flow of sensitive knowledge to China in academic research settings between 2017 and 2020. The first is the extent to which the universities and research centers worked along with every arm of the federal government and the intelligence community to mitigate the risk of IP flowing to China. FBI director Christopher Wray's call for a whole-of-society response to the threat posed by China was heeded. This response was driven in part by the fear that the spigot that feeds federal research dollars into the academic research system would be shut off if universities and research centers did not comply. But less pragmatic concerns were also at play, notably the political conviction that China posed a genuine challenge to America's economic competitiveness and, by extension, to its global leadership.

This leads to the second observation. While there was widespread political consensus that measures had to be taken to restrict international collaboration with China and to monitor closely the research activities of people who engaged in it, there was also considerable push-back against some of the policies adopted by the Trump administration between 2017 and 2020. As we have seen, leaders of major universities called for a more balanced and transparent approach to deal with the risks posed by the open nature of university research. Accusations of racial profiling abounded among the Chinese American community. Major employers of highly qualified foreign nationals in STEM fields, firms like Google and Apple, filed an amicus brief in November 2020 against the federal government's tightening of criteria for getting an H-1B visa to work in the United States.[151] In March 2021 forty academics across the country signed an open letter in support of Harvard's Charles Lieber claiming that he was the "target of a tragically misguided government campaign" that was discouraging international scientific collaboration with China, in particular, to the detriment of the United States' position as world scientific leader, and to science itself.[152] Shortly before, more than one hundred faculty members at MIT signed an open letter to their president attacking the arrest of one of their nanotechnology professors for allegedly failing to disclose his ties to the Beijing government.[153]

As we have stressed time and again in this book, export controls and other modes of regulating the transnational flow of knowledge come up against deeply ingrained American values of social and academic openness, freedom of expression, and faith in free markets, all subtended by a conviction that the government should "interfere" as little as possible in the "marketplace" of goods and services and of academic ideas. The enactment of the protection of national security, extended in the 1990s to embrace economic security, sought to limit the expression of these cardinal values. The intersection between the two occurred on a constantly contested terrain. It was not simply that the benefits of the one had to be weighed against the costs of the other. It was also that the very concept of national security, along with the associated reach of its application, was part of an ongoing political and contested project, a demand by the government to suspend basic freedoms that always had to be justified. We wrote this epilogue in the midst of a major extension of controls over the transnational circulation of money in the form of FDI, and of scientific and technological knowledge embedded in people, ideas, and things, between the United States and the People's Republic of China. The evolution of the tensions between the regulating national security state and the corporate and academic actors that we have described in this book, and the trace that their engagement leaves on history, is a topic for future study, not present speculation.

Notes

Chapter 1

1. Department of Commerce, *Export Control, 62nd Quarterly Report* (4th Quarter 1962) (Washington, DC: Government Printing Office), 1963, 24–25.
2. Department of Commerce, "Hydrocarbon Research, Inc., et al."
3. Department of Commerce, *Export Control, 62nd Quarterly Report*, 25.
4. "Hydrocarbon Research, Inc., et al.: Consent Denial and Probation Order," 12488.
5. "Hydrocarbon Research, Inc., et al.: Consent Denial and Probation Order," 12488
6. Department of Commerce, *Export Control, 62nd Quarterly Report*, 25–26.
7. Department of Commerce, 26.
8. "Hydrocarbon Research, Inc., et al.: Consent Denial and Probation Order," 12487 (emphasis added).
9. "Hydrocarbon Research, Inc., et al.: Consent Denial and Probation Order," 12487. See also J. Forrester Davison, "Exports of Technical Data and the Export Control Act: Hearing Examiners and Consent Decrees," *George Washington Law Review* 33 (1964), 209–39, at 237.
10. Department of Commerce, *Export Control, 62nd Quarterly Report*, 26.
11. "Hydrocarbon Research, Inc., et al.: Consent Denial and Probation Order," 12487. The very same argument was invoked to control the circulation of unpublished data that American engineers had shared with their British counterparts in a collaborative project to develop gas centrifuges for uranium enrichment in the early 1960s; see John Krige, "Hybrid Knowledge: The Transnational Coproduction of the Gas Centrifuge for Uranium Production in the 1960s," *British Journal for the History of Science* 45:3 (2012), 337–57. The same determination to control unclassified knowledge in movement informed export controls by the Department of Commerce, nuclear cooperation by the Atomic Energy Commission, and the granting of visas by the State Department.

12. Harold J. Berman and John R. Garson, "United States Export Controls—Past, Present, and Future," *Columbia Law Review* 67:5 (1967), 791–890, at 860.

13. Department of Commerce, *Export Control, 62nd Quarterly Report*, 24; "Hydrocarbon Research, Inc., et al.: Consent Denial and Probation Order," 12488.

14. "US Claims NY Firm Made Illegal Red Deal," *Washington Post*, December 12, 1962, A16. "Leak to Rumania Disclosed by US," *New York Times*, December 14, 1962, 11. The serious effects of blacklisting on business companies are also stressed by Davison, "Exports of Technical Data and the Export Control Act," 235.

15. Department of Commerce, "Hydrocarbon Inc.: Temporary Consent Order," *Federal Register* 26:118 (June 21, 1961), 5535; Department of Commerce, "Hydrocarbon Inc.: Temporary Order Denying Certain Export Privileges," *Federal Register* 26:241 (December 15, 1961), 12044; Department of Commerce, "Hydrocarbon Inc.: Extension of Temporary Order Denying Certain Export Privileges," *Federal Register* 27:17 (January 25, 1962), 757.

16. Davison, "Exports of Technical Data and the Export Control Act, 227.

17. Berman and Garson, "United States Export Controls," 862 (emphasis added).

18. Davison, "Exports of Technical Data and the Export Control Act," 236.

19. Frank E. Samuel, "Technical Data Export Regulations," *Harvard International Law Club Journal* 6:2 (1965), 125–65, at 134, 155, 164.

20. Besides the articles already quoted, see J. N. Behrman, "US Government Controls over Export of Technical Data," *Patent, Trademark and Copyright Journal of Research and Education* 8:3 (1964), 303–15; Teruaki Inada, "International Know-How Licensing and Territorial Restraint Clauses," *Harvard International Law Journal* 8:2 (1967), 241–79, at 245–47; James T. Haight, "US Regulation of East-West Trade," *Business Lawyer* 19:4 (1964), 875–86, at 880.

21. When the questions of data controls and the extraterritorial reach of export controls became virulent again in the 1970s and 1980s, the Hydrocarbon precedent showed up once more, albeit hidden in the footnotes of law papers. See, for example, Ruth Greenstein, "National Security Controls on Scientific Information," *Jurimetrics* 23:1 (1982), 40–88, at 176n141 (misunderstanding the case as irrelevant to the control over scientific data); Kenneth Propp, "Export Controls: Restrictions on the Export of Critical Technologies," *Harvard International Law Journal* 22:2 (1981), 411–18, at 417n33; Janet Lunine, "High Technology Warfare: The Export Administration Act Amendments 1985 and the Problem of Foreign Reexport," *New York University Journal of International Law and Politics* 18:2 (1986), 663–702, at 688n145.

22. Michael Mastanduno, *Economic Containment: CoCom and the Politics of East-West Trade* (Ithaca, NY: Cornell University Press, 1992). Alan P. Dobson, *US Economic Statecraft for Survival 1933–1991: Of Sanctions, Embargoes and Economic Warfare* (London: Routledge, 2002); Philip J. Fungiello, *American-Soviet Trade in the Cold War* (Chapel Hill: University of North Carolina Press, 1988).

23. Hugo Meijer, *Trading with the Enemy: The Making of US Export Control Policy toward the People's Republic of China* (Oxford: Oxford University Press, 2016).

24. Still one of the best introductions to thinking about economic sanctions is David A. Baldwin, *Economic Statecraft* (Princeton, NJ: Princeton University Press), 1985.

25. Savita Pande, "The Challenge of Nuclear Export Controls," *Strategic Analysis* 23:4 (1999), 575–99; Danielle Peterson et al., "Export Controls and International Safeguards: Strengthening Nonproliferation through Interdisciplinary Integration," *Nonproliferation Review* 15:3 (2008), 515–27.

26. Australia Group, "20 Years of Australia Group Cooperation," April 2005, http://www.australiagroup.net/en/20years.pdf (accessed March 10, 2017).

27. Dinshaw Mistry, "Technological Containment: The MTCR and Missile Proliferation," *Security Studies* 11:3 (2002), 91–122; Wyn Q. Bowen, "US Policy on Ballistic Missile Proliferation: The MTCR's First Decade (1987–1997)," *Nonproliferation Review* 5:1 (1997), 21–39; Micah Zenko and Sarah Kreps, *Limiting Armed Drone Proliferation* (New York: Council on Foreign Relations, 2014).

28. See the reflections on the definition of "strategic goods" in Tor Egil Førland, "'Economic Warfare' and 'Strategic Goods': A Conceptional Framework for Analyzing COCOM," *Journal of Peace Research* 28:2 (1991), 191–204, at 197–200.

29. Hence, the US intelligence community plays a crucial researched dimension of export controls. The research done in this field is insufficient with the exception of nuclear nonproliferation intelligence. See, for example, Tanya Ogilvie-White, "The IAEA and the International Politics of Nuclear Intelligence," *Intelligence and National Security* 29:3 (2014), 323–40; Kristen A. Lau and Kevin C. Desouza, "Intelligence and Nuclear Non-proliferation Programs: The Achilles Heel," *Intelligence and National Security* 29:3 (2014), 387–431. But for a study on British intelligence and export controls that goes beyond the nuclear field, see Huw Dylan, *Defense Intelligence and the Cold War: Britain's Joint Intelligence Bureau 1945–1964* (Oxford: Oxford University Press, 2014), esp. 69–106.

30. *National Security Decision Memorandum 247: US Policy on the Export of Computers to Communist Countries*, March 14, 1974, 1, https://fas.org/irp/offdocs/nsdm-nixon/nsdm_247.pdf (accessed January 26, 2020).

31. Berman and Garson, "United States Export Controls," 820–23.

32. Berman and Garson, 814.

33. The literature on secrecy is vast. Good starting points are Peter Galison, "Removing Knowledge," *Critical Inquiry* 31:1 (2004), 229–43; Alex Wellerstein, "Knowledge and the Bomb: Nuclear Secrecy in the United States, 1939–2008," PhD diss., Harvard University, 2010; Alex Wellerstein, *Restricted Data: The History of Nuclear Secrecy in the United States* (Chicago: University of Chicago Press, 2021); Harold C. Relyea, "Information, Secrecy, and Atomic Energy," *NYU Review of Law and Social Change* 10:2 (1980/81), 265–86; Michael Aaron Dennis, "Secrecy and Science Revisited: From Politics to Historical Practice and Back,"

in *Secrecy and Knowledge Production*, ed. Judith Reppy, Cornell University Peace Studies Program, Occasional Paper No. 23, 1999, 1–16; Walter Gellhorn, *Security, Loyalty, and Science* (Ithaca, NY: Cornell University Press), 1950; Herbert N. Foerstel, *Secret Science: Federal Control of American Science and Technology* (Westport, CT: Praeger, 1993), 21–95.

34. For an introduction, see Genevieve Knezo, *"Sensitive but Unclassified" and Other Federal Security Controls on Scientific and Technical Information: History and Current Controversy* (Washington, DC: Congressional Research Service), 2003.

35. Joseph Masco, "'Sensitive but Unclassified': Secrecy and the Counterterrorist State," *Public Culture* 22:3 (2010), 433–63; David E. Pozen, "The Mosaic Theory, National Security, and the Freedom of Information Act," *Yale Law Journal* 115:3 (2005), 628–79.

36. Mario Daniels and John Krige, "Beyond the Reach of Regulation? 'Basic' and 'Applied' Research in the Early Cold War United States," *Technology and Culture* 59:2 (2018), 226–50; Lisa A. Shay, "The Great Debate over Unclassified Information: National Security versus Scientific Freedom," *IEEE Transactions on Professional Communication* 32:3 (1989), 139–48; John Shattuck, "Federal Restrictions on Free Flow of Academic Information and Ideas," *Government Information Quarterly* 3:1 (1986), 5–29; Kenneth R. Foster and Irving A. Lurch, "Collateral Damage: American Science and the War on Terrorism," *IEEE Technology and Science Magazine* 24:3 (2005), 45–52; Samuel A. W. Evans and Walter D. Valdivia, "Export Controls and the Tensions between Academic Freedom and National Security," *Minerva* 50 (2012), 169–90.

37. In this regard not much has changed since Tibor Kremic, "Why the Lack of Academic Literature on Export Controls?," NASA/TM-2001-210982 (July 2001), https://ntrs.nasa.gov/archive/nasa/casi.ntrs.nasa.gov/20010090818.pdf (accessed May 10, 2020).

38. This is obviously simplifying the sophisticated globalization debates. A good starting point for a more balanced view is Manfred B. Steger, *Globalization: A Very Short Introduction* (Oxford: Oxford University Press, 2003), esp. chapter 4.

39. Mastanduno *Economic Containment*, 64–108; James K. Libbey, "Cocom, Comecon, and the Economic Cold War," *Russian History* 37 (2010), 133–52.

40. Wendy Brown, *Walled States, Waning Sovereignty* (Cambridge, MA: MIT Press, 2010).

41. Michael Mastanduno, "Economics and Security in Statecraft and Scholarship," *International Organization* 52:4 (1998), 825–54, at 834–37; Susan Strange, "International Economics and International Relations: A Case of Mutual Neglect," *International Affairs* 46:2 (1970), 304–15; Shahar Hameiri and Lee Jones, "Probing the Links between Political Economy and Non-traditional Security: Themes, Approaches and Instruments," *International Politics* 52:4 (2015), 371–88.

42. David C. Engerman, "American Knowledge and Global Power," *Diplomatic History* 31:4 (2007), 599–632; John Krige, ed., *How Knowledge Moves: Writing the Transnational History of Science and Technology* (Chicago: University of

Chicago Press, 2019); John Krige, "Techno-diplomacy: A Concept and Its Application to US-France Nuclear Cooperation in the Nixon-Kissinger Era," *Federal Register* 12 (2020), 99–116; Walter LaFeber, "Technology and US Foreign Relations," *Diplomatic History* 24:1 (2000), 1–19; Odd Arne Westad, "The New International History of the Cold War: Three (Possible) Paradigms," *Diplomatic History* 24:4 (2000), 551–65.

43. Berman and Garson, "United States Export Controls"; Richard T. Cupitt, *Reluctant Champions: US Presidential Policy and Strategic Export Controls; Truman, Eisenhower, Bush, and Clinton* (New York: Routledge, 2000); Alan P. Dobson, *US Economic Statecraft for Survival 1933–1991: Of Sanctions, Embargoes and Economic Warfare* (London: Routledge, 2002); Claus Hofhansel, *Commercial Competition and National Security: Comparing US and German Export Control Policies* (Westport: Praeger, 1996); Mastanduno, *Economic Containment*; Hugo Meijer, *Trading with the Enemy: The Making of US Export Control Policy toward the People's Republic of China* (Oxford: Oxford University Press, 2016).

44. The home page of the Association and a brief survey of its history is at http://aueco.org/about_aueco/ (accessed May 14, 2021).

45. Tor Egil Førland, "The History of Economic Warfare: International Law, Effectiveness, Strategies," *Journal of Peace Research* 30:2 (1993), 151–62, at 156.

46. For a good introduction to the history of blockades, see Lance E. Davis and Stanley L. Engerman, *Naval Blockades in Peace and War: An Economic History since 1750* (Cambridge: Cambridge University Press, 2006). See also Nicholas A. Lambert, *Planning Armageddon: British Economic Warfare and the First World War* (Cambridge, MA: Harvard University Press, 2012).

47. Dietrich Beyrau, Michael Hochgeschwender, and Dieter Langewiesche, eds., *Formen des Krieges: Von der Antike bis zur Gegenwart* (Paderborn: Schöningh, 2007); Maurice Pearton, *Diplomacy, War and Technology since 1830* (Lawrence: University Press of Kansas, 1984).

48. Berman and Garson, "United States Export Controls," 791–92.

49. "Trading with the Enemy Act, October 6, 1917," in *Trading with the Enemy* (New York: Guaranty Trust, 1917), 39–64.

50. Berman and Garson, "United States Export Controls," 792n1.

51. Edward S. Miller, *Bankrupting the Enemy: The US Financial Siege of Japan before Pearl Harbor* (Annapolis, MD: Naval Institute Press, 2007), 77–78; Michael A. Barnhart, *Japan Prepares for Total War: The Search for Economic Security, 1919–1941* (Ithaca, NY: Cornell University Press, 1987), 129–30, 178–82; "'Moral' Embargo Helps to Conserve Raw Materials," *Wall Street Journal*, December 21, 1939, 2; "US Acts to Curb Further Sales of Planes to Japan," *New York Herald Tribune*, June 12, 1938, 1; "The 'Moral' Embargo," *New York Herald Tribune*, December 4, 1939, 22.

52. Section 6 of this law pertaining to export controls can be found in *Report to Congress on Operations of the Foreign Economic Administration* (Washington, DC: Government Printing Office), 1944, 47, https://fraser.stlouisfed.org/docs/historical

/martin/54_01_19440925.pdf (accessed March 6, 2017); Berman and Garson, "United States Export Controls," 792n2.

53. 81st Congress, 1st Session, S. 548: A Bill to Provide Continuation of Authority for the Regulation of Exports, and for Other Purposes, January 18, 1949, 2. This bill became the Export Control Act of 1949.

54. Berman and Garson, "United States Export Controls," 792.

55. Melvyn P. Leffler, "The American Conception of National Security and the Beginnings of the Cold War, 1945–48," *American Historical Review* 89:2 (1984), 346–81; Emily S. Rosenberg, "The Cold War and the Discourse of National Security," *Diplomatic History* 17:2 (1993), 277–84; Andrew Preston, "Monsters Everywhere: A Genealogy of National Security," *Diplomatic History* 38:3 (2014), 477–500.

56. For a much broader discussion, see Robert A. Pollard, *Economic Security and the Origins of the Cold War, 1945–1950* (New York: Columbia University Press, 1985).

57. Ian Fergusson, *The Export Administration Act: Evolution, Provisions and Debate*, CRS Report for Congress, RL31832, July 15, 2009, 3.

58. These conflicts are one of the main topics in the literature on the history of Cocom as quoted above that is the basis of this entire subsection.

59. Mastanduno, *Economic Containment*, 72–78; Alan P. Dobson, "From Instrumental to Excessive: The Changing Goals of the US Cold War Strategic Embargo," *Journal of Cold War Studies* 21:1 (2010), 98–119, at 99–108.

60. Angela Stent, *From Embargo to Ostpolitik: The Political Economy of West German–Soviet Relations, 1955–1980* (Cambridge: Cambridge University Press, 1981).

61. Timothy Aeppel, "The Evolution of Multinational Export Controls: A Critical Study of the COCOM Regime," *Fletcher Forum* 12 (1985), 105–24; Hofhansel, *Commercial Competition and National Security*.

62. Rosa Rosanelli, *US Export Control Regulations Explained to the European Exporter: A Handbook* (Université de Liège, European Studies Unit, 2014), 26–27.

63. Bruce W. Jentleson, *Pipeline Politics: The Complex Political Economy of East-West Trade* (Ithaca, NY: Cornell University Press, 1986), 172–214; Patrick B. Fazzone, "Business Effects of the Extraterritorial Reach of US Export Control Laws," *New York University Journal of International Law and Politics* 15:3 (1983), 545–94; Alan P. Dobson, "The Reagan Administration, Economic Warfare, and Starting to Close Down the Cold War," *Diplomatic History* 29:3 (2005), 531–56. Another interesting example, in this case from the 1960s, of the effects of US re-export on the trade of allies is offered by Jeffrey A. Engel, "The Surly Bonds: American Cold War Constraints on British Aviation," *Enterprise and Society* 6:1 (2005), 1–44, at 31–39.

64. Since the late 1940s, foreign visitors have been screened on the basis of export control lists in order to assess the risks of them acquiring technology in the

United States. If the danger of technology loss was greater than the benefits of, for example, academic exchanges, the State Department denied visas to the individual; see Mario Daniels, "Restricting the Transnational Movement of 'Knowledgeable Bodies': The Interplay of US Visa Restrictions and Export Controls in the Cold War," in John Krige, ed., *How Knowledge Moves: Writing the Transnational History of Science and Technology* (Chicago: University of Chicago Press, 2019), 35–61.

65. For discussions of the notion of "decline," see, for example, Aaron L. Friedberg, "The Strategic Implications of Relative Economic Decline," *Political Science Quarterly* 104:3 (1989), 401–31; Henry R. Nau, *The Myth of America's Decline: Leading the World Economy into the 1990s* (Oxford: Oxford University Press, 1990).

66. The criticism is very well summarized in several reports of the National Academy of Sciences: *Analysis of the Effects of US National Security Controls on US-Headquartered Industrial Firms* (Washington, DC: National Academy Press, 1986); *Balancing the National Interest: US National Security Export Controls and Global Economic Competition* (Washington, DC: National Academy Press, 1987); *Finding Common Ground: US Export Controls in a Changed Global Environment* (Washington, DC: National Academy Press, 1991).

67. Reinsch, Testimony at Hearings before the Committee on Commerce, Science, and Transportation, September 17, 1998, 22.

68. Testimony, Gary Milhollin, Executive Director, Wisconsin Project for Nuclear Arms Control, *US Export Control and Nonproliferation Policy and the Role and Responsibility of the Department of Defense*, Hearings before the Committee on Armed Services, US Senate, 105th Congress, July 9, 1998, 29.

69. Samuel A. Evans, "Technological Ambiguity and the Wassenaar Arrangement," PhD diss., Oxford University, 2009; Michael Lipson, "The Reincarnation of Cocom: Explaining Post–Cold War Export Controls," *Non-proliferation Review* 6:2 (1999), 33–51. Additionally, the United States embedded export controls in many bilateral economic, technical aid, and military agreements with allies as well as neutral states like Sweden and Switzerland; see Michael Mastanduno, "Trade as a Strategic Weapon: American and Alliance Export Control Policy in the Early Postwar Period," *International Organization* 42:1 (1988), 121–50; Tor Egil Førland, *Cold Economic Warfare: CoCom and the Forging of Strategic Export Controls, 1948–1954* (Dordrecht: Republic of Letters, 2009); Yoko Yasuhara, "The Myth of Free Trade: The Origins of COCOM, 1945–1950," *Japanese Journal of American Studies* 4 (1991), 127–48; Timothy Aeppel, "The Evolution of Multinational Export Controls: A Critical Study of the COCOM Regime," *Fletcher Forum* 12 (1985), 105–24; Vibeke Sørensen, "Economic Recovery versus Containment: The Anglo-American Controversy over East-West Trade, 1947–1951," *Cooperation and Conflict* 24 (1989), 69–97; Mikael Nilsson, "Aligning the Non-aligned: A Reinterpretation of Why and How Sweden Was Granted Access to US Military

Materiel in the Early Cold War, 1948–1952," *Scandinavian Journal of History* 35:3 (2010), 290–309.

70. President of the United States, *National Security Strategy of the United States of America*, December 2017, 17, https://www.whitehouse.gov/wp-content/uploads/2017/12/NSS-Final-12-18-2017-0905.pdf (accessed September 28, 2019).

71. Brandt J. C. Pasco, "United States National Security Reviews of Foreign Direct Investment: From Classified Programs to Critical Infrastructure, This Is What the Committee on Foreign Investment in the United States Cares About," *ICSID Review* 29:2 (2014), 350–71.

72. Michael J. Hogan, *A Cross of Iron: Harry S. Truman and the Origins of the National Security State, 1945–1954* (Cambridge: Cambridge University Press, 1998); Douglas T. Stuart, *Creating the National Security State: A History of the Law that Transformed America* (Princeton, NJ: Princeton University Press, 2008); Stephen Wertheim, *Tomorrow the World: The Birth of US Global Supremacy* (Cambridge, MA: Belknap Press of Harvard University Press, 2020).

73. As an example, see the essay by Reagan's secretary of defense for research and engineering Richard D. DeLauer, "The Force Multiplier: Advanced Technology, and Particularly Electronics, Provides Leverage in Modern Military Defense Systems," *IEEE Spectrum* 19:10 (1982), 36–37. For a broader overview, see Robert R. Tomes, *US Defense Strategy from Vietnam to Operation Iraqi Freedom: Military Innovation and the New American Way of War, 1973–2003* (London: Routledge, 2007). See also Robert R. Tomes, "The Cold War Offset Strategy: Origins and Relevance," November 6, 2016, https://warontherocks.com/2014/11/the-cold-war-offset-strategy-origins-and-relevance/ (accessed May 10, 2020).

74. Mastanduno, *Economic Containment*, 113–26; Aaron L. Friedberg, *In the Shadow of the Garrison State: America's Anti-statism and Its Cold War Grand Strategy* (Princeton, NJ: University Press, 2000), chapter 8; R. Scott Kemp, "The Nonproliferation Emperor Has No Clothes: The Gas Centrifuge, Supply-Side Controls, and the Future of Nuclear Proliferation," *International Security* 38:4 (2014), 39–78, at 75–76; Mara Drogan, "The Nuclear Imperative: Atoms for Peace and the Development of US Policy on Exporting Nuclear Power, 1953–1955," *Diplomatic History* 40:5 (2016), 948–74, at 949; Matthew Evangelista, *Innovation and the Arms Race: How the United States and the Soviet Union Develop New Military Technologies* (Ithaca, NY: Cornell University Press, 1988).

75. Vannevar Bush, *Science: The Endless Frontier; A Report to the President on a Program for Postwar Scientific Research* (July 1945; rpt. Washington, DC: National Science Foundation, 1960). For the history of the development of the US innovation system, see David M. Hart, *Forged Consensus: Science, Technology, and Economic Policy in the United States, 1921–1953* (Princeton, NJ: Princeton University Press, 1998); Alfred K. Mann, *For Better or for Worse: The Marriage between Science and Government in the United States* (New York: Columbia University Press, 2000).

76. For a broader discussion of the role of American technological leadership for US international hegemony after 1945, see John Krige, *American Hegemony and the Postwar Reconstruction of Science in Europe* (Cambridge, MA: MIT Press, 2006).

77. For a critical discussion of the concept of "techno-nationalism," see David E. H. Edgerton, "The Contradictions of Techno-nationalism and Techno-globalism: A Historical Perspective," *New Global Studies* 1:1 (2007), 1–32.

78. Kenneth Flamm, "Controlling the Uncontrollable: Reforming US Export Controls on Computers," *Brookings Review* 14:1 (1996), 22–25; Dong Jung Kim, "Trading with the Enemy? The Futility of US Economic Countermeasures against the Chinese Challenge," *Pacific Review* 30:3 (2017), 289–308.

79. The same happened with the British gas centrifuge isotope enrichment program in the late 1960s and early 1970s; see John Krige, "US Technological Superiority and the Special Relationship: Contrasting British and American Policies for Controlling the Proliferation of Gas Centrifuge Enrichment," *International History Review* 36:2 (2014), 230–51; Krige, "Hybrid Knowledge."

80. For the US discussion about how far behind the Soviets were in their efforts to catch up with the American nuclear program, see Michael D. Gordin, *Red Cloud at Dawn: Truman, Stalin, and the End of the Atomic Monopoly* (New York: Farrar, Straus and Giroux, 2009), 63–88.

81. On the debate about the "American challenge" and the "technological gap," see Jean Jacques Servan-Schreiber, *The American Challenge* (New York: Atheneum, 1968); Bernard Nossiter, "Europe's Technology Gap," parts 1–5, *Washington Post*, February 12–16, 1967; Robert Gilpin, "European Disunion and the Technology Gap," *Public Interest* 10 (1968), 43–54.

82. These observations are informed by the broad literature on the security implications of economic asymmetric dependencies and interdependencies in international relations. See, for example, Dale C. Copeland, "Economic Interdependence and War: A Theory of Trade Expectations," *International Security* 20:4 (1996), 5–41; Robert Gilpin, "Economic Interdependence and National Security in Historical Perspective," in *Economic Issues and National Security*, ed. Klaus Knorr and Frank N. Trager (Lawrence: Published for the National Security Education Program by the Regents Press of Kansas, 1977), 19–66; Mastanduno, "Economics and Security in Statecraft and Scholarship." Stimulating, but lacking historical depth: Henry Farrell and Abraham Newman, "Weaponized Interdependence: How Global Networks Shape State Coercion," *International Security* 44:1 (2019), 42–79. See also John Krige, *Sharing Knowledge, Shaping Europe: US Technological Collaboration and Nonproliferation* (Cambridge, MA: MIT Press, 2016).

83. Our overview of the export control system is a pragmatic simplification of a very complicated system. For arguably the best introduction to the bureaucratic and conceptual complexities of the US export control system, see Berman and Garson, "United States Export Controls." For a much shorter, less sophisticated

but up-to-date introduction, see Congressional Research Service, *The US Export Control System and the Export Control Reform Initiative*, Report R41916, April 5, 2019, 1–9. A great overview is also offered by Rosanelli, *US Export Control Regulations*, 18–23. http://local.droit.ulg.ac.be/jcms/service/file/20140108134656_Handbook-RR-0801.pdf (accessed May 11, 2020). For the institutional dynamics, see, for example, John R. McIntyre, "The Distribution of Power and the Interagency Politics of Licensing East-West High-Technology Transfer," in *Controlling East-West Trade and Technology Transfer: Power, Politics, and Policies*, ed. Gary K. Bertsch (Durham, NC: Duke University Press, 1988), 97–133.

84. Congress is the main actor in the analysis of Gernot Stenger, "The Development of American Export Control Legislation after World War II," *Wisconsin International Law Journal* 6:1 (1987), 1–42. See also William J. Long, "The Executive, Congress, and Interest Groups in US Export Control Policy: The National Organization of Power," in Bertsch, *Controlling East-West Trade and Technology Transfer*, 27–62.

85. Hugo Meijer, "Actors, Coalitions, and the Making of Foreign Security Policy: US Strategic Trade with the People's Republic of China," *International Relations of the Asia-Pacific* 15 (2015), 433–75; *US National Security and Military/Commercial Concerns with the People's Republic of China: Report of the Select Committee on National Security and Military/Commercial Concerns with the People's Republic of China* (Cox Report) (Washington, DC: Government Printing Office, 1999).

86. James A. Glasgow, Elina Teplinsky, and Stephen L. Marcus, "Nuclear Export Controls: A Comparative Analysis of National Regimes for the Control of Nuclear Materials, Components, and Technology" (Washington, DC, 2012), https://www.pillsburylaw.com/images/content/3/3/v2/332/NuclearExportControls.pdf (accessed March 9, 2017).

87. William J. Long, *US Export Control Policy: Executive Autonomy vs. Congressional Reform* (New York: Columbia University Press, 1989); Steven Elliott, "The Distribution of Power in the US Politics of East-West Energy Trade Controls," in Bertsch, *Controlling East-West Trade and Technology Transfer*, 63–96.

88. It seems indeed quite odd that even the most thorough histories of the computer industry (and its relations to national governments) virtually ignore national security export controls. See, for example, the classic studies by Kenneth Flamm, *Targeting the Computer: Government Support and International Competition* (Washington, DC: Brookings Institution, 1987); Kenneth Flamm, *Creating the Computer: Government, Industry, and High Technology* (Washington, DC: Brookings Institution, 1988); Paul E. Ceruzzi, *A History of Modern Computing*, 2nd ed. (Cambridge, MA: MIT Press, 2003). During the Cold War, oil drilling equipment and the know-how to produce large-diameter pipelines could not be traded freely, nor could a large variety of machine tools or certain chemical processes. For the energy sector as still the best starting point, see Jentleson, *Pipeline Politics*.

89. The central role of knowledge for the US export control policy is also emphasized by John Krige, "Regulating the Academic 'Marketplace of Ideas': Commercialization, Export Controls, and Counterintelligence," *Engaging Science, Technology, and Science* 1 (2015), 1–24.

90. Deemed Export Advisory Committee, "The Deemed Export Rule in the Era of Globalization, Report to the Secretary of Commerce" (Augustine Report), December 2007, https://fas.org/sgp/library/deemedexports.pdf (accessed January 27, 2017); John Krige, "National Security and Academia: Regulating the International Circulation of Knowledge," *Bulletin of the Atomic Scientists* 70:2 (2014), 42–52; Benjamin Carter Findley, "Revisions to the United States Deemed-Export Regulations: Implications for Universities, University Research, and Foreign Faculty, Staff, and Students," *Wisconsin Law Review* 2006, 1223–74.

91. The White House, *National Strategy for Critical and Emerging Technologies*, October 2020; Edgerton, "Contradictions of Techno-Nationalism and Techno-Globalism"; Paul Evans, "Techno-nationalism in US-China Relations: Implications for US Universities," *East Asia Policy* 12:2 (2020), 80–92.

Chapter 2

1. Peter Galison, "Secrecy in Three Acts," *Social Research* 77 (2010), 941–74.

2. "An Act to Punish Acts of Interference with the Foreign Relations, the Neutrality, and the Foreign Commerce of the United States, to Punish Espionage, and Better to Enforce the Criminal Laws of the United States, and for other Purposes" (Espionage Act), *American Journal of International Law* 11:4 (1917), Supplement: Official Documents, 178–98, at 179, section 1 (d).

3. See *Control of Travel from and into the United States: Hearings before the House Committee on Foreign Affairs*, February 1918 (Washington, DC: Government Printing Office 1918); Craig Robertson, *The Passport in America: The History of a Document* (Oxford: Oxford University Press, 2010), 187–88. For a more detailed analysis, see Mario Daniels, "Controlling Knowledge, Controlling People: Travel Restrictions of US Scientists and National Security," *Diplomatic History* 43:1 (2019), 57–82.

4. *Report of the Commission on Government Security Pursuant to Public Law 304, 84th Congress, as Amended* (Washington, DC: Government Printing Office, 1957), 446–47.

5. John Torpey, "The Great War and the Birth of the Modern Passport System," in *Documenting Individual Identity: The Development of State Practices in the Modern World*, ed. Jane Caplan and John Torpey (Princeton, NJ: Princeton University Press, 2001), 256–70, at 257.

6. Espionage Act, 191–92.

7. Robertson, *Passport*, 190.

8. Espionage Act, 181–90, at 189.

9. "Trading with the Enemy Act, October 6, 1917," in *Trading with the Enemy* (New York: Guaranty Trust, 1917), 40–41; Harold Relyea, *Silencing Science: National Security Controls and Scientific Communication* (Norwood: Ablex, 1994), 73–74.

10. Katherine C. Epstein has described some of the measures taken by the British Admiralty to restrict the circulation of new naval technologies in the run up to World War I in "Harnessing Invention: The British Admiralty and the Political Economy of Knowledge in the World War I Era," in John Krige, ed., *Writing the Transnational History of Knowledge Flows in a Global Era* (Chicago: University of Chicago Press, forthcoming).

11. Relyea, *Silencing Science*, 68–72.

12. "Trading with the Enemy Act," section 10 (i), 57.

13. Ch. 95, 40 Stat. 394 (1917), "An Act to Prevent the Publication of Inventions by the Grant of Patents That Might Be Detrimental to the Public Safety or Convey Useful Information to the Enemy, to Stimulate Invention, and Provide Adequate Protection to Owners of Patents, and for Other Purposes." See also Sabing H. Lee, "Protecting the Private Inventor under the Peacetime Provisions of the Invention Secrecy Act," *Berkeley Technology Law Journal* 12:2 (1997), 345–411.

14. Kathryn Steen, *The American Synthetic Organic Chemicals Industry: War and Politics, 1910–1930* (Chapel Hill: University of North Carolina Press, 2014); Kathryn Steen, "Patents, Patriotism, and 'Skilled in the Art': USA v. The Chemical Foundation, Inc., 1923–1926," *Isis* 92:1 (2001), 91–122; Peter J. Hugill and Veit Bachmann, "The Route to the Techno-industrial World Economy and the Transfer of German Organic Chemistry to America before, during, and Immediately after World War I," *Comparative Technology Transfer and Society* 3:2 (2005), 158–86.

15. Richard T. Cupitt, *Reluctant Champions: US Presidential Policy and Strategic Export Controls; Truman, Eisenhower, Bush, and Clinton* (New York: Routledge, 2000), 38–39.

16. Relyea, *Silencing Science*, 75, claims that "policywise," export controls "were probably the most effective deterrents to enemy acquisition of especially useful American science and technology." It is unclear what this statement is based on.

17. Committee on Public Information, "What the Government Asks the Press," in *Complete Report of the Committee on Public Information, 1917–1919* (Washington, DC: Government Printing Office, 1920), 10–12, at 12.

18. Relyea, *Silencing Science*, 74–75.

19. Cupitt, *Reluctant Champions*, 41.

20. Cupitt, 43–45; Edward S. Miller, *Bankrupting the Enemy: The US Financial Siege of Japan before Pearl Harbor* (Annapolis: Naval Institute Press, 2007), 77–79; Michael A. Barnhart, *Japan Prepares for Total War: The Search for Economic Security, 1919–1941* (Ithaca, NY: Cornell University Press, 1987), 129–30, 178–82; "The 'Moral' Embargo," *New York Times*, December 22, 1939, 18; "'Moral' Embargo Helps to Conserve Raw Materials," *Wall Street Journal*, December 2, 1939, 2; "US Acts to Curb Further Sales of Planes to Japan," *New York Herald Tribune*,

June 12, 1938, 1; "US to Stop Plane Sales," *North-China Herald and Supreme Court and Consular Gazette*, June 15, 1938, 469; "The 'Moral' Embargo," *New York Herald Tribune*, December 4, 1939, 22.

21. Department of State, *Peace and War: United States Foreign Policy, 1931–1941* (Washington, DC: Government Printing Office, 1943), 65–67; the Neutrality Acts of 1935 and 1937 are printed on pp. 266–71 and 355–65.

22. Message to Congress on Appropriations for National Defense, May 16, 1940, http://www.presidency.ucsb.edu/ws/index.php?pid=15954 (accessed September 1, 2017). For the context, see also *Report to Congress of the Operations of the Foreign Economic Administration, September 25, 1944* (Washington, DC: Government Printing Office, 1944), 35.

23. Ch. 508, 54 Stat. 712, "Act to Expedite the Strengthening of the National Defense," July 2, 1940. The act does not use the name "Export Control Act," but other documents from the 1940s do, for example, Department of State, *Peace and War*, 97; *Report to Congress of the Operations of the Foreign Economic Administration*, 47.

24. *Report to Congress of the Operations of the Foreign Economic Administration*, 35.

25. 50 Stat 1834, Proclamation No. 2337, "Enumeration of Arms, Ammunition and Implements of War by the President of the United States of America," May 1, 1937.

26. Proclamation No. 2413, "Administration of Section 6 of the Act Entitled 'An Act to Expedite the Strengthening of the National Defense,'" July 2, 1940, in Administrator of Export Control, *Export Control Regulations and Export Control Schedule No. 1* (Washington, DC: Government Printing Office, 1941), 35–37.

27. Proclamations No. 2417, 2428, 2449, 2453, "Administration of Section 6 of the Act Entitled 'An Act to Expedite the Strengthening of the National Defense,'" July 26, September 30, December 10, 1940; January 10, 1941, in *Export Control Regulations and Export Control Schedule No. 1*, 37–42.

28. *Report to Congress of the Operations of the Foreign Economic Administration*, 36.

29. Proclamation No. 2423, September 12, 1940, in *Export Control Regulations and Export Control Schedule No. 1*, 38–39.

30. Proclamation No. 2451, December 20, 1940, in *Export Control Regulations and Export Control Schedule No. 1*, 41.

31. Department of State, Chief of the Office of Arms and Munitions Control, Joseph C. Green, to 148 Persons and Companies Manufacturing Airplane Parts, July 1, 1938, in Department of State, *Peace and War*, 109.

32. Department of State, Press Release, December 20, 1939, in *Papers Relating to the Foreign Relations of the United States, Japan: 1931–1941*, vol. 2 (Washington, DC: Government Printing Office, 1943), 203–4.

33. Franklin Delano Roosevelt, "The Great Arsenal Democracy," radio speech, December 29, 1940, http://www.americanrhetoric.com/speeches/fdrarsenalofdemoc

racy.html (accessed September 3, 2017). This speech includes another reference to air warfare.

34. Proclamation No. 2465, March 4, 1941, in *Export Control Regulations and Export Control Schedule No. 1*, 46–47.

35. Administrator of Export Control, *Export Schedule A*, Effective April 15, 1941, Washington, DC, 1941, 7.

36. Administrator of Export Control, 7.

37. Department of Commerce, Office of Industry and Commerce, "Report to the National Resources Board on Wartime Procedures for Export Control of Technical Data," June 30, 1950, 3, NARA, RG 40, UD Entry 59, box 3. These destinations were Canada, Great Britain and Northern Ireland, Curacao, Australia, Bahamas, Barbados, Bermuda, Newfoundland, New Zealand, Union of South Africa, Greenland Island, and the Philippines (5).

38. Administrator of Export Control, *Export Schedule A*, 10.

39. Department of Commerce, Office of Industry and Commerce, "Report to the National Resources Board," 3–4, 13.

40. Lee, "Protecting the Private Inventor," 349–50. The Act of July 1, 1940, can be found printed in "Secrecy of Applications for Defense Inventions," *Journal of the Patent Office Society* 22:9 (1940), 712–14.

41. Quoted in *The Government's Classification of Private Ideas: Thirty-Fourth Report by the Committee on Government Operations*, House Report No. 96-1540 (Washington, DC: Government Printing Office, 1980), 40–41.

42. Lee, "Protecting the Private Inventor," 349.

43. *Government's Classification*, 37, 46.

44. Department of Commerce, *Historical Statistics of the United States, 1789–1945: A Supplement to the Statistical Abstract of the United States; Prepared by the Bureau of the Census with Cooperation of the Social Science Research Council* (Washington, DC: Government Printing Office, 1949), 312.

45. *Government's Classification*, 33, 35, 40.

46. *Government's Classification*, 37.

47. Alex Wellerstein, "Patenting the Bomb: Nuclear Weapons, Intellectual Property, and Technological Control," *Isis* 99:1 (2008), 57–87, at 69–76.

48. Department of Commerce, Office of Industry and Commerce, "Report to the National Resources Board," 27.

49. Office of Censorship, US Postal Censorship Regulations, Edition of April 13, 1942 (Washington, DC: Government Printing Office, 1942), sec. 3 (f), sec. 4 (b) (1), (2), (9), sec. 5.

50. Department of Commerce, Office of Industry and Commerce, "Report to the National Resources Board," 19–20. For a general history of the Office of Censorship, see Michael S. Sweeney, *Secrets of Victory: The Office of Censorship and the American Press and Radio in World War II* (Chapel Hill: University of North Carolina Press, 2001).

51. Graham H. Stuart, "Safeguarding the State through Passport Control," *State Department Bulletin* 12:311 (1945), 1066-70, at 1067-68.

52. *Government's Classification*, 447.

53. Department of Commerce, Office of Industry and Commerce, "Report to the National Resources Board," 21.

54. DoS, A. Richard, Memorandum, "Policy for the Guidance of American Business Organizations in Their Dealings with Foreign Countries, June 24, 1944, NARA, RG 59, Entry A1 1494D, box 35.

55. DoS, A. Richard, Memorandum.

56. DoS, Munitions Control Section, Memorandum of conversation ... on December 30, 1944, concerning business relations between American citizens and organizations and representatives of the Soviet Union, January 6, 1945, NARA, RG 59, Entry A1 1494D, box 36.

57. DoS memorandum on technical aid contracts, June 10, 1946, NARA, RG 59, Entry A1 1494D, box 36. See also Kendall E. Bailes, "The American Connection: Ideology and the Transfer of American Technology to the Soviet Union, 1917–1941," *Comparative Studies in Society and History* 23:3 (1981), 421–48, at 429, 433 (quote); Timothy W. Luke, "Technology and Soviet Foreign Trade: On the Political Economy of an Underdeveloped Superpower," *International Studies Quarterly* 29:3 (1985), 327–53; Stefan J. Link, *Forging Global Fordism: Nazi Germany, Soviet Russia, and the Contest over the Industrial Order* (Princeton, NJ: Princeton University Press, 2020), 110–15.

58. Katherine S. Sibley, "Soviet Military-Industrial Espionage in the United States and the Emergence of an Espionage Paradigm in US-Soviet Relations, 1941–45," *American Communist History* 2:1 (2003), 21–61.

59. "How to Do Business with the USSR," text draft, without date, Haley to Mason (both DoS), April 17, 1945; Munitions Control Section, "Comments on 'How to Do Business with the USSR,'" March 23, 1945, NARA, RG 59, Entry A1 1494D, box 36.

60. DoS, Memorandum, "Export Licenses for Technical Data," November 7, 1945; Memorandum, Wilcox, Schaetzel, Terrill, Anderson, October 5, 1945, NARA, RG 59, Entry A1 1494D, box 35.

61. DoS, Munitions Control Section, Memorandum of conversation ... on December 30, 1944, concerning business relations between American citizens and organizations and representatives of the Soviet Union, January 6, 1945, NARA, RG 59, Entry A1 1494D, box 36.

62. Munitions Control Section, "Memorandum of Conference" held with officers of the Military Intelligence Service, G2 (War Department) and Office of Naval Intelligence, February 19, 1945, NARA, RG 59 Entry A1 1494D, box 36.

63. DoS, Stinebower to Anderson, "A Proposed Committee on USSR/US Trade," February 21, 1945, NARA, RG 59, Entry A1 1494D, box 36.

64. "How to Do Business with the USSR," 9, NARA, RG 59, Entry A1 1494D, box 36.

65. List of Technical Aid Contracts, May 14, 1945 (with later additions); DoS, Memorandum of conversation, "Standard Oil—USSR Synthetic Rubber Agreement, November 21, 1945; Memorandum of conversation, "Soviet—Standard Rubber Agreement," November 9, 1945; Memorandum of conversation (telephone), "Soviet—Standard Rubber Agreement," November 9, 1945; DoS, Memorandum, Dubrow, August 1, 1945, NARA, RG 59, Entry A1 1494D, box 36.

66. A similar, less systematic campaign began after V-J Day in Japan.

67. There is now a considerable corpus of literature on the forced technology transfers from Germany to the United States and the other wartime allies. Foundational is John Gimbel, *Science, Technology, and Reparations: Exploitation and Plunder in Postwar Germany* (Stanford, CA: Stanford University Press, 1990). But see also Matthias Judt and Burghard Ciesla, eds., *Technology Transfer Out of Germany after 1945* (Amsterdam: Harwood Academic), 1996; Douglas O'Reagan, *Taking Nazi Technology: Allied Exploitation of German Science after the Second World War* (Baltimore: Johns Hopkins University Press, 2019); Michael J. Neufeld, *Von Braun: Dreamer of Space, Engineer of War* (New York: Alfred A. Knopf, 2007).

68. Vannevar Bush to Henry L. Stimson, Secretary of War, August 28, 1944, NARA, RG 59, Entry A1 1494D, box 35.

69. DoS, Office of War Mobilization and Office of Scientific Research and Development, Memorandum, "Dissemination of German Technological Information," April 26, 1945. For an early discussion of the clash of the policy of dissemination with international patent law, see also "Memorandum regarding Dissemination of Technological Information Collected in Foreign Countries," March 29, 1945, NARA, RG 59, Entry A1 1494D, box 35.

70. Memorandum, "Dissemination of Technological Information Gathered in Germany by CIOS or a Successor Organization," February 27, 1945, NARA, RG 59, Entry A1 1494D, box 35. See also Stephen C. Schlesinger, *Act of Creation: The Founding of the United Nations; A Story of Superpowers, Secret Agents, Wartime Allies and Enemies and Their Quest for a Peaceful World* (Boulder, CO: Westview, 2003), 53–72.

71. Memorandum, "Some Further Preliminary Considerations concerning the Dissemination of German Technological Information," January 12, 1945, NARA, RG 59, Entry A1 1494D, box 35.

72. EOs 9568 of June 5, 1945, and 9604 of August 25, 1945, http://www.presidency.ucsb.edu/ws/index.php?pid=60663 and http://www.presidency.ucsb.edu/ws/index.php?pid=60669 (accessed September 15, 2017).

73. CORSI was chaired by the commissioner of patents and consisted of representatives of the State, Justice, War, and Navy Departments and the OSRD. NARA, RG 59, Entry A 1 1494D, box 35, Anderson to Thorp ("The Work of the Committee on the Release of Scientific Information"), December 19, 1945; Memorandum to the members of CORSI, November 15, 1945; Minutes of CORSI meeting, September 12, 1945. On CORSI and the Office of Technical Services, see Rob-

ert K. Stewart, "The Office of Technical Services: A New Deal Idea in the Cold War," *Knowledge: Creation, Diffusion, Utilization* 15:1 (1993), 44–77; Robert K. Stewart, "Merging Foreign and Domestic Information Policy Goals: The US Government's Office of Technical Services (1946–1950)," *Electronic Reference Materials* 8 (1989), 1–32; Gimbel, *Science*, 26–27.

74. Another $23.5 million were spent for technology to scramble the enemy's radar. See Arthur A. Bright and John Exter, "War, Radar, and the Radio Industry," *Harvard Business Review* 25:2 (1947), 255–72; James Phinney Baxter, *Scientist against Time* (Boston: Little, Brown, 1946), 134–69. For a modern account, see Robert Buderi, *The Invention That Changed the World: The Story of Radar from War to Peace* (London: Little, Brown, 1997).

75. *Government's Classification*, 33–37, at 35.

76. Minutes of meeting of the Committee on Release of Scientific Information, September 12, 1945, and Minutes of CORSI meeting, October 17, 1945, NARA, RG 59, Entry A1 1494D, box 35. On the reconversion problems of the radar industry, see Bright and Exter, "War," 270–71.

77. Henry DeWolf Smyth, *Atomic Energy for Military Purposes: The Official Report on the Development of the Atomic Bomb under the Auspices of the United States Government, 1940–1945, Written at the Request of Maj. Gen. Leslie R. Groves* (Princeton, NJ: Princeton University Press), 1945.

78. Rebecca Press Schwartz, "The Making of the History of the Atomic Bomb: Henry DeWolf Smyth and the Historiography of the Manhattan Project," PhD diss., Princeton University, 2008, 56–78, 93–94.

79. Alex Wellerstein "Knowledge and the Bomb: Nuclear Secrecy in the United States, 1939–2008," PhD. diss., Harvard University, 2010, 193–213.

80. The McMahon Bill of December 20, 1945, is printed in *The American Atom: A Documentary History of Nuclear Policies from the Discovery of Fission to the Present*, ed. Philip L. Cantelon, Richard G. Hewlett, and Robert C. Williams (University Park: University of Pennsylvania Press, 1991), 77–91, at 78. The liberality of the information policy envisioned in the bill is also stressed by Richard G. Hewlett and Oscar E. Anderson Jr., *The New World, 1939/1946: A History of the United States Energy Commission* (University Park: University of Pennsylvania Press, 1962), 483; and Wellerstein, "Knowledge," 182.

81. "Report of the Meeting of Foreign Ministers of the United States, the United Kingdom, and the Union of Soviet Republic in Moscow," December 27, 1945, http://avalon.law.yale.edu/20th_century/decade19.asp (accessed September 16, 2017); Joseph M. Siracusa, *Nuclear Weapons: A Very Short Introduction* (Oxford: Oxford University Press, 2008), 27–29.

82. *Government's Classification*, 47; Lee, "Protecting," 351.

83. Department of Commerce, Office of Industry and Commerce, "Report to the National Resources Board on Wartime Procedures for Export Control of Technical Data," June 30, 1950, 17, NARA RG 40, UD Entry 59, box 3; DoS, Terrill to Wilcox, December 4, 1945, NARA, RG 59, Entry A1 1494D, box 35.

84. Exton to Green, December 13, 1945, NARA, RG 59, Entry A 1 1494D, box 35.

85. Minutes of the Eighteenth Meeting to Consider a Tentative Draft Treaty of Friendship, Commerce and Navigation Between the United States and the USSR, September 24, 1945, NARA, RG 59, Entry A1 1494D, box 36.

86. DoS, Munitions Control Section, Memorandum of conversation . . . on December 30, 1944, concerning business relations between American citizens and organizations and representatives of the Soviet Union, January 6, 1945, NARA, RG 59, Entry A1 1494D, box 36.

87. See DoS, Memorandum, "Some Reflections on a Memorandum Entitled 'Some Preliminary Considerations concerning the Transfer of Technology to the USSR,'" January 24, 1945; Memorandum, "Exchange of Technology between the United States and the USSR," n.d., NARA, RG 59, Entry A1 1494D, box 36.

88. This is not least a critique of Douglas O'Reagan, "Know-How in Postwar Business Law," *Technology and Culture* 58:1 (2017), 121–53. While an important pioneering endeavor toward a rich *Begriffsgeschichte* of "know-how," O'Reagan's study touches only very superficially on the national security history of the concept, and his time line focuses on the postwar era, basically ignoring his own empirical finding that the term began to quickly enter the common language exactly with the start of World War II.

89. DoS, "Memorandum for United States Negotiators on Patent and Technology Agreement with the USSR," April 15, 1946, NARA, RG 59, Entry A1 1494D, box 36.

90. Memorandum, Wilcox, Schaetzel, Terrill, Anderson, October 5, 1945, NARA, RG 59, Entry A1 1494D, box 35.

91. "Second Draft of Report to the President on US Policies and Programs in the Economic Field Which May Affect the War Potential of the Soviet Bloc," January 1951, 58, NARA, CREST files, CIA-RDP79-01143A000400110002-5.

92. Office of Intelligence Research, Division of Research for the USSR and Eastern Europe, "Vulnerability of the Soviet Bloc to Existing and Tightened Western Economic Controls," OIR Report No. 5447, January 26, 1951, 65, NARA, CREST files, CIA-RDP79R01012A000500030013-1.

93. Memorandum of conversation on a meeting between representatives of the Department of State and the OSRD, June 30, 1944, NARA, RG 59, Entry A1 1494D, box 35.

94. DoS, Memorandum, "US-USSR Technological Relations," December 13, 1945, NARA, RG 59, Entry A1 1494D, box 36.

95. Minutes of the Eighteenth Meeting to Consider a Tentative Draft Treaty of Friendship, Commerce and Navigation between the United States and the USSR, September 24, 1945; DoS, Memorandum, "Exchange of Technology and Protection of Technical and Artistic Property," August 20, [1945], NARA, RG 59, Entry A1 1494D, box 36.

96. DoS to Secretary of Commerce, Wallace, October 2, 1945, NARA, RG 59, Entry A 1 1494D, box 36.

97. DoS (Wilcox), Memorandum, "Policy and Information Statement for the USSR Technology," November 23, 1945, NARA, RG 59, Entry A 1 1494D, box 35.

98. Exton to Green, December 13, 1945, NARA, RG 59, Entry A 1 1494D, box 35.

99. Exton to Green, December 12, 1945, NARA, RG 59, Entry A 1 1494D, box 35.

100. Igor Gouzenko was a cipher clerk in the Soviet embassy in Ottawa during World War II. He defected to the West late in 1945 with documents that revealed the presence of a Soviet spy ring in Canada that sought information on North American technology, including the bomb. See, for example, "Canada Seizes 22 as Spies: Atom Secrets Believed Aim," *New York Times*, February 16, 1946, 1, 6; Neal Stanford, "US Says Atom Spies Are 'Under Control,'" *Christian Science Monitor*, February 19, 1946, 1, 16; Paul W. Ward, "Only US Has Bomb Secret, Byrnes Says," *Baltimore Sun*, February 20, 1946, 1. For the standard assessment of an immediate strong reaction of the public, see Hewlett and Anderson, *New World*, 501; Gregg Herken, "'A Most Deadly Illusion': The Atomic Secret and American Nuclear Weapons Policy, 1945–1950," *Pacific Historical Review* 49:1 (1980), 51–76, at 64. On the Gouzenko case, see Amy Knight, *How the Cold War Began: The Gouzenko Affair and the Hunt for Soviet Spies* (Toronto: McClelland and Stewart, 2005).

101. Campbell Craig and Fredrik Logevall, America's Cold War: The Politics of Insecurity (Cambridge, MA: Harvard University Press, 2009), 66, speak, without a proper calendar but nevertheless fittingly, of "five fateful weeks."

102. John Lewis Gaddis, *The United States and the Origins of the Cold War, 1941–1947* (New York: Columbia University Press, 2000), 299–301. Craig and Logevall, *America's Cold War*, 68–69.

103. Melvyn P. Leffler, *A Preponderance of Power: National Security, the Truman Administration, and the Cold War* (Stanford, CA: Stanford University Press, 1992), 109; Gaddis, *United States*, 304.

104. For the text of the speech, originally titled "Sinews of Peace," see https://www.winstonchurchill.org/resources/speeches/1946-1963-elder-statesman/the-sinews-of-peace (accessed September 23, 2017). For the context of the speech, see Patrick Wright, *Iron Curtain: From Stage to Cold War* (Oxford: Oxford University Press, 2007), 21–61.

Chapter 3

1. John R. Steelman (Chairman of the President's Scientific Research Board), *Science and Public Policy: A Report to the President*, vol. 1, *A Program for the Nation* (Washington, DC: Government Printing Office, 1947), 10.

2. Letter of Roosevelt to Vannevar Bush, November 17, 1944, in Vannevar Bush, *Science—the Endless Frontier: A Report to the President on a Program for*

Postwar Scientific Research (1945; rpt. Washington, DC: Government Printing Office, 1960), 3–4, at 3. OSRD's role as a postwar model is also emphasized by Michael J. Hogan, *A Cross of Iron: Harry S. Truman and the Origins of the National Security State, 1945–1954* (Cambridge: Cambridge University Press, 1998), 224.

3. Bush, *Science—the Endless Frontier*, 5. On Bush, see the biography by G. Pascal Zachary, *Endless Frontier: Vannevar Bush, Engineer of the American Century* (New York: Free Press, 1997). On the later significance of distinguishing basic from applied science, see Mario Daniels and John Krige, "Beyond the Reach of Regulation? 'Basic' and 'Applied' Research in the Early Cold War United States," *Technology and Culture* 59:2 (2018), 226–50.

4. On the concept and the institutional history of the national security state, see Hogan, *Cross*; Douglas T. Stuart, *Creating the National Security State: A History of the Law that Transformed America* (Princeton, NJ: Princeton University Press, 2008); Aaron L. Friedberg, *In the Shadow of the Garrison State: America's Anti-statism and Its Cold War Grand Strategy* (Princeton, NJ: Princeton University Press, 2000); Anna Kasten Nelson, "The Evolution of the National Security State: Ubiquitous and Endless," in *The Long War: A New History of US National Security Policy since World War II*, ed. Andrew J. Bacevich (New York: Columbia University Press, 2007), 265–301.

5. DoS, Memorandum of conversation, "Exchange of Technology—USSR," February 27, 1946, NARA, RG 59, Entry A 1 1494D, box 36.

6. DoS, Minutes of Meeting of the USSR Committee, March 25, 1946, 2, RG 40, Entry UD 76, box 3. Among the participants of this meeting was Alexander Gerschenkron.

7. DoS, Memorandum for United States Negotiators on Patent and Technology Agreement with the USSR, April 15, 1946, 11, NARA, RG 59, Entry A 1 1494D, box 36.

8. DoS, Minutes of Meeting of the USSR Committee, March 25, 1946, 3, RG 40, Entry UD 76, box 3.

9. DoS, Memorandum for United States Negotiators on Patent and Technology Agreement with the USSR, April 15, 1946, 17, NARA, RG 59, Entry A 1 1494D, box 36.

10. DoS, Memorandum, "Proposed Legislation to Protect Industrial 'Know-How' and Scientific Knowledge and to Provide for the Secrecy of Designated Patents," April 23, 1946, RG 59, Entry A 1 1494D, box 35.

11. Friedberg, *In the Shadow*, 297.

12. Steelman, *Science*, 3–4, 6. For the concept of scientists as a war resource and reserve, see also John Krige, *American Hegemony and the Postwar Reconstruction of Science in Europe* (Cambridge, MA: MIT Press, 2006), 193–94; David Kaiser, "Cold War Requisitions, Scientific Manpower, and the Production of American Physicists after World War II," *Historical Studies in the Physical and Biological Sciences* 33:1 (2002), 131–59, at 137–38.

13. David M. Hart, *Forged Consensus: Science, Technology, and Economic Policy in the United States, 1921–1953* (Princeton, NJ: Princeton University Press, 1998), 184–92.

14. David C. Mowery and Nathan Rosenberg, *Technology and the Pursuit of Economic Growth* (Cambridge: Cambridge University Press, 1989), 129; Hart, *Forged Consensus*, 173.

15. Hart, *Forged Consensus*, 195, 197; Mowery and Rosenberg, *Technology*, 129; Friedberg, *In the Shadow*, 298–99 (figs. 8.1 and 8.2).

16. DoS, Minutes of Meeting of the USSR Committee, March 25, 1946, 4, RG 40, Entry UD 76, box 3.

17. DoS, Memorandum for United States Negotiators on Patent and Technology Agreement with the USSR, April 15, 1946, 17, NARA, RG 59, Entry A 1 1494D, box 36.

18. DoS, Minutes of Meeting of the USSR Committee, March 25, 1946, 3, RG 40, Entry UD 76, box 3; DoS, Terrill, to Senator Burton K. Wheeler (draft), March 23, 1946, NARA, RG 59, Entry A 1 1494D, box 36.

19. See, for example, "Learn US Urges Sale of Secret War Devices to Russia," *Chicago Daily Tribune*, April 29, 1946, 1; "US War Secrets Deal Reported," *Los Angeles Times*, April 29, 1946, 1; "Senate Group Sifts War Secret's Sale," *New York Times*, April 29, 1946, 1, 7.

20. DoS, Minutes of Meeting of the USSR Committee, March 25, 1946, 4–5, 7, RG 40, Entry UD 76, box 3.

21. The McMahon Bill v. 20.12.1945, in *The American Atom: A Documentary History of Nuclear Policies from the Discovery of Fission to the Present*, ed. Philip L. Cantelon, Richard G. Hewlett, and Robert C. Williams, 2nd ed. (University Park: University of Pennsylvania Press, 1991), 77–91, quote at 78. The liberality of the information policy envisioned by the bill is also stressed by Richard G. Hewlett and Oscar E. Anderson Jr., *The New World, 1939/1946: A History of the United States Energy Commission* (University Park: University of Pennsylvania Press, 1962); Alex Wellerstein, *Restricted Data: The History of Nuclear Secrecy in the United States* (Chicago: University of Chicago Press, 2021), 151.

22. Hewlett and Anderson, *World*, 512.

23. Atomic Energy Act of 1946, PL 585, 79th Congress, 13 (sec. 10b, 1), http://science.energy.gov/~/media/bes/pdf/Atomic_Energy_Act_of_1946.pdf (accessed September 30, 2017).

24. Wellerstein, *Restricted Data*, 154–55; Richard G. Hewlett, "A Historian's View: 'Born Classified' in the AEC," *Bulletin of the Atomic Scientists* 37:10 (1981), 20–27; Galison, "Secrecy in Three Acts," 952, 961; Howard Morland, "Born Secret," *Cardozo Law Review* 26:4 (2005), 1401–8, at 1402.

25. Atomic Energy Act of 1946, 1.

26. US Senate, *Hearing Held before Committee on the Judiciary, S. 1953 to Amend the Espionage Act*, Executive Session, April 29, 1946, after 1.

27. US Senate, *Hearing Held before Committee on the Judiciary, S. 1953 to Amend the Espionage Act*, Executive Session, May 6, 1946, 33, 49–50.

28. DoS, Memoranda, "Proposed Legislation to Protect Industrial 'Know-How' and Scientific Knowledge and to Provide for the Secrecy of Designated Patents," April 23 and April 29, 1946; "A Bill to Amend Section 6 of the Act of July 2, 1940, . . . Relating to the Exportation of Certain Commodities," Undated; "A Bill to Amend the Act Relating to Preventing the Publication of Inventions in the National Interest . . . ," RG 59, Entry A 1 1494D, box 35.

29. DoS, Memorandum, "Disclosure of Technical Information to a Foreign Government," May 17, 1945, RG 59, Entry A 1 1494D, box 35.

30. DoS, Memorandum (Robert G. Hooker Jr.), "Proposed Legislation to Protect Industrial 'Know-How' and Scientific Knowledge and to Provide for the Secrecy of Designated Patents," April 29, 1946, RG 59, Entry A 1 1494D, box 35.

31. James R. Newman, "Control of Information Relating to Atomic Energy," *Yale Law Journal* 56 (1947), 769–802, at 783. Newman, who had been counsel to the Senate Special Committee on Atomic Energy, criticized the Atomic Energy Act in terms similar to those of Hooker's reflections, stating that "the terror at the loss of the 'secret' is a tribal fear which, once gaining ascendancy in our minds, must inevitably weaken rather than strengthen our defensive power as a nation. Preoccupation with the 'secret,' instead of with the thing itself will stifle the scientific research from which our real strength is derived" (782–83).

32. See William L. Thorp to Secretary of State, February 3, 1948, RG 40 Entry UD 76, box 2.

33. In mid-1948 the UTI was replaced by the Interdepartmental Committee on Industrial Security (CIS), which was a division of the State-Army-Navy-Air Force Coordinating Committee (SANACC). Because of the reorganization of the National Security system, the CIS, renamed Interdepartmental Committee on Internal Security (ICIS), became in summer 1949 part of the National Security Council. Its deliberations led to the directive NSC 5427, which established in 1954 the Office of Strategic Information (OSI) within the Department of Commerce. The OSI was abolished in 1958 after fierce public critique of its information policy. *Science and Foreign Relations: International Flow of Scientific and Technological Information*, Department of State Publication 3860 (Berkner Report), May 1950, 82; Mitchel B. Wallerstein, "The Office of Strategic Information, Department of Commerce, 1954–1957," in National Academy of Sciences, National Academy of Engineering, and Institute of Medicine, *Working Papers of the Panel on Scientific Communication and National Security* (Washington, DC: National Academy Press, 1982), 84–88.

34. Unclassified Technological Information Committee (UTI), Minutes, January 20, 1947, 2, 5, 10, NARA, RG 40, Entry UD 76, box 2.

35. The shift of emphasis of export control policy from short supply to national security concerns is stressed by Harold J. Berman and John R. Garson. "United

States Export Controls—Past, Present, and Future," *Columbia Law Review* 67:5 (1967), 791–890, at 795–96.

36. Hogan, *Cross*; Melvyn P. Leffler, "The American Conception of National Security and the Beginnings of the Cold War, 1945–48," *American Historical Review* 89:2 (1984), 346–81; Emily S. Rosenberg, "The Cold War and the Discourse of National Security," *Diplomatic History* 17:2 (1993), 277–84.

37. Stuart, *Creating*.

38. Hogan, *Cross*, 12–14. For US science, a similar point is made by Dominique Pestre, "Scientists in Time of War: World War II, the Cold War, and Science in the United States and France," *French Politics, Culture and Society* 24:1 (2006), 27–39, at 30.

39. This is the main argument of Friedberg, *In the Shadow*.

40. Robert A. Pollard, *Economic Security and the Origins of the Cold War, 1945–1950* (New York: Columbia University Press, 1985).

41. Yoko Yasuhara, "The Myth of Free Trade: The Origins of COCOM, 1945–1950," *Japanese Journal of American Studies* 4 (1991), 127–48; James K. Libbey, "CoCom, Comecon, and the Economic Cold War," *Russian History* 37 (2010), 133–52.

42. Berman and Garson, "United States Export Controls," 792.

43. Export Control Act, February 28, 1949, ch. 11, 63 Stat. 7, sec. 1 and 2 (emphasis added).

44. Berman and Garson, "United States Export Controls," 792.

45. "Voluntary Controls Put over Export of Knowhow," *Washington Post*, November 11, 1949, 2.

46. Department of Commerce, *Export Control and Allocation Powers: Eleventh Quarterly Report* (Washington, DC: Government Printing Office, 1950), 2.

47. "'Voluntary' Plan Bars Data Export," *New York Times*, November 11, 1949, 11.

48. Department of Commerce, "Plan to Control Export of Advanced Technical Data," *Foreign Commerce Weekly* 37:9 (November 28, 1949), 43. See also Department of Commerce, *Export Control and Allocation Powers, Ninth Quarterly Report the President, the Senate and House of Representatives* (Washington, DC: Government Printing Office, 1949), 3.

49. Executive Proclamation 2776, March 26, 1948, "Enumeration of Arms, Ammunition, and Implements of War," http://www.presidency.ucsb.edu/ws/index.php?pid=87144 (accessed March 3, 2018); Department of State, "New Requirements Relating to the Licensing for Export and Import of Articles Defined of Arms, Ammunition and Implements of War," MD Bulletin No. 1, April 1, 1948, https://www.cia.gov/library/readingroom/docs/CIA-RDP75-00662R000300070013-1.pdf (accessed March 3, 2018).

50. Jeffrey A. Engel, "The Surly Bonds: American Cold War Constraints on British Aviation," *Enterprise and Society* 6:1 (2005), 1–44, at 20.

51. Department of Commerce, "Plan to Control Export of Advanced Technical Data," *Foreign Commerce Weekly* 37:9 (November 28, 1949), 43.

52. Department of Commerce, 43.

53. Robert A. Bowman to John C. Green, Summary of Activities of the Program for Voluntary Protection of Technical Information to April 31, 1951, NARA, RG 40, Entry UD 65, box 1.

54. "A Program for Voluntary Protection of Technical Information," Brochure, January 1951, 1–2, NARA, RG 40, Entry UD 65, box 1.

55. "Program for Voluntary Protection," 3.

56. Department of Commerce, *Export Control and Allocation Powers, Tenth Quarterly Report* (Washington, DC, 1950), 3. Department of Commerce, Office of Industry and Commerce, "Report to the National Security Resources Board on Wartime Procedures for Export Control of Technical Data," June 30, 1950, 25–26, NARA, RG 40, entry UD 59, box 3.

57. Berkner Report, 7. For some background on the Berkner Report, see Allan A. Needell, *Science, Cold War and the American State: Lloyd V. Berkner and the Balance of Professional Ideals* (Amsterdam: Harwood Academic, 2000), 141–48.

58. Berkner Report, 2.

59. Berkner Report, 82–84.

60. Jessica Wang, *American Science in an Age of Anxiety: Scientists, Anticommunism, and the Cold War* (Chapel Hill: University of North Carolina Press, 1999).

61. Second Draft of Report to the President on US Policies and Programs in the Economic Field Which May Affect the War Potential of the Soviet Bloc, January 1951, 61, NARA, CREST files, CIA-RDP79-01143A000400110002-5.

62. Craig Robertson, "The Documentary Regime of Verification: The Emergence of the US Passport and the Archival Problematization of Identity," *Cultural Studies* 23:3 (2009), 329–54.

63. Interdepartmental Committee on Industrial Security, Basic Problems of Industrial Security, October 14, 1948, NARA, RG 40, Entry UD 59, box 8.

64. *Report of the Commission on Government Security Pursuant to Public Law 304, 84th Congress, as Amended* (Washington, DC: Government Printing Office, 1957), 467.

65. Office of Strategic Information to Joint Operating Committee, March 10, 1955, NARA, RG 40, UD entry 56, box 1.

66. For a thorough analysis, see Mario Daniels, "Controlling Knowledge, Controlling People: Travel Restrictions of US Scientists and National Security," *Diplomatic History* 43:1 (2019), 57–82.

67. Mario Daniels, "Restricting the Transnational Movement of 'Knowledgeable Bodies': The Interplay of US Visa Restrictions and Export Controls in the Cold War," in *How Knowledge Moves: Writing the Transnational History of Science and Technology*, ed. John Krige (Chicago: University of Chicago Press, 2019), 35–61.

68. Berkner Report, 77–79.

69. Michael J. Ybarra, *Washington Gone Crazy: Senator Pat McCarran and the Great American Communist Hunt* (Hanover, NH: Steerforth), 2004.

70. The Immigration and Nationality Act of 1952 amended and extended section 212, pertaining to visa denials, of the Internal Security Act of 1950. Both versions are reprinted in *Bulletin of the Atomic Scientists* 8:7 (1952), 257–58.

71. Victor F. Weisskopf, "Report on the Visa Situation," *Bulletin of the Atomic Scientists* 8:7 (1952), 221–22, at 221.

72. Edward Shils, "Editorial: America's Paper Curtain," *Bulletin of the Atomic Scientists* 8:7 (1952), 210–17, at 212. See also *Whom We Should Welcome: Report of the President's Commission on Immigration and Naturalization* (Washington, DC: Government Printing Office, 1953), 67.

73. *Whom We Should Welcome*, 66; Weisskopf, "Report on the Visa Situation," 222.

74. Director of Office of Defense Mobilization to Secretary of Commerce, December 10, 1954, attachment, Control of Visits to Industrial Facilities in the Interest of Internal Security, NARA, RG 40, UD entry 56, box 2.

75. Federal Bureau of Investigation, *Soviet Intelligence Travel and Entry Techniques*, April 1953, i–iii, v, 1–2, 22–23. Even though the FBI focuses on Soviet citizens, it uses the scientist Alan Nunn May, born in the UK, as one example. NARA, CREST files, CIA-RDP65-0076R000400080001-9.

76. Berkner Report, 79–80.

77. Shils, "America's Paper Curtain," 213. Shils also writes, along these lines, "The transformation of science into a subject of crucial importance to national defense has changed patronizing distrust into active and harassing suspicion— recently exaggerated and aggravated by the Fuchs, Nunn May, and Pontecorvo cases" (213).

78. The term was also used to criticize the spread of government secrecy in the early Cold War. In this context the "paper curtain" was between the US government and the people. *Availability of Information from Federal Departments and Agencies: Twenty-Fifth Intermediate Report of House Committee on Government Operations* (Washington, DC: Government Printing Office, 1956), 3.

79. *Whom We Should Welcome*, 67.

80. Wellerstein, *Restricted Data*, section 4.2.

81. Berkner Report, 82.

82. "Position Taken by CIA Representative on Mandatory Export Control of Technical Data," February 23, 1951, NARA, CREST files, CIA-RDP75-00662R000 200150068-3.

83. Department of Defense Memorandum, "Mandatory Export Control of Technical Data," January 17, 1951, NARA, RG 40, Entry UD 59, box 8.

84. Department of Commerce, *Export Control and Allocation Powers, Fifteenth Quarterly Report* (Washington, DC: Government Printing Office, 1951), 1.

85. "Chronology of Technical Data Licenses," John Donovan to Robert A. Bowman, March 1, 1956, NARA, RG 40, Entry UD 59, box 3.

86. See Advisory Committee on Export Policy Document No. 109.1, Export Control of Technical Data, August 17, 1954, NARA, RG 40, Entry UD 65, box 1.

87. Department of Commerce, *Export Control and Allocation Powers, Tenth Quarterly Report* (Washington, DC: Government Printing Office, 1950), 3.

88. Minutes Task Group Meeting on Technical Data Exports, June 16, 1954, 1, NARA, RG 40, Entry UD 65, box 1.

89. On this committee, see Cupitt, *Reluctant Champions*, 70.

90. See Minutes Task Group Meeting on Technical Data Exports, June 16, 1954; Advisory Committee on Export Policy Document No. 109.1, Export Control of Technical Data, August 17, 1954, NARA, RG 40, Entry UD 65, box 1.

91. On the shift in the Cocom allies' approach to export controls in 1954, see Michael Mastanduno, "Trade as a Strategic Weapon: American and Alliance Export Control Policy in the Early Postwar Period," *International Organization* 42:1 (1988), 121–50, at 142–46.

92. Department of Commerce, *Export Control: Thirtieth Quarterly Report* (Washington, DC: Government Printing Office, 1955), 4–5.

93. Department of Defense Memorandum, "Mandatory Export Control of Technical Data," January 17, 1951, NARA, RG 40, Entry UD 59, box 8.

94. Department of Commerce, *Export Control*.

95. Advisory Committee on Export Policy Document No. 109.1, Export Control of Technical Data, August 17, 1954, 4–5, NARA, RG 40, Entry UD 65, box 1.

96. Advisory Committee on Export Policy Document No. 109.1, 4.

97. "Export Regulations, Part 385: Exportations of Technical Data," *Federal Register* 19:253 (December 31, 1954), 9384–86, at 9384 (sec. 385.1c) (emphasis added).

98. "Export Regulations," 9385 (sec. 385.4 d iii and vi).

99. "Export Regulations," 9384 (sec. 385.1a) (emphasis added).

100. Department of State, "International Traffic in Arms, Ammunition, and Implements of War, Part 75: Exportation of Technical Data," *Federal Register* 20:73 (April 14, 1955), 2461–63. For sake of not making a complex story even more complicated, our analysis focuses on the Department of Commerce strand of export controls. For a discussion of central aspects of arms data controls, see Berman and Garson, "United States Export Controls," 808–10, 823–30.

101. For the texts of EO 10290 and EO 1051, see http://www.presidency.ucsb.edu/ws/?pid=78426 and http://www.presidency.ucsb.edu/ws/index.php?pid=485 (accessed March 8, 2018).

102. EO 10501, November 5, 1953, Safeguarding Official Information in the Interest of the Defense of the United States, NARA, RG 40, UD 56, box 1; Atomic Energy Act of 1954, 68 Stat. 919.

103. *The Government's Classification of Private Ideas: Thirty-Fourth Report by*

the Committee on Government Operations, House Report No. 96-1540 (Washington, DC: Government Printing Office 1980), 3, 12–13, 53–62.

104. Department of Commerce, *Export Control: Thirtieth Quarterly Report* (Washington, DC: Government Printing Office, 1955), 4; "Chronology of Technical Data Licenses," John Donovan to Robert A. Bowman, March 1, 1956, NARA, RG 40, Entry UD 59, box 3; "Export Regulations, Part 385," 9386 (sec. 385.51, table).

105. Engineering College Research Council of the American Society for Engineering Education, Recommendation to the US Department of Commerce to Exempt Colleges and Universities from Export Control of Technical Data, February 23, 1955, NARA, RG 40, Entry UD 59, box 3.

106. Department of Commerce Press Release, April 16, 1955, NARA, RG 40, Entry UD 59, box 3.

107. Department of Commerce Press Release.

108. For problems of definition, see Frank E. Samuel, "Technical Data Export Regulations," *Harvard International Law Club Journal* 6:2 (1965), 125–65, esp. 135–36; J. N. Behrman, "US Government Controls over Export of Technical Data," *Patent, Trademark, and Copyright Journal of Research and Education* 8 (1964), 303–15, esp. 304–5.

109. *Availability of Information from Federal Departments and Agencies: Hearings before a Subcommittee of the Committee on Government Operations, House of Representatives, Part 6*, Statement, Donald J. Hughes (Chairman of the Federation of American Scientists; Physiker am Brookhaven National Laboratory), 1448; Department of Commerce, *Export Control: Thirty-Ninth Quarterly Report* (Washington, DC: Government Printing Office, 1957), 15–16.

110. See, for example, Mastanduno, "Trade as a Strategic Weapon"; Tor Egil Førland, *Cold Economic Warfare: CoCom and the Forging of Export Controls, 1948–1954* (Dordrecht: Republic of Letters, 2009); Tor Egil Førland, "'Selling Firearms to the Indians': Eisenhower's Export Control Policy, 1953–54," *Diplomatic History* 15:2 (1991), 221–44; Vibeke Sørensen, "Economic Recovery versus Containment: The Anglo-American Controversy over East-West Trade, 1947–1951," *Cooperation and Conflict* 24 (1989), 69–97; Ian Jackson, *The Economic Cold War: America, Britain, and East-West Trade, 1948–1963* (Houndsmills: St. Martin's Press, 2001); Alan P. Dobson, *US Economic Statecraft for Survival 1933–1991: Of Sanctions, Embargoes and Economic Warfare* (London: Routledge, 2002), chapters 5 and 6.

111. NSC 104, "US Policies and Programs in the Economic Field Which May Affect the War Potential of the Soviet Bloc," Memorandum by the Executive Secretary of the National Security Council, James S. Lay Jr., to the National Security Council, February 12, 1951, Enclosure 1: President Truman to the Secretary of State, December 28, 1950; *Foreign Relations of the United States, 1951, National Security Affairs, Foreign Economic Policy*, vol. 1 (Washington, DC: Government Printing Office, 1979), 1024–25.

112. NSC 104, Subenclosure, "Report to the President on United States Policies and Programs in the Economic Field Which May Affect the War Potential of the Soviet Bloc," February 1951, 1027–34. The report added, "These program should be devised in such a way to create the least possible impediment to the exchange of such information among the nations of the free world."

113. Office of Intelligence Research, Division of Research for the USSR and Eastern Europe, "Vulnerability of the Soviet Bloc to Existing and Tightened Western Economic Controls," OIR Report No. 5447, January 26, 1951, 62, NARA, CREST files, CIA-RDP79R01012A000500030013-1.

114. NSC 104, Subenclosure, "Report to the President." The CIA apparently studied how much technological information was lost the Soviet bloc through Western Europe. See CIA, Office of Research and Reports, "Project Initiation Memorandum: Flow of Strategic Resources and Technological Information into the USSR via Western Europe," June 29, 1951, https://www.cia.gov/library/reading room/docs/CIA-RDP79T01049A000300100001-6.pdf (accessed March 6, 2018).

115. Council on Foreign Economic Policy, Economic Defense Policy Review, Summary of Staff Study No. 8, "Differentials between US and Multilateral Controls," June 23, 1955, https://www.cia.gov/library/readingroom/docs/CIA-RDP63-00084A000100090001-6.pdf (accessed March 6, 2018).

116. Council on Foreign Economic Policy.

117. Mastanduno, "Trade as a Strategic Weapon," 146–47, 149; "Chronology of Technical Data Licenses," John Donovan to Robert A. Bowman, March 1, 1956, NARA, RG 40, Entry UD 59, box 3.

118. Cocom, Record of Discussion on Revision of the Strategic Export Controls—Administrative Principles, Cocom Document No. 2869.76, December 17, 1958, and January 8, 1959, https://www.cia.gov/library/readingroom/docs/CIA-RDP62-00647A000100150034-9.pdf (accessed March 6, 2018).

119. Cocom, Record of Statements by the United Kingdom and United States Delegates on the Review of the Strategic Export Controls—Administrative Principle No. 3, Cocom Document No. 2849.82, February 5, 1959, https://www.cia.gov/library/readingroom/docs/CIA-RDP62-00647A000100150026-8.pdf (accessed March 6, 2018).

120. Cocom, Record of Discussion on Review of Strategic Export Controls—Administrative Principle No. 3, Cocom Document 2869.84, February 16, 1959, https://www.cia.gov/library/readingroom/docs/CIA-RDP62-00647A000100150024-0.pdf (accessed March 6, 2018). On the French position, see Cocom, Record of Discussion on Revision of the Strategic Export Controls—Administrative Principles, Cocom Document No. 2869.76, December 17, 1958, and January 8, 1959.

121. See Cocom, Record of Discussion on Revision of the Strategic Export Controls—Administrative Principles, Cocom Document No. 2869.76, December 17, 1958, and January 8, 1959; Cocom, Record of Discussion on Revision of the Strategic Export Controls—Administrative Principles, Cocom Document No. 2869.80,

January 26, 1959, https://www.cia.gov/library/readingroom/docs/CIA-RDP62-006 47A000100150028-6.pdf (accessed March 6, 2018).

122. Cocom, Record of Discussion on Review of Strategic Export Controls—Administrative Principle No. 3, Cocom Document No. 2869.87, March 9, 1959, https://www.cia.gov/library/readingroom/docs/CIA-RDP62-00647A000100 150021-3.pdf (accessed March 6, 2018).

123. Cocom, Record of Discussion on the Installation of Embargoed Equipment in Civil Aircraft Exported to the Soviet Bloc, Cocom Document No. 3513, May 4, 1959, https://www.cia.gov/library/readingroom/docs/CIA-RDP62-00647 A000100180088-7.pdf (accessed March 6, 2018).

124. Cocom, A Further Memorandum by the United Kingdom Delegation on Their Proposal to Export Infra-red Detector Cells to Poland, Cocom Document No. 3995, May 11, 1960, https://www.cia.gov/library/readingroom/docs/CIA-RDP62-00647A000200020005-4.pdf (accessed March 6, 2018).

125. *Investigation and Study of the Administration, Operation, and Enforcement of the Export Control Act of 1949, and Related Acts, Report of the House Select Committee on Export Controls* (Washington, DC: Government Printing Office, 1962), 11, 14.

126. On the history of the East-West exchange programs, see Yale Richmond, *US-Soviet Cultural Exchanges, 1958–1986: Who Wins?* (Boulder, CO: Westview, 1987); Yale Richmond, *Cultural Exchange and the Cold War: Raising the Iron Curtain* (University Park: Pennsylvania State University Press, 2003); Glenn E. Schweitzer, *Scientists, Engineers, and Two-Track Diplomacy: Half a Century of US-Russian Interacademy Cooperation* (Washington, DC: National Academy Press, 2004); Robert F. Byrnes, *Soviet-American Academic Exchanges, 1958–1975* (Bloomington: Indiana University Press, 1976); Herbert Kupferberg, *The Raised Curtain: Report of the Twentieth Century Fund Task Force on Soviet-American Scholarly and Cultural Exchanges* (New York: Twentieth Century Fund, 1977). For the Atoms for Peace conference, see John Krige, "Atoms for Peace, Scientific Internationalism, and Scientific Intelligence," *Osiris* 21:1 (2006), 161–81. For the general development of exchange relations as well as for a reprint of NSC 5706 and the accompanying Policy Statement, see Richmond, *US-Soviet Cultural Exchanges*, 1–9, 133–37.

127. NSC 5607, in Richmond, *US-Soviet Cultural Exchanges*, 134.

128. John C. Borton (DoC) to Henry Kearns (DoS), Change in Export Control Regulations to Facilitate Visits to US Plants under East-West Contacts Program, December 20, 1957, NARA, RG 489, A1 entry 7, box 7.

129. Loring E. Nacy (DoC) to Henry Kearns (DoS), Visits by Soviet bloc Nationals to US Plants under the East-West Contacts Program, November 14, 1957, with attachment: Proposed Amendment to PD 1192. Draft letter Secretary of Commerce to Secretary of State, November 15, 1957, with attachment: Proposed Amendment to PD 1192 [different version]. The need of the DoC of the visa

application information for meeting its export control responsibilities is spelled out in Report on Meeting held by Representatives of State and Commerce on October 14, 1957. All documents in NARA RG 489, entry 7, box 7.

130. John C. Borton (DoC) to Henry Kearns (DoS), Change in Export Control Regulations to Facilitate Visits to US Plants under East-West Contacts Program, December 20, 1957, NARA, RG 489, A1 entry 7, box 7.

131. Statement, Meyer (DoC), Our position re: enforcement of the Export Control Program on unclassified technical data as it relates to the East/West Exchange Program and Industrial Conferences and Trade Shows, November 21, 1960, NARA, RG 489, A1 entry 7, box 7.

132. "US Blocks Trade of Oil Drill Data," *New York Times*, May 17, 1956, 1. See also Joseph R. Slevin, "Dresser-Red Deal Weighed by Weeks," *New York Herald Tribune*, April 11, 1956, A3.

133. Department of Commerce, *Export Control: Thirty-Sixth Quarterly Report* (Washington, DC: Government Printing Office, 1956), 9.

134. Department of Commerce, *Export Control: Thirty-Fourth Quarterly Report* (Washington, DC: Government Printing Office, 1956), 42; *Availability of Information from Federal Departments and Agencies: Hearings before the House Subcommittee on Government Operations, Part 6, Department of Commerce* (Washington, DC: Government Printing Office, 1956), 1496, 1514 (Statement, Newton H. Foster, Finished Products Division, Department of Commerce, Office of the Assistant Director of Export Supply); Department of Commerce, *Export Control: Forty-Second Quarterly Report* (Washington, DC: Government Printing Office, 1958), 35.

135. Newton H. Foster, Finished Products Division, Department of Commerce, Office of the Assistant Director of Export Supply, to DuPont Laboratories, March 7, 1956, NARA, RG 40, Entry UD 59, box 1.

136. NARA, RG 489, A1 Entry 7, box 7, Frank Sheaffer (DoC) to F. D. Hockersmith (DoC), January 11, 1961, Draft letter to the International Conference of the American Nuclear Society and the International Conference on Strong Magnetic Fields at MIT, May 15, 1961.

137. NARA, RG 489, A1 Entry 7, box 7, Loring E. Nacy (DoC) to Henry Kearns (DoS), Visits by Soviet Bloc Nationals to US Plants under the East-West Contacts Program, November 14, 1957; quote in Draft letter Secretary of Commerce to Secretary of State, November 27, 1957.

138. [CIA] Informal History: Intelligence Involvement in the East-West Exchanges Program [ca. 1974], 3, https://www.cia.gov/library/readingroom/docs/DOC_0001495225.pdf (accessed January 15, 2017); Standing Committee on Exchanges, July 16, 1956. https://www.cia.gov/library/readingroom/docs/CIA-RDP85 S00362R000600030005-7.pdf (accessed January 15, 2017).

139. Director of Central Intelligence, Directive No. 2/6, Committee on Exchanges (Coordination and Exploitation of East-West Exchange Program), April 3,

1963 (showing the redactions of the version of 1959), attachment to United States Intelligence Board Memorandum, Proposed Amendments to Director of Central Intelligence Directives, April 3, 1963, https://www.cia.gov/library/readingroom/docs/CIA-RDP86B00269R000200060080-5.pdf (accessed January 15, 2017).

140. NARA, CREST files, IAC Standing Committee on Exchanges, First Semi-annual Report, October 4, 1956, quote 1, CIA-RDP61-00459R000300050005-3. For the close connection between scientific exchanges and scientific intelligence, see also Krige, "Atoms for Peace." From a CIA perspective: Guy E. Coriden, "The Intelligence Hand in East-West Exchange Visits," *Studies in Intelligence* 2:3 (1958), 63–71; NARA, CREST files, CIA-RDP78-03921A000300210001-1.

141. Secretary of Commerce, Lewis L. Strauss, to Allen Dulles, June 11, 1959, Strauss to Secretary of State, May 8, 1959, with attachment, Suggestions for modifying present procedures for implementation of the East-West Exchange Program, quote in the attachment, https://www.cia.gov/library/readingroom/docs/CIA-RDP80B01676R000800010018-6.pdf (accessed January 15, 2017).

142. IAC Standing Committee on Exchanges, First Semi-annual Report, October 4, 1956, appendix A.

143. IAC Standing Committee on Exchanges Third Semi-annual Report, February 11, 1958, 4, and annex A to this report, Interim Evaluation of the Intelligence Aspects of the East-West Exchange Program, February 11, 1958, 5, NARA, CREST files, CIA-RDP61-00549R000300050002-6.

144. James McGrath, "The Scientific and Cultural Exchange," *Studies in Intelligence* 7:1 (1963), 25–30, at 27–28; NARA, CREST files, CIA-RDP78T03194A000200010001-2.

145. Ralph Jones, DoS, Soviet and Eastern European Exchanges Staff, to IBM, February 21, 1963, NARA, RG 489, A1 Entry 1, box 4. A similar letter mentioned the involvement of the DoD in making decisions about exchanges: NARA, RG 489, A1 Entry 1, box 4, Frank G. Siscoe, Director, Soviet and Eastern European Exchanges Staff, to the President of the System Development Corporation, March 29, 1963.

146. Daniels and Krige, "Beyond the Reach of Regulation?"

147. McGrath, "Scientific and Cultural Exchange," 30.

148. *Investigation and Study of the Administration, Operation, and Enforcement of the Export Control Act of 1949, and Related Acts, Report of the House Select Committee on Export Controls* (Washington, DC: Government Printing Office, 1962), 2, 16.

149. Department of Commerce, *Export Control and Allocation Powers: Tenth Quarterly Report* (Washington, DC: Government Printing Office, 1950), 1.

150. Department of Commerce, *Export Control: Twelfth Quarterly Report* (Washington, DC: Government Printing Office, 1950), 2, 16; Department of Commerce, *Export Control: Twenty-Fourth Quarterly Report* (Washington, DC; Government Printing Office, 1953), 9–11.

151. Berman and Garson, "United States Export Controls," 811.

152. *Legislative Reference Service of the Library of Congress: A Background Study on East-West Trade, Prepared for the Committee on Foreign Relations United States Senate* (Washington, DC: Government Printing Office, 1965), 6; Department of Commerce, *Export Control: Thirty-Second Quarterly Report* (Washington, DC: Government Printing Office, 1955), 6, 8; Department of Commerce, *Export Control: Forty-Eighth Quarterly Report* (Washington, DC: Government Printing Office, 1959), 6.

153. Lutz Frühbrodt, "American and German Trade Relations," in *The United States and Germany in the Era of the Cold War, 1945–1990: A Handbook*, vol. 1, *1945–1968*, ed. Detlef Junker (Cambridge: Cambridge University Press, 2004), 317–25, at 318, table 1. For purpose of comparison, the West German figures were 1.5 and 2.0 percent respectively (319, table 2).

154. Berman and Garson, "United States Export Controls," 878.

155. Availability of Information from Federal Departments and Agencies, Hearings, part 6, Statement, Borton, 1510.

156. Berman and Garson, "United States Export Controls," 808, 843.

157. Availability of Information from Federal Departments and Agencies, Hearings, part 6, Statement, Donald J. Hughes, 1448.

158. Berman and Garson, "United States Export Controls," 851, 854, 859.

Chapter 4

1. "An Analysis of Export Control of US Technology—a DOD Perspective," in *A Report of the Defense Science Board Task Force on Export of US Technology, 4 February 1976* (Washington, DC: Office of the Director of Defense Research Engineering, 1976) (Bucy Report). See also J. Fred Bucy, "On Strategic Technology Transfer to the Soviet Union," *International Security* 1:4 (1977), 25–43; J. Fred Bucy, "Technology Transfer and East-West Trade: A Reappraisal," *International Security* 5:3 (1980–81), 132–51.

2. "Détente: A Trade Giveaway?," *Business Week*, January 12, 1974, 64, 66.

3. Bucy was a member of the Defense Science Board of the Department of Defense (DoD) and had chaired DSB task forces on design-to-cost and on the export of US technology. He was also a member of the Technology Assessment Advisory Council of the Office of Technology Assessment (OTA) and served as a member of the OTA's advisory panel to the technology and world trade program; see *Transfer of Technology to the Soviet Union and Eastern Europe: Hearing before the Permanent Subcommittee on Investigations of the Committee on Governmental Affairs*, US Senate, 95th Congress, May 25, 1977, 1.

4. Michael Mastanduno, *Economic Containment: CoCom and the Politics of East-West Trade* (Ithaca, NY: Cornell University Press, 1992), 187.

5. Stuart Macdonald, *Technology and the Tyranny of Export Controls: Whisper Who Dares* (New York: St. Martin's, 1990), 66.

6. Bucy, "On Strategic Technology Transfer to the Soviet Union," 28.

7. The Munitions Control Regulations, by contrast, did control the sharing of know-how of military systems with Communist countries.

8. Harold J. Berman and John R. Garson, "United States Export Controls—Past, Present, and Future," *Columbia Law Review* 67:5 (1967), 791–890, at 828.

9. Stephen Linde, "Arms Control—State Department Regulation of Exports of Technical Data Relating to Munitions Held to Encompass General Knowledge and Experience," *NYU Journal of International Law and Politics* 9:91 (1976–77), 91–112, at 99. The same argument was invoked by the US Joint Committee on Atomic Energy to restrict British engineers sharing gas centrifuge technology with continental colleagues. See John Krige, "Hybrid Knowledge: The Transnational Co-production of the Gas Centrifuge for Uranium Enrichment in the 1960s," *British Journal for the History of Science* 45:3 (2102), 337–57.

10. Berman and Garson, "United States Export Controls," 862.

11. Steven Lazarus, quoted in "Détente: A Trade Giveaway," 66.

12. Bruce Seely has pointed out that a search of the OCLS WorldCat database using the phrase "technology transfer" produced 138 books between 1961 and 1970, and 1,497 books between 1971 and 1980, a more than tenfold increase between the decades: Bruce E. Seely, "Historical Patterns in the Scholarship of Technology Transfer," *Comparative Technology Transfer and Society* 1:1 (April 2003), 7–48.

13. CIA National Foreign Assessment Center, *US Export Competitiveness: A Review and Evaluation*, Research Paper ER81-10044, February 1981.

14. US Department of Commerce, International Trade Administration, *An Assessment of US Competitiveness in High Technology Industry*, Washington, DC, February 1983, 46.

15. Richard R. Nelson and Gavin Wright, "The Rise and Fall of American Technological Leadership: The Postwar Era in Historical Perspective," *Journal of Economic Literature* 40:4 (1992), 1931–64, 1932.

16. Nelson and Wright, 1952.

17. Jean-Jacques Servan-Schreiber, *The American Challenge* (New York: Avon, 1971). The original French version was published in 1967.

18. Céline Pessis, Sezin Topçu, and Christophe Bonneuil, eds., *Une autre histoire des "Trente glorieuses": Modernisation, contestations, et pollutions dans la France de l'après-guerre* (Paris: La Découverte, 2013).

19. Nelson and Wright, "Rise and Fall," figure 10.

20. Nelson and Wright, "Rise and Fall," 1959.

21. Herbert Meyer, "Those Worrisome Technology Exports," *Fortune* 97 (February 1978), 106–9.

22. Daniel Lloyd Spencer, *Technology Gap in Perspective: Strategy of International Technology Transfer* (New York: Spartan Books, 1970), 14.

23. Paul Kennedy, *The Rise and Fall of the Great Powers: Economic Change and Military Conflict from 1500 to 2000* (1987; New York: Vintage Books, 1989), 405. For the economic implications, see 434–35.

24. Jeremy Adelman, "International Finance and Political Legitimacy: A Latin American View of the Global Shock," in Niall Ferguson, Charles S. Maier, Erez Manela, and Daniel Sargent, eds., *The Shock of the Global: The 1970s in Perspective* (Cambridge, MA: Harvard University Press, 2010), 113–27.

25. Japan reevaluated the yen against the dollar by 41 percent between 1976 and 1978, reversing the trend of US loss in market shares. CIA National Foreign Assessment Center, *US Export Competitiveness: A Review and Evaluation*, Research Paper ER81-10044, February 1981.

26. Kennedy, *Rise and Fall of the Great Powers*, 434–35.

27. Charles S. Maier, *Among Empires: American Ascendancy and Its Predecessors* (Cambridge, MA: Harvard University Press, 2016). The quote is from Stephen Wertheim, "The Final Frontiers? Charles Maier's Breakthrough *Among Empires*," *Harvard International Review*, November 22, 2007, http://hir.harvard.edu/the-final-frontiers. See also Victoria de Grazia, "The Crisis of Hyper-consumerism: Capitalism's Latest Forward Lurch," in Jürgen Kocka and Marcel van der Linden, eds., *Capitalism: The Emergence of a Historical Concept* (London: Bloomsbury Academic, 2016), 71–106.

28. Daniel J. Sargent, "The United States and Globalization in the 1970s," in Ferguson et al., *Shock of the Global*, 49–64, 50 for both quotes.

29. Marshall D. Shulman, "Toward a Western Philosophy of Coexistence," *Foreign Affairs* 52:1 (October 1973), 35–58, at 39.

30. A. Köves, "The Impact of Western Trade Restrictions on East-West Trade after World War II," *Acta Oeconomica* 19:1 (1977), 67–76.

31. George D. Holliday, *Technology Transfer to the USSR, 1928–1937 and 1966–1975: The Role of Western Technology in Soviet Economic Development* (Boulder CO: Westview 1979), 61.

32. Shulman, "Western Philosophy of Coexistence," 41.

33. As stated in the Nixon-Brezhnev Joint US-USSR Communique, San Clemente, June 24, 1973, http://www.washingtonpost.com/wp-srv/inatl/longterm/summit/archive/com1973-1.htm (accessed December 7, 2018). This relaxation of tension included, too, a major shift in relationships between Western Europe and the Soviet bloc led by Chancellor Willy Brand's "Ostpolitik, which, through recognizing—if not formalizing—the de facto division of Germany, paved the way for much closer cooperation between the Federal Republic and the European Economic Community with Eastern Europe and the Soviet Union. See M. E. Sarotte, *Dealing with the Devil: East Germany, Détente, and Ostpolitik, 1969–1973* (Chapel Hill: University of North Carolina Press, 2001); Carole Fink and Bernd Schaefer, eds., *Ostpolitik, 1969–1974: European and Global Responses* (Cambridge: Cambridge University Press, 2009).

34. Jeremi Suri, "Henry Kissinger and the Geopolitics of Globalization," in Ferguson et al., *Shock of the Global*, 173–88.

35. Mastanduno, *Economic Containment*, 112.

36. Bucy, "On Strategic Technology Transfer to the Soviet Union," 28.

37. Board on International Scientific Exchange, Commission on International Relations, National Research Council, *Review of the US/USSR Agreements on Cooperation in the Fields of Science and Technology* (Washington, DC: National Academy of Sciences, 1977); see also *East-West Technology Transfer: A Congressional Dialog with the Reagan Administration Prepared for the Use of Joint Economic Committee, Congress of the United States*, December 19, 1984 (Washington DC: Government Printing Office, 1984), 117. See also NSSM 176, *Review of US-Soviet Bilateral Issues*, March 13, 1973, https://fas.org/irp/offdocs/nssm-nixon/nssm_176.pdf; and NSDM 215, *US-Soviet Bilateral Issues*, May 3, 1973, https://fas.org/irp/offdocs/nsdm-nixon/nsdm_215.pdf); both signed by Henry Kissinger.

38. Sections 5, 6, Joint Communiqué.

39. Quoted in Reinhard Rode and Hanns-Dieter Jacobsen, eds., *Economic Warfare or Détente: An Assessment of East-West Economic Relations in the 1980s* (Boulder CO: Westview, 1985), 3.

40. Office of Technology Assessment, *Technology and East-West Trade* (Washington, DC: Government Printing Office, 1979), 119–20.

41. Samuel P. Huntington, "Trade, Technology and Leverage: Economic Diplomacy," *Foreign Policy* 32 (Autumn 1978), 63–80.

42. The Export-Import Bank provided $469 million in credits in a fifteen-month period in 1973–74 for about sixteen different projects, which also received an equal amount of private credit; Huntington, "Economic Diplomacy," 77. The Jackson-Vanik Amendment to the 1974 Trade Act stopped the practice by denying most-favored-nation status to certain countries with nonmarket economies that restricted emigration.

43. National Foreign Assessment Center of the CIA, *The Soviet Economy in 1978–79 and Prospects for 1980* (Washington, DC: Director of Public Affairs, Central Intelligence Agency, 1980), table 7.

44. CIA, *Soviet Economy in 1978–79*, 16.

45. Testimony of William Perry, *Technology Exports: Department of Defense Organization and Performance: Hearing before the Subcommittee on International Economic Policy and Trade of the Committee on Foreign Affairs, House of Representatives, 96th Congress, October 30, 1979* (Washington, DC: Government Printing Office, 1980), 2. Corporate investment in R&D has to be added to the US figure to get a better sense of relative weights.

46. Huntington, "Trade, Technology and Leverage: Economic Diplomacy," 64.

47. Fred J. Bucy, Full Statement, *Transfer of Technology to the Soviet Union and Eastern Europe, Hearing before the Permanent Subcommittee on Investigations*

of the Committee on Governmental Affairs of the US Senate, 95th Congress, May 25, 1977 (Washington, DC: Government Printing Office, 1977), 15.

48. Dr. Ellen Frost, Statement, *Extension and Revision of the Export Administration Act of 1969: Hearings and Markup before the Subcommittee on International Economic Policy and Trade of the Committee on Foreign Affairs, House of Representatives, 96th Congress, Part I, February 15, 22; March 7, 8, 14, 15, 21, 22, 26, 27, 28; April 3, 4, 24, 25 and 26, 1979* (Washington, DC: Government Printing Office, 1979), 159.

49. Bucy, Full Statement, *Transfer of Technology*, 16. The Soviets had enormous difficulties maintaining their production targets in high technology goods. Typically, the 1972 Ninth Five-Year Plan called for the production of forty-five to forty-eight thousand computers by 1975, a nine- to tenfold increase over the numbers at hand in 1970. The system failed miserably to reach this target, producing only two thousand to three thousand annually in 1974 and 1975. This was not for lack of expertise but perhaps because the authorities decided it was far better to adopt IBM solutions than to channel further resources into successful indigenous efforts like those of S. A. Lebedev, who built an institute in Kiev, and who was one of the most accomplished computer scientists and engineers of the twentieth century. Seymour A. Goodman, "The Origins of Digital Computing in Europe: Retracing the Paths of Influential, but Often Isolated, Computer Pioneers," *Communications of the ACM* 46:9 (2003), 21–25.

50. "Détente: A Trade Giveaway?," 66.

51. OTA, *Technology and East-West Trade*, 112.

52. Berman and Garson, "United States Export Controls," 800 (emphasis added).

53. For Mastanduno, "economic warfare may be considered the waging of hostilities against an adversary by economic rather than military instruments. It's immediate or proximate objective is to weaken the economy or economic potential of a target state by denying it the benefits of international economic exchange." It can have the ultimate goal to also "weaken its *military* capabilities or potential." Mastanduno, *Economic Containment*, 40.

54. OTA, *Technology and East-West Trade*, 115.

55. OTA, 117.

56. Christopher J. Donovan, "The Export Administration Act of 1979: Refining United States Export Control Machinery," *Boston College International and Comparative Law Review* 4:1 (1981), 77–114, 84.

57. Mastanduno, *Economic Containment*, 114.

58. Gary K. Bertsch, "Technology Transfers and Technology Controls: A Synthesis of the Western-Soviet Relationship," in Ronald Amann and Julian Cooper, eds., *Technical Progress and Soviet Economic Development* (Blackwell: Oxford, 1986), table 6.2. For later data, see OTA, *Technology and East-West Trade*.

59. Testimony of Edith W. Martin, Deputy Undersecretary of Defense for Research and Engineering, *Scientific Communications and National Security*, Hear-

ing before the Subcommittee on Science, Research and Technology and the Subcommittee on Investigations and Oversight of the Committee on Science and Technology, House of Representatives, 98th Congress, 2nd Session, May 24, 1984, 135.

60. Bucy, "Technology Transfer and East-West Trade," 150 (emphasis added).

61. Bucy Statement, *Transfer of Technology to the Soviet Union and Eastern Europe*, 17, 18.

62. Perry, *Technology Exports*, 34.

63. John Krige, *Sharing Knowledge, Shaping Europe: US Technological Cooperation and Nonproliferation* (Cambridge, MA: MIT Press, 2016), 149–67.

64. On high-performance computer safeguards, see Mario Daniels, "Dangerous Calculations: The Origins of the US High Performance Computer Export Safeguards Regime (1968–1974)," in John Krige, ed., *Writing the Transnational History of Knowledge Flows in a Global Age* (Chicago: Chicago University Press, forthcoming).

65. Bucy Report, 25.

66. Bucy Report, vi.

67. "Détente: A Trade Giveaway?," 64.

68. "Détente: A Trade Giveaway?," 64.

69. Bucy Report, v.

70. See, for example, *Report of the Technology Transfer Panel of the Committee on Armed Services*, House of Representatives, 98th Congress, June 13, 1984.

71. Daniel Lloyd Spencer, *Technology Gap in Perspective: Strategy of International Technology Transfer* (New York: Spartan Books, 1970), 29.

72. Spencer, 28.

73. Spencer, 32.

74. OTA, *Technology and East-West Trade*, 89, quoting a report by the National Academies.

75. Thane Gustafson, *Selling the Russians the Rope? Soviet Technology Policy and US Export Controls*, Report R-2649-ARPA (Santa Monica, CA: RAND, April 1981), 1 (emphasis added).

76. Bucy, "Strategic Technology Transfer," 28.

77. *Report of the Technology Transfer Panel*, June 13, 1984, 20.

78. Bucy Report, 12.

79. OTA, *Technology and East-West Trade*, 119–20.

80. Bucy, "Strategic Technology Transfer," 28.

81. Bucy defined science as being "devoted to the collection and expansion of knowledge." In contrast to technology, he said, "science has no boundaries and the free and open exchange of science benefits all those engaged in research." *Transfer of Technology to the Soviet Union*, 3. He thus aligned the (basic) science/technology boundary with the freely circulating/regulated knowledge boundary, which was developed in 1958. See Mario Daniels and John Krige, "Beyond the Reach of Regulation? 'Basic' and 'Applied' Research in the Early Cold War United States," *Technology and Culture* 59:2 (2018), 226–50.

82. Bucy Report, xiii.

83. Bucy Report, 2.

84. Bucy Report, 2.

85. Bucy Report, 3 (emphasis in the original).

86. As noted by Maurice J. Mountain, "Technology Exports and National Security," in his contribution at 95–103 to Huntington et al., "Trade, Technology and Leverage: Economic Diplomacy," 98.

87. Bucy Report, 5.

88. Bucy Report, 34.

89. Bucy, "*Strategic Technology Transfer*," 37, for both quotes.

90. Bucy Report, 39.

91. Bucy, "Strategic Technology Transfer," 32–33.

92. Bucy Report, 38. On the use of export and visa controls to regulate US-Soviet scientific exchanges in the early Cold War, see Mario Daniels, "Restricting the Transnational Movement of 'Knowledgeable Bodies': The Interplay of US Visa Restrictions and Export Controls in the Cold War," in *How Knowledge Moves: Writing the Transnational History of Science and Technology*, ed. John Krige (Chicago: University of Chicago Press, 2019), 35–61.

93. Bucy Report, iv.

94. See also Patrick B. Fazzone, "Business Effects of the Extraterritorial Reach of the US Export Control Laws," *NYU Journal of International Law and Politics* 15 (1982–83), 545–94.

95. Bucy Report, 19–22.

96. Comptroller General, *Report to the Congress of the United States: Export Controls; Need to Clarify Policy and Simplify Administration*, Report ID-79-16, March 1, 1979. For a discussion of Cocom that one-sidedly emphasizes European failings, see Jonathan B. Bingham and Victor C. Johnson, "A Rational Approach to Export Controls," *Foreign Affairs* 57:4 (Spring 1979), 894–920. Bingham chaired a House committee that had jurisdiction over export controls when he wrote the article.

97. OTA, *Technology and East-West Trade*, 158. This report includes a useful short survey of Cocom activities. The extent of exceptions agreed to by Cocom tended to be exaggerated. Between 1971 and 1975 Cocom countries exported $86 billion worth of goods to the Eastern bloc. Exceptions amounted to $600 million worth of business—less than 1 percent. The United States was the major "offender" throughout the 1970s. *International Transfer of Technology: Report of the President to the Congress together with Assessment . . .* , Prepared for the Subcommittee on International Security and Scientific Affairs of the Committee on International Relations, December 1978 (Washington, DC: Government Printing Office, 1979), 21.

98. OTA, *Technology and East-West Trade*, 157.

99. Mastanduno, *Economic Containment*, 211.

100. A validated license authorizes a specific export in response to an application by an exporter. A general license authorizes exports without application by an exporter.

101. Mastanduno, *Economic Containment*, 198 (emphasis in the original).

102. J. Fred Bucy, "Going, Going Goooonnnnne," *New York Times*, September 11, 1976, https://www.nytimes.com/1976/09/11/archives/going-going-goooonn nnne.html (accessed May 6, 2021).

103. Cited in Meyer, "Those Worrisome Technology Exports," 106.

104. *International Transfer of Technology: Report of the President to the Congress*, December 1978.

105. Bucy Report, 28.

106. Bucy Report, xv, 37.

107. Memo from the Secretary of Defense, "Interim DoD Policy Statement on Export Control of United States Technology," August 26, 1977, Exhibit No. 5 in *Transfer of Technology in the Dresser Industries Export Licensing Actions*, hearing before the Permanent Subcommittee on Investigations of the Committee on Governmental Affairs, United States Senate, 95th Congress, 2nd Session, October 3, 1978 (Washington, DC: Government Printing Office, 1979), 90–92.

108. Memo from the Secretary of Defense, 91 (emphasis added).

109. Memo from the Secretary of Defense, 91.

110. Memo from the Secretary of Defense, 91.

111. Memo from the Secretary of Defense, 92.

112. OTA, *Technology and East-West Trade*, 121.

113. *Dresser Industries Export Licensing Actions*, 27.

114. *Dresser Industries Export Licensing Actions*, 31.

115. *Dresser Industries Export Licensing Actions*, 28.

116. OTA, *Technology and East-West Trade*, 122 (emphasis added).

117. *Technology Exports: Department of Defense Organization and Performance: Hearing before the Subcommittee on International Economic Policy and Trade of the Committee on Foreign Affairs*, 96th Congress, October 30, 1979 (Washington, DC: Government Printing Office, 1980), 21.

118. United States, Export Administration Act of 1979, *International Legal Materials* 18:6 (November 1979), 1508–24, section 5 (d).

119. Donovan, "Export Administration Act of 1979."

120. Luther Karl Branting, "Reconciliation of Conflicting Goals in the Export Administration Act of 1979—a Delicate Balance," *Law and Policy in International Business* 12 (1980), 415–60, at 431.

121. *Hearings and Markup of the Export Administration Act of 1979 Held before Subcommittee on International Economic Policy and Trade, House of Representatives*, February 15, 22; March 7, 8, 14, 15, 21, 22, 26, 27, 28; April 3, 4, 24, 25, and 26, 1979.

122. Frost, *Hearings and Markup*, 168.

123. Davis, *Hearings and Markup*, 410. Davis did not complain about the low level of resources: she saws this as "not atypical of formative, innovative process design activities."

124. Davis, *Hearings and Markup*, 418.

125. Bucy, *Hearings and Markup*, appendix 23, 1020.

126. Frost, *Hearings and Markup*, 169–72.

127. *Technology Exports: Department of Defense Organization and Performance: Hearing before the Subcommittee on International Economic Policy and Trade of the Committee on Foreign Affairs*, 96th Congress, October 30, 1979 (Washington, DC: Government Printing Office, 1980), 7.

128. Davis, *Hearings and Markup*, 411.

129. Davis, *Hearings and Markup*, 428, 430.

130. Comptroller General, *Report to the Congress of the United States: Export Controls*, 61.

131. *Federal Register* 45 (October 1, 1980), 65014–19, 65152–67.

132. The US Munitions List was used by the Department of State in implementing the ITAR (International Traffic in Arms Regulations).

133. *East-West Technology Transfer: A Congressional Dialog*, 16, 91–93

134. *East-West Technology Transfer: A Congressional Dialog*, 113.

135. *Technology Transfer: Hearings before the Technology Transfer Panel of the Committee on Armed Services*, 98th Congress, June 9, 21, 23; July 13, 14, 1983 (Washington, DC: Government Printing Office, 1984), 187, 199.

136. *Technology Transfer Panel*, 218.

137. Shirley Miller Dvorin, "The Export Administration Act of 1979: An Examination of Foreign Availability of Controlled Goods and Technologies," *Northwestern Journal of International Law and Business* 2:1 (1980), 179–99, at 187.

138. *Technology Transfer Panel*, 202.

139. *Technology Transfer Panel*, 179.

140. *Technology Transfer Panel*, 202.

141. Mastanduno, *Economic Containment*, 217.

142. Mastanduno, *Economic Containment*, 216–19 for this paragraph; 219 for the quote.

143. Comptroller General, *Report to the Congress of the United States: Export Controls*, March 1979, 63.

144. Panel on the Impact of National Security Controls on International Technology Transfer, Committee on Science, Engineering and Public Policy, *Balancing the National Interest: US National Security Export Controls and Global Economic Competition* (Washington, DC: National Academy Press, 1987), 129.

Chapter 5

1. CIA, "Soviet Acquisition of Western Technology," included in *East-West Trade and Technology Transfer, Hearing before the Subcommittee on International Finance and Monetary Policy of the Committee on Banking, Housing, and Urban Affairs, United States Senate, Ninety-Seventh Congress, Second Session, on the Extent of Technology Transfers from the West to the Soviet Union during the Past*

Decade and the Contributions such Transfers Have Made to Strengthen the Soviet Military-Industrial Base, April 1982 (Washington, DC: Government Printing Office, 1982), 23–37. See also Director of Central Intelligence, *The Technology Acquisition Efforts of the Soviet Intelligence Services (U)*, Interagency Intelligence Memorandum, NI IIM 82-10006, June 1982, https://www.cia.gov/library/reading room/docs/CIA-RDP82M00786R000104810001-5.pdf (accessed March 8, 2019). These documents will be referred to as Soviet Acquisition, Congress; and Soviet Acquisition, RROOM, respectively.

2. Gus W. Weiss, "The Farewell Dossier: Duping the Soviets," [*CIA*] *Studies in Intelligence* 39:5 (1996), 121–26, at 124, https://www.cia.gov/static/887689795bd91e d08ca926a2f6278ee4/The-Farewell-Dossier.pdf (accessed May 5, 2021).

3. Stuart Mcdonald, "Controlling the Flow of High-Technology Information from the United States to the Soviet Union: A Labour of Sisyphus?," *Minerva* 24:1 (1986), 39–73, at 66.

4. As mentioned by a James Buckley, State Department official who chaired the American delegation in his congressional testimony, *Transfer of United States High Technology to the Soviet Union and Soviet Bloc Nations, Hearings before the Permanent Subcommittee on Investigations of the Committee on Governmental Affairs*, US Senate, 97th Congress, May 4, 5, 6, 11, and 12, 1982, 158.

5. Other interpretations about the impact and consequences of technology transfer were possible. In 1981 Gustafson stressed that meaningful technology transfer would be possible only if the Soviets developed their own capacities for technological innovation, generation and diffusion, which many thought was most unlikely. Autio-Sarasmo argued in 2016 that "the importance of transferred technology for Soviet economic performance was modest. There were some domestic innovations, and some cases of successful reverse engineering based on technology transfer, but no important technological breakthroughs emerged." Hanson locates what innovation there was in the military-industrial complex, with little spin off into the civilian sector. Bertsch wisely emphasized in 1981 that most analyses of impact and consequences of technology transfer made over the previous decade were "tentative and ambiguous," leaving policy makers considerable "freedom of action" in dealing with a contentious issue. Thane Gustafson, *Selling the Russians the Rope? Soviet Technology Policy and US Export Controls*, RAND Report R-2649-ARPA, for the Defense Advanced Research Projects Agency (Santa Monica, CA: RAND, April 1981), vii, https://www.rand.org/pubs/reports/R2649.html (accessed October 29, 2018); Sari Autio-Sarasmo, "Technological Modernisation in the Soviet Union and Post-Soviet Russia: Practices and Continuities," *Europe-Asia Studies* 68:1 (2016), 79–96, at 92; P. Hanson, *The Rise and Fall of the Soviet Economy: An Economic History of the USSR from 1945* (London: Longman, 2003), 21; Gary K. Bertsch, "Technology Transfers and Technology Controls: A Synthesis of the Western-Soviet Relationship," 115, in Ronald Amann and Julian Cooper, eds., *Technical Progress and Soviet Economic Development* (New York: B. Blackwell, 1986).

6. John Shattuck, Vice President for Government and Community Affairs, Harvard University, Testimony to the hearings, *1984: Civil Liberties and the National Security State*, Congressional Subcommittee on Courts, Civil Liberties, and the Administration of Justice of the Committee on the Judiciary, 98th Congress, November 2, 3, 1983, and January 24, April 5, and September 26, 1984 (Washington, DC: Government Printing Office, 1984), 333.

7. CIA, Directorate of Intelligence, *Truck Production at the Soviet Kama River Plant—Western Technology in Action*, Secret Report SOV 85-10147X, August 1985, 1, https://www.cia.gov/library/readingroom/docs/CIA-RDP86T00591R000300400003-5.pdf (accessed September 24, 2018).

8. CIA, 7; Gustafson, *Selling the Russians the Rope?*, vii.

9. Gustafson, *Selling the Russians the Rope?*, 24.

10. Soviet Acquisition, Congress; and Soviet Acquisition, Interagency Intelligence Memorandum, NI IIM 82-10006, June 1982.

11. Technology was defined more broadly than by Bucy's panel, through it incorporated his ideas. For the CIA "technology" for weapons (specifically) meant "the application of scientific knowledge, technical information, know-how, critical materials, keystone manufacturing and test equipment, and end products, that are essential to the research and development, as well as the series manufacture of high-quality weapons and military equipment." Soviet Acquisition, Congress, 23.

12. Congressional Hearings, *East-West Trade and Technology Transfer: Hearing before the Subcommittee on International Finance and Monetary Policy of the Committee on Banking, Housing and Urban Affairs, US Senate, 97th Congress, 2nd Session*, April 14, 1982. President Kennedy had rejected a request by Bryant Chucking Grinder Corporation in Vermont to sell high-precision ball-bearing grinding machines (that were necessary to build inertial guidance systems for MIRVed ICBMs). *East-West Trade and Technology Transfer*, April 1982, 10.

13. *East-West Trade and Technology Transfer*, 9.

14. Soviet Acquisition, RROOM, 3.

15. Soviet Acquisition, RROOM, 10.

16. Cited in National Academy of Engineering, *Scientific Communication and National Security* (Washington, DC: National Academies Press, 1982) (Corson Report), 9–10.

17. Weiss, "Farewell Dossier," 124.

18. Soviet Acquisition, RROOM, table 2, 14.

19. The Department of State did not endorse this position. It wrote that bilateral exchanges in culture and in science and technology, between the United States and the USSR and its allies "call for mutual benefit, equality and reciprocity." Those that result in unacceptable technology loss are altered or disapproved. "Studies conducted on these exchange programs have found no evidence that sensitive technologies have been transferred; they have also found that reciprocity has been maintained. Most importantly, though, these exchange programs allow US

scientists and officials access to Soviet science institutions, which could otherwise be closed, and provide the US Government with leverage to acquire a consistent, meaningful pattern of reciprocity." *East-West Trade and Technology Transfer*, 118.

20. Frank Carlucci, "Scientific Exchanges and National Security," Reply to William D. Carey, *Science* 215:4529 (January 8, 1982), 140–41, at 140.

21. Cited by Lara Baker Jr., in congressional testimony, *Transfer of United States High Technology to the Soviet Union and Union and Soviet Bloc Nations*, 55.

22. *Transfer of United States High Technology to the Soviet Union and Union and Soviet Bloc Nations*, 24.

23. Carlucci, "Scientific Exchanges," 141.

24. Department of Defense, *Soviet Military Power*, Government Printing Office, 1981, 81, available at http://edocs.nps.edu/2014/May/SovietMilPower1981.pdf (accessed October 20, 2018).

25. Soviet Acquisition, RROOM, 19.

26. Soviet Acquisition, RROOM, 25, 26.

27. Phillip Mirowksi, *Science-Mart: Privatizing American Science* (Cambridge, MA: Harvard University Press, 2011); Shelia Slaughter and Gary Rhoades, *Academic Capitalism and the New Economy: Markets, State and Higher Education* (Baltimore: Johns Hopkins University Press, 2004).

28. Corson Report, 11.

29. Mitchel B. Wallerstein, "Historical Context of National Security Concerns about Science and Technology," appendix B, 126, Corson Report.

30. Wallerstein, "Historical Context."

31. *East-West Trade and Technology Transfer*, 85; see also 68 for Brady's statement.

32. Nicholas Wade, "Science Meetings Catch the US—Soviet Chill," *Science* 207:4435 (March 7, 1980), 1056, 1058.

33. Gloria B. Lubkin, "Government Bars Soviets from AVS and OSA Meetings," *Physics Today* 33:4 (April 1980), 81–83.

34. Lubkin, 82.

35. Kenneth Kalivoda, "The Export Administration Act's Technical Data Regulations: Do They Violate the First Amendment?," *Georgia Journal of International and Comparative Law* 11:3 (1981), 563–87, at 569.

36. Kalivoda, 570.

37. Wade, "Science Meetings," 1058.

38. Quoted in Wade, 1058.

39. Cited by Constance Holden, "Feds Defend Bubble Meddle," *Science* 208:4444 (May 9, 1980), 577.

40. Holden, "Feds Defend Bubble Meddle."

41. For the quotes from Press, see Lubkin, "Government Bars Soviets," 81.

42. Lubkin, 82.

43. Corson Report, 182–88, for the Gololobov correspondence.

44. Corson Report, 103.

45. Corson Report, 172.

46. These items were also used to build a questionnaire that covered the campus activities of foreign students authorized to study in the United States by the Department of State. It is reproduced in appendix B of John Shattuck, "Federal Restrictions on the Free Flow of Academic Information and Ideas," *Government Information Quarterly* 3:1 (1986), 5–29, at 26.

47. Corson Report, 172.

48. Bruce B. Weyhrauch, "Operation Exodus: The United States Government's Program to Intercept Illegal Exports of High Technology," *Computer/Law Journal* 7:2 (1986), 203–25, 211.

49. Harold C. Relyea, *Silencing Science: National Security Controls and Scientific Communication* (New Jersey: Ablex, 1994), 120.

50. Letter from Five University Presidents, Corson Report, 136–39. All quotations that follow in the next two paragraphs are from this letter.

51. Corson Report, 136–39.

52. Corson Report, 137.

53. Corson Report, 175, 181.

54. Corson Report, 188.

55. Paul E. Gray, "Technology Transfer at Issue: The Academic Viewpoint," *IEEE Spectrum*, May 1982, 64–68. Gray, the president of MIT, noted that the line between basic and applied research was very difficult to draw in the program, while drawing a line between US citizens and immigrant aliens disregarded the international character of American universities.

56. Letter from Five University Presidents, 137.

57. Corson Report, 139.

58. David Wilson, "National Security Control of Technological Information," *Jurimetrics* 25:2 (1989), 109–29, at 120.

59. Unless specified otherwise, this section is based on the *Report of the Public Cryptography Study Group*, and its minority report, by George I. Davida, *The Case against Restraints on Non-governmental Research in Cryptography*, both included in the documentary evidence presented to the Congressional Subcommittee on Courts, Civil Liberties, 561–75, and 576–82.

60. Mitchel. B. Wallerstein, "Voluntary Restraints on Research with National Security Implications: The Case of Cryptography, 1975–1982," Corson Report, appendix E, 120–25.

61. Mary M. Cheh, "Government Controls of Private Ideas: Striking a Balance between Scientific Freedom and National Security," *Jurimetrics* 23:1 (1982), 1–32, at 13–14.

62. In the early 1980s there were only a couple of dozen researchers in cryptography in the United States producing about one hundred papers a year, most of which had no direct impact on the NSA's mission (of the first forty-six papers

voluntarily submitted to the NSA for review using the study group's process, only two were deemed to have national security implications). Wallerstein, "Case of Cryptography," Corson Report, appendix E, 120–25.

63. The NSF-NSA agreement was tested in summer 1980, while the study group was still deliberating. Inman informed the NSF that the probable results in two of the grant proposals that had been sent to it for review would, if openly published, seriously impugn national security. The NSF approved the awards, informing one of the grantees that he was responsible for immediately notifying the NSF program official of "any data, information or materials developed under this grant which may require classification." The grantee would "allow the NSF the option to review such materials" and would "defer dissemination or publication pending the review and determination that the results are not classified"—providing the determination was completed within sixty days of the NSF's receipt of the proposal. Wallerstein, "Case of Cryptography," 122. NSF sponsorship was thus conditional on the researcher accepting responsibility for identifying potentially classifiable material produced by the research, and on accepting prepublication review by the program official of any such results before they were disseminated. It should be noted that, in defining this policy, the NSF was simply following its responsibilities under EO 12065. This directive stipulated that if any employee or contractor of an agency not having classification authority (here the NSF) originated information that may require classification, that information had to be submitted to the relevant agency (here the NSA) for review.

64. *Report of the Public Cryptography Study Group*, 4.

65. The Department of Justice told Frank Press, then science adviser to President Carter, that the use of ITAR to restrict the dissemination of cryptographic information developed by scientists and mathematicians in the private sector, independent of government supervision or support, amounted to unconstitutional prior restraint. US Department of Justice, Memorandum Larry L. Simms to Davis R. Robinson, "Revised Proposed International Traffic in Arms Regulations (ITAR)," July 5, 1984, www.eff.org/files/filenode/bernstein/exhibit.a.tien.html (accessed November 10, 2018).

66. *Report of the Public Cryptography Study Group*, 4.

67. Wallerstein, "Case of Cryptography," 122; Paul E. Gray, "Technology Transfer at Issue: The Academic Viewpoint," *IEEE Spectrum*, May 1982, 64–68. Reagan's EO 12356 of April 1982, to be discussed later, placed the responsibility for the initial judgment about the sensitivity of the research results squarely on the grantee, as in the NSF directive.

68. *Report of the Public Cryptography Study Group*, 14.

69. Davida, *Case against Restraints*, 582.

70. Relyea, *Silencing Science*, 79–80, for both comments.

71. Office of the Undersecretary of Defense for Research and Engineering, *Report of the Defense Science Board Task Force on University Responsiveness to National Security Requirements*, Washington, DC, January 1982.

72. Genevieve J. Kenzo, *Defense Basic Research Priorities: Funding and Policy Issues*, CRS Report, Washington, DC, October 24, 1990, notes that a subsequent amendment allowed for exceptions to the restrictive mandate.

73. *University Responsiveness to National Security*, 1-2.

74. *University Responsiveness to National Security*, 1-1.

75. *University Responsiveness to National Security*, 5-1.

76. *University Responsiveness to National Security*, 5-5, 2-8.

77. *University Responsiveness to National Security*, 4-2.

78. *University Responsiveness to National Security*, vi.

79. *University Responsiveness to National Security*, 4-4.

80. *University Responsiveness to National Security*, 4-3.

81. *University Responsiveness to National Security*, 4-4

82. *University Responsiveness to National Security*, 4-7.

83. *University Responsiveness to National Security*, 4-5.

84. The DoD suggested that the same procedure could eventually be invoked by other federal agencies that sponsored research (NSF, NASA, DoE, etc.). DoD would take the burden of determining military criticality off the shoulders of the other agencies (since the MCTL was classified), establish the pertinence of ITAR and EAR to the research being supported, and help draft the initial contract guidelines between the government and the university. There were likely to be far fewer contracts affected, percentage-wise, since non-DoD-sponsored research was generally not militarily critical and not as directly linked to process or utilization technology. These controls also applied to non–federally funded research, of course. If need be, the DoD could perhaps also serve as a consultant and adviser to private industry, foundations, and so on, just as it would to a government agency, though here the government lacked the lever provided by having contractual ties with the researcher.

85. *University Responsiveness to National Security*, vii.

86. Adm. Bobby Inman, "Classifying Science: A Government Proposal," *Aviation Week and Space Technology*, February 8, 1982, 10, 11, 82, for this and all quotes from Inman's speech.

87. Christopher Paine (Federation of Atomic Scientists), "Admiral Inman's Tidal Wave," *Bulletin of the Atomic Scientists* 38:3 (1982), 3–6. See also Rosemary Chalk, "Security and Scientific Communication," *Bulletin of the Atomic Scientists* 39:7 (1983), 19–23, at 19.

88. Corson Report, 10.

89. William D. Carey, "Classifying Science: A Scientist's Objection," *Aviation Week and Space Technology*, February 8, 1982, 10, 11, 82, for this and all quotes from Careys's speech.

90. Multiple authors will be cited in what follows. Of crucial relevance is James R. Ferguson, "National Security Controls on Technological Knowledge: A Constitutional Perspective," *Science, Technology and Human Values* 10:2 (1985), 87–98. See also James R. Ferguson. "Scientific Freedom, National Security and the

First Amendment," *Science* 221 (August 12, 1983), 622; James R. Ferguson, "Scientific and Technological Expression: A Problem in First Amendment Theory," *Harvard Civil Rights—Civil Liberties Law Review* 16:2 (1981), 519–60, at 525. In 1985 Ferguson was an attorney with the Department of Justice in Chicago. His analyses of these issues stand out for their lucidity, brevity, and coherence.

91. Robert F. Ladenson, "Scientific and Technical Information, National Security and the First Amendment: A Jurisprudential Inquiry," *Public Affairs Quarterly* 1:2 (1987), 1–20, at 3–4.

92. Ferguson, "National Security Controls on Technological Knowledge," 90.

93. Ferguson, 90.

94. Ferguson, "Scientific and Technological Expression," 519; Ladenson, "National Security and the First Amendment," 4–5. For more detailed discussions of the "Progressive" case, see Ian M. Dumain, "No Secret, No Defense: United States v. Progressive," *Cardozo Law Review* 26:4 (2005), 1323–36. Janet M. Nesse, "United States v. Progressive, Inc.: The National Security and Free Speech Conflict," *William and Mary Law Review* 22:1 (1980), 141–60. The trigger of the case, the November 1979 issue of the *Progressive*, can be found at https://progressive.org/downloads/2722/download/1179.pdf (accessed June 8, 2020). See also the court rulings on this case: *United States v. Progressive Inc.*, 467 F. Supp. 990 (W.D. Wis. 1979), and 486 F. Supp. 5 (W.D. Wis. 1979).

95. Ferguson argues that all other types of technological knowledge promoted the three values that the First Amendment was intended to protect: the social interest in the free circulation of ideas, an individual interest in self-expression (because speech is a mode of self-fulfillment), and the political interest of providing the information that citizens in a democratic society need to form opinions on matters of public importance. Note that he is staking out a legal position that could of course be challenged as being too generous to the risks posed by the dissemination of technological knowledge.

96. Thomas I. Emerson, "First Amendment Doctrine and the Burger Court," *California Law Review* 68:3 (1980), 422–89, at 454.

97. For more insights on this, see the discussion of *United States v. Edler* in Cheh, "Government Control of Private Ideas," 7–8; Roger Funk, "National Security Controls on the Dissemination of Privately Generated Scientific Information," *UCLA Law Review* 30 (1982–83), 405–54, at 427–28; Congress of the US, Office of Technology Assessment, *Science, Technology and the First Amendment*, Special Report (Washington, DC: Government Printing Office, 1987).

98. The Defense University forum comprised eight university presidents, three heads of educational organizations, and nine members from defense agencies. Harold C. Relyea, *National Security Controls and Scientific Information, Updated 02/08/85*, Congressional Research Service Issue Brief, 1985, CRS-10.

99. David A. Wilson, "National Security Control of Technological Information," *Jurimetrics* 25:2 (1985), 109–29, at 116.

100. Edward Gerjuoy, "Embargo on Ideas: The Reagan Isolationism," *Bulletin of the Atomic Scientists* 38:9 (1982), 31–37, at 31.

101. Edward Gerjuoy, "Controls on Scientific Information Exports," *Yale Law and Policy Review* 3:2 (1985), 447–78, at 472.

102. The Association of University Professors rejected this analogy, arguing that it was one thing voluntarily to submit to prepublication restraints to protect a patentable invention and to file it on a timely basis. It was quite another to do so because the government wanted to determine whether an article should be published, in whole or in part. The former was acceptable, the latter "should be unacceptable to a scientist and a university." See Robert W. Rosenzweig, "Research as Intellectual Property: Influences within the University," *Science, Technology and Human Values* 10:2 (1985), 41–48, at 48.

103. *United States v. Edler Industries, Inc.*, 579 F. 2d 516 (1978), https://law.re source.org/pub/us/case/reporter/F2/579/579.F2d.516.76-3370.html (accessed June 8, 2020).

104. For the following, see Cheh, "Government Control of Private Ideas," 7–8; Funk, "National Security Controls," 427–28; Office of Technology Assessment, *Science, Technology and the First Amendment*, 49–53; Allen M. Shinn, "The First Amendment and the Export Laws: Free Speech on Scientific and Technical Matters," *George Washington Law Review* 58 (1989–90), 368–403.

105. Shinn, "First Amendment," 381.

106. Office of Technology Assessment, *Science, Technology, and the First Amendment*, 53.

107. Testimony to the hearings, *1984: Civil Liberties and the National Security State*, 334.

108. *Snepp v. United States*, 444 US 507 (1980), https://caselaw.findlaw.com/us -supreme-court/444/507.html (accessed June 8, 2020).

109. The "weight" of the burden of proof imposed on the government if it wanted to restrict the dissemination of technical data differed immensely between information in which the state had a proprietary stake and that in which it had a nonproprietary interest. A considerable effort would be needed to persuade the Supreme Court that the state had a "compelling interest" in controlling the dissemination of sensitive information in which it had a *nonproprietary* interest. It was not enough simply to appeal to the national interest. For example, the government would need to confirm that the information was not available from a third-party source, it would need to assess the gravity of the alleged harm that the state sought to avert, it would have to establish the likelihood that the recipient nation could use the data to develop the technology in question, and it would have to argue that the acquired technical data was subject to "near-term application" (cf. Ferguson, "National Security Controls on Technological Knowledge").

110. Ferguson, "National Security Controls on Technological Knowledge," 96.

Chapter 6

1. David A. Wilson, testimony to *Technology Transfer, Hearings before the Technology Transfer Panel of the Committee on Armed Services, House of Representatives*, 98th Congress, 1st Session, June 9, 21, 23, July 13, 14, 1983 (Washington, DC, Government Printing Office, 1984), 255–56.

2. Office of the Undersecretary of Defense for Research and Engineering, *Report of the Defense Science Board Task Force on University Responsiveness to National Security Requirements*, Washington, DC, January 1982, xi. The three were the American Council on Education, the Association of American Universities, and the National Association of State Universities and Land Grant Colleges; see David Wilson, "National Security Control of Technological Information," *Jurimetrics* 25:2 (1989), 121.

3. Wilson, "National Security Control," 121. It was not the only audience, of course. That included, on the government's side, the intelligence community, the State and Commerce Departments, and the Justice Department, and on the scientific community's side, in addition to those already involved in the forum and the National Academy complex, there was the AAAS, the American Physical Society, and the IEEE, all of whom had taken explicit positions on the issues involved.

4. This paragraph is based on the written testimony of F. Karl Willenbrock, Chairman of the IEEE Technology Transfer Committee, to the hearings, *1984: Civil Liberties and the National Security State*, Congressional Subcommittee on Courts, Civil Liberties, and the Administration of Justice of the Committee on the Judiciary, 98th Congress, November 2, 3, 1983, and January 24, April 5, and September 26, 1984 (Washington, DC: Government Printing Office, 1984), 105–17.

5. The possible abuse of academic programs by people from China who were not bona fide students was increasingly causing concern. As we mentioned earlier, five people were told to leave their seats just before their plane left New York for China, suspected of exporting sensitive technical data. Relationships with China are analyzed in depth in chapter 10.

6. National Academy of Engineering, *Scientific Communication and National Security* (Washington, DC: National Academies Press, 1982) (Corson Report), 14, 39.

7. Statement of Admiral B. R. Inman for the May 11, 1982, Senate Governmental Affairs Subcommittee on Investigations, *Hearing on Technology Transfer*, appendix H, Corson Report, 140, and its summary in the report itself, at 17.

8. Corson Report, 19. Even then many of the "recent episodes" reported by the intelligence community that made up this "small percentage" were no clear threat to national security: thus "the visitor's technical activities and studies went beyond his or her agreed field of study; (b) the visitor's time was poorly accounted for, including reports of excessive time spent collecting information (e.g. in the library) not related to his or her field of study; (c) the visitor ... attempted to evade visa or

exchange restrictions imposed on his or her itinerary." In fact there were only one or two incidents of "clearly illegal activities of an intelligence nature" by a foreign visitor (Corson Report, 17, 18).

9. Corson Report, 15.

10. Corson Report, 50.

11. Lara Baker, Testimony before the Permanent Subcommittee on Investigations of the Senate Committee on Government Affairs, *Hearing on Transfer of United States High Technology to the Soviet Union and Soviet Bloc Nations*, 97th Congress, May 4, 5, 6, 11, and 12, 1982 (Washington, DC: Government Printing Office, 1982), 340.

12. Corson Report, 41 (emphasis in the original).

13. Testimony at *Scientific Communications and National Security*, Hearings before the Subcommittee on Science, Research and Technology of the Subcommittee on Investigations and Oversight of the Congressional Committee on Science and Technology, 98th Congress, May 24, 1984, 96.

14. Statement of Admiral B. R. Inman, appendix H, Corson Report, 141.

15. Corson Report, 42.

16. Corson Report, 4.

17. Corson Report, 33.

18. Corson Report, 34.

19. Corson Report. 35.

20. Corson Report, 6.

21. General licenses were extended automatically without the need to actually apply to the government for a license or to acquire an official document authorizing the export general licenses are to be contrasted with validated licenses, issued by the Office of Export Administration in the Department of Commerce, which were granted on a case-by-case basis after considering a formal application and the justification for the "shipment" by the exporter. One variant, a General License GTDA allowed for data exports to any destination; the GTDR variant was more restrictive and took into consideration the nature of the technical data being exported and its destination, distinguishing between different categories of "countries of concern."

22. Rosemary Chalk, "Commentary on the NAS Report," *Science, Technology and Human Values* 8:1 (1983), 21–24, at 24.

23. David Dickson, *The New Politics of Science* (Chicago: University of Chicago Press, 1988), 112.

24. Memo for the Acting Director of Central Intelligence from the Chairman, Technology Transfer Intelligence Committee, September 27, 1982. CIA-RDR83 M00914R002200240005-1, https://www.cia.gov/library/readingroom/document/cia-rdp83m00914r002200240005-1 (accessed June 8, 2020).

25. Ronald Reagan, National Security Study Directive 14-82, "Scientific Communication and National Security," December 23, 1982, CIA-RDP90301013R000

300510002-7, https://www.cia.gov/library/readingroom/document/cia-rdp90b01013 r000300510002-7 (accessed June 8, 2020).

26. Mitchel B. Wallerstein, "Scientific Communication and National Security in 1984," *Science* 224:4648 (1984), 460–66, at 461.

27. John Prados, "The Strategic Defense Initiative: Between Strategy, Diplomacy and US Intelligence Estimates," in Leopoldo Nuti, ed., *The Crisis of Détente in Europe: From Helsinki to Gorbachev, 1975–1985* (London: Routledge, 2008), 86–98.

28. EO 12356, https://www.archives.gov/federal-register/codification/executive-order/12356.html (accessed June 8, 2020).

29. John Shattuck, "Federal Restrictions on the Free Flow of Academic Information and Ideas," *Government Information Quarterly* 3:1 (1986), 5–29, at 9–11.

30. Robert A. Rosenbaum, Morton J. Tenzer, Stephen H. Unger, William Van Alstyne, and Jonathan Knight, "Academic Freedom and the Classified Information System," *Science* 219:4582 (1983), 257–59, at 258.

31. Sissela Bok, Testimony at *1984, Civil Liberties and the National Security State*, 247.

32. John Shattuck, "Federal Restrictions on the Free Flow of Academic Information and Ideas," *Government Information Quarterly* 3:1 (1986), 5–29, includes the text of NSDD 84 and the agreement.

33. Robert A. Rosebaum, Morton J. Tenzer, Stephen H. Unger, William Van Alstyne, and Jonathan Knight, "Government Censorship and Academic Freedom," *Academe*, November–December 1983, 15a–17a, at 16a.

34. Testimony at, *Scientific Communications and National Security*, 63–67.

35. Ehrlich, Testimony at *Scientific Communications and National Security*, 66.

36. Aerospace Industries Association of America, Inc., *Trade and R&D Policies: An Aerospace Industries Association Proposal*, January 1984, 13.

37. Wallerstein, "National Security in 1984," 461.

38. Wilson, "National Security Control," 124.

39. Paul Mann, "Export Policy Triggers Dispute," *Aviation Week and Space Technology*, December 19, 1983, 11–12. The issues were far broader than those isolated here, including relations with Cocom.

40. Wilson, "National Security Controls," 125.

41. Caspar Weinberger, Department of Defense Directive Number 2040.2, January 17, 1984, *International Transfers of Technology, Goods, Services and Munitions*, https://biotech.law.lsu.edu/blaw/dodd/corres/pdf2/d20402p.pdf (accessed March 6, 2019). See also Statement of Dr. Edith W. Martin, *Scientific Communications and National Security*, 135–50.

42. It would resolve DoD differences "regarding technical standards and definitions in the dissemination and exchange of technical information," DoD Directive 2040.2, 17.

43. Martin, Testimony at *Scientific Communications and National Security*, 149.

44. Irwin Goodwin, "Pentagon Lowers Heat on Science Secrecy," *Physics Today* 37:7 (July 1984), 57–59, 58.

45. Colin Norman, "Universities Gag on Research Controls," *Science* 224:4645 (April 13, 1984), 134.

46. Norman, 134.

47. Colin Norman, "DeLauer Questions DOD Censorship," *Science* 224:4648 (May 4, 1984), 471.

48. Stephen Unger, Testimony to the hearings, *1984: Civil Liberties and the National Security State*, 120–21.

49. Dale Corson, Testimony at, *Scientific Communications and National Security*, 48–52, 50.

50. Corson, 50 (emphasis added).

51. Goodwin, "Pentagon Lowers Heat," 58.

52. Martin, Testimony at *Scientific Communications and National Security*, 138.

53. Martin, 138.

54. Letter, George Keyworth to Robert C. MacFarlane, June 29, 1984, CIA-RDP87M00220R0001000800097.4.

55. Colin Norman, "Universities Prevail on Secrecy," *Science* 226 (October 26, 1984), 418. See also Letter, Caspar Weinberger to Richard Gowen, January 4, 1985, CIA-RDP87M00220R000100080009-4.

56. See also Gerjuoy, "Scientific Export Controls," 478.

57. NSDD 189, *National Policy on the Transfer of Scientific, Technical and Engineering Information*, September 21, 1985, https://fas.org/irp/offdocs/nsdd/nsdd-189.htm (accessed March 28, 2017).

58. David A. Wilson, "Federal Control of Information in Academic Science," *Jurimetrics* (1987), 283–96, at 295.

59. Mario Daniels and John Krige, "Beyond the Reach of Regulation? 'Basic' and 'Applied' Research in the Early Cold War United States," *Technology and Culture* 59:2 (2018), 226–50.

60. Office of Technology Assessment, *Science, Technology and the First Amendment: Special Report* (Washington, DC: Government Printing Office, 1987).

61. Barbara J. Culliton, "Pajaro Dunes: A Search for Consensus," *Science* 216 (4542), 155, 156, 158.

62. Dickson, *New Politics of Science*, 60. See also Mirowski, *Science-Mart*; and Slaughter and Rhoades, *Academic Capitalism and the New Economy*.

63. John J. Young, "Memorandum for Secretaries of the Military Departments: Contracted Fundamental Research," June 26, 2008; Ashton B. Carter, "Memorandum for Secretaries of the Military Departments: Fundamental Research," May 24, 2010, https://research.uci.edu/policy-library/export-control-policies/govt-fundamental-research-policy.pdf (accessed October 29, 2020). One reviewer remarked that Young's memo was contested, and that his text on fundamental research was removed, only to be reinstated by Carter.

64. Deemed Export Advisory Committee, *The Deemed Export Rule in the Era of Globalization*, December 20, 2007, https://fas.org/sgp/library/deemedexports.pdf (accessed June 12, 2020) (DEAC hereafter).

65. See John Krige, "Export Controls to Regulate Knowledge Acquisition in a Globalizing Economy," in John Krige, ed., *How Knowledge Moves: Writing the Transnational History of Science and Technology* (Chicago: University of Chicago Press, 2016), 62–92.

66. DEAC, 1.

67. DEAC, 56–57.

68. DEAC, 86.

69. Jessica Wang, "A State of Rumor: Low Knowledge, Nuclear Fear, and the Scientist as Security Risk," *Journal of Policy History* 28:3 (2016), 401–46, 420–21.

Chapter 7

1. *American Economic Power: Redefining National Security for the 1990's*, Hearings before the Joint Committee, part 1, November 9, 15, and 16, 1989 (Washington, DC: Government Printing Office, 1991), Opening Statement, Stephen J. Solarz, 1.

2. *American Economic Power*, 2.

3. *American Economic Power*, 2.

4. Beverly Crawford, "The New Security Dilemma under International Economic Interdependence," *Millennium: Journal of International Studies* 23:1 (1994), 25–55.

5. For an excellent overview, see Robert Gilpin, "Economic Interdependence and National Security in Historical Perspective," in *Economic Issues and National Security*, ed. Klaus Knorr and Frank N. Trager (Lawrence: Published for the National Security Education Program by the Regents Press of Kansas, 1977), 19–66.

6. John Krige, *Sharing Knowledge, Shaping Europe: US Technological Collaboration and Nonproliferation* (Cambridge, MA: MIT Press, 2016).

7. Michael Mastanduno, *Economic Containment: CoCom and the Politics of East-West Trade* (Ithaca, NY: Cornell University Press, 1992), 266–309, writes perceptively about the growing impact of export controls on West-West trade relations. In his interpretation, this development is, however, mainly understood as an effect of the tightening up of controls against Soviet legal and illegal acquisition of Western technology. Mastanduno briefly alludes to the "blurring of . . . national security concerns" and "economic (i.e. competitiveness) objectives in a fundamental way" but does not explore this route. Moreover, he does not at all talk about FDI regulations and their new and growing role in the export control system.

8. Matthew J. Baltz, "Institutionalizing Neoliberalism: CFIUS and the Governance of Inwards Foreign Direct Investment in the United States since 1975," *Review of International Political Economy* 24:5 (2017), 859–80; Aaron L. Friedberg,

"The End of Autonomy: The United States after Five Decades," *Daedalus* 120:4 (1991), 69–90.

9. The concept of "securitization" has been developed since the 1980s by the so-called Copenhagen school of international relations. See Barry Buzan, *People, States, and Fear: The National Security Problem in International Relations* (Brighton: Wheatsheaf Books, 1983); Barry Buzan and Lene Hansen, *The Evolution of International Security Studies* (Cambridge: Cambridge University Press, 2009); Barry Buzan, Ole Wæver, and Jaap de Wilde, *Security: A New Framework for Analysis* (Boulder: Lynne Rienner, 1998). Recently, the concept found favorable reception among historians, not the least in Germany, who stress its usefulness for the analysis of the development of security policies over time. Christopher Daase, "Die Historisierung der Sicherheit: Anmerkungen zur historischen Sicherheitsforschung aus politikwissenschaftlicher Sicht," *Geschichte und Gesellschaft* 38 (2012), 387–405; Eckart Conze, "Securitization: Gegenwartsdiagnose oder historischer Analyseansatz?," *Geschichte und Gesellschaft* 38 (2012), 453–67. Historical security studies focus on the "change of security terms, security concepts and security perceptions" (Conze, "Securitization," 455), on the changes of "what is deemed in need of protection" and the development of "strategies which are deemed appropriate for this safeguard" (Daase, "Historisierung der Sicherheit," 396). Our approach follows these lines and the idea that "securitization can be analyzed by a look at political discourses and practices." What security is supposed to mean is always controversial because it usually causes a clash between different moral and ideological concepts (Conze, "Securitization," 456–57).

10. Testimony, Paul Kennedy, at Hearing, *American Economic Power: Redefining National Security for the 1990's*, 23–111.

11. An excellent exception is Joseph. J. Romm, *Defining National Security: The Nonmilitary Aspects* (New York: Council of Foreign Relations Press, 1993) (which, however, is part of the debate analyzed here). Without reference to "economic security" but nevertheless very helpful: Andrew Preston, "Monsters Everywhere: A Genealogy of National Security," *Diplomatic History* 38:3 (2014), 477–500; Emily S. Rosenberg, "The Cold War and the Discourse of National Security," *Diplomatic History* 17:2 (1993), 277–84; Melvyn P. Leffler, "The American Conception of National Security and the Beginnings of the Cold War, 1945–48," *American Historical Review* 89:2 (1984), 346–81; David Jablonsky, "The State of the National Security State," *Parameters* 32 (2002/3), 4–20.

12. Michael Borrus, Wayne Sandholtz, Steve Weber, and John Zysman, "Prologue," in Wayne Sandholtz, Michael Borrus, John Zysman, Ken Conca, Jan Stowsky, Steven Vogel, and Steve Weber, *The Highest Stakes: The Economic Foundations of the Next Security System* (Oxford: Oxford University Press, 1992), 3–5.

13. Edward N. Luttwak, "From Geopolitics to Geo-economics: Logic of Conflict, Grammar of Commerce," *National Interest* 20 (1990), 17–23, at 17, 19.

14. Crawford, "New Security Dilemma," 25.

15. James R. Golden, *Economics and National Strategy in the Information Age: Global Networks, Technology Policy, and Cooperative Competition* (Westport, CT: Praeger, 1994), 3; for Golden's arguments against "economic security" policies, see esp. 7–8.

16. For a critical view, see Vincent Cable, "What Is International Economic Security?," *International Affairs* 71:2 (1995), 305–24.

17. Samuel P. Huntington, "Why International Primacy Matters," *International Security* 17:4 (1993), 68–83, at 72.

18. Golden, *Economics*, 7.

19. Samuel P. Huntington, "Advice for a Democratic President: The Economic Renewal of America," *National Interest* 27 (1992), 14–19, at 16.

20. Michael Borrus and John Zysman, "Industrial Competitiveness and American National Security," in Wayne Sandholtz et al., *Highest Stakes*, 7–52, at 9. For some more reflection on this changing perspective on markets, see Crawford, "New Security Dilemma," 28–29.

21. Huntington, "Why International Primacy Matters," 68–60. For a broader, more balanced discussion, see Michael Mastanduno, "Do Relative Gains Matter? America's Response to Japanese Industrial Policy," *International Security* 16:1 (1991), 73–113.

22. Differentiated introductions to these topics are offered by Aaron L. Friedberg, "The Strategic Implications of Relative Economic Decline," *Political Science Quarterly* 104:3 (1989), 401–31; Richard R. Nelson and Gavin Wright, "The Rise and Fall of American Technological Leadership: The Postwar Era in Historical Perspective," *Journal of Economic Literature* 30:4 (1992), 1931–64.

23. Romm, *Defining National Security*, 61.

24. Stephen E. Haynes, Michael M. Hutchison, and Raymond F. Mikesell, "Japanese Financial Policies and the US Trade Deficit," *Essays in International Finance* 162 (1986), 3 (table 1).

25. Paul Kennedy, *The Rise and Fall of the Great Powers: Economic Change and Military Conflict from 1500 to 2000* (New York: Random House, 1987), xv–xvi (emphasis in the original).

26. Kennedy, 515, 532–33 (emphasis in the original).

27. Kennedy, 519. Ironically, one of the numerous authors who criticized Kennedy's obituary for American hegemony as premature was Samuel P. Huntington. In 1988, he was much more optimistic about the state of US power and much less worried about Japan's rise than a mere three years later. See his antideclinist "The US: Decline or Renewal?," *Foreign Affairs* 67:2 (1988), 76–96. See also (published prior to Kennedy's book) Susan Strange, "The Persistent Myth of Lost Hegemony," *International Organization* 51:4 (1987), 551–74.

28. Kent H. Hughes, *Building the Next American Century: The Past and Future of American Economic Competitiveness* (Washington, DC: Woodrow Wilson Center Press, 2005).

29. Borrus and Zysman, "Industrial Competitiveness and American National Security," 7.

30. See, e.g., Clyde V. Prestowitz Jr., *Trading Places: How We Allowed Japan to Take the Lead* (New York: Basic Books, 1988).

31. Huntington, "Why International Primacy Matters," 75–76.

32. Huntington, 72.

33. Richard J. Samuels, "Reinventing Security: Japan since Meiji," *Daedalus* 120:4 (1991), 47–68, at 47–48.

34. Samuels, 61.

35. Samuels, 53, 60–61. Borrus and Zysman, "Industrial Competitiveness and American National Security," 20, also describe Japan as a "learning economy."

36. Huntington, "Advice for a Democratic President," 16–17.

37. J. David Richardson, "The Political Economy of Strategic Trade Policy," *International Organization* 44:1 (1990), 107–35; Paul R. Krugman, ed., *Strategic Trade Policy and the New International Economics* (Cambridge, MA: MIT Press, 1986).

38. For a good overview of many aspects of this debate, see Theodore H. Moran, *American Economic Policy and National Security* (New York: Council of Foreign Relations Press, 1993).

39. A good example from a mercantilist, albeit nonrealist, perspective is Robert Kuttner, *The End of Laissez-Faire: National Purpose and the Global Economy after the Cold War* (New York: Alfred A. Knopf, 1991). For the larger context of this discussion, see his chapter 6, "Security: Military or Economic?"

40. Aaron L. Friedberg, "The End of Autonomy: The United States after Five Decades," *Daedalus* 120:4 (1991), 69–90, at 69–71; Aaron L. Friedberg, "The Changing Relationship between Economics and National Security," *Political Science Quarterly* 106:2 (1991), 265–76, at 268–71.

41. For an introduction to the impact of globalization since the 1970s, see Niall Ferguson, Charles S. Maier, Erez Manela, and Daniel J. Sargent, eds., *The Shock of the Global: The 1970s in Perspective* (Cambridge, MA: Harvard University Press, 2010).

42. A similar conclusion is reached in the very thorough study by Erik R. Pages, *Responding to Defense Dependence: Policy Ideas and the American Defense Industrial Base* (Westport, CT: Praeger, 1996), 145. Pages criticizes realism on a theoretical level but does not historicize its role in shaping the discourses and policies he analyzes.

43. Souvik Saha, "CFIUS Now Made in China: Dueling National Security Frameworks as a Countermeasure to Economic Espionage in the Age of Globalization," *Northwestern Journal of International Law and Business* 33:199 (2012), 199–234, at 209; James K. Jackson, *Committee on Foreign Investment (CFIUS)*, Congressional Research Service, January 16, 2018, 3–5.

44. The amendment was named after its main congressional sponsors, Senator J. James Exon (D-WV) and Representative James Florio (D-NJ).

45. Exon-Florio Amendment, Omnibus Trade and Competitiveness Act of 1988, 5021, PL 100-418, *US Statutes at Large* 102 (1988), 1107.

46. Bill Emmott, *Japanophobia: The Myth of the Invincible Japanese* (New York, 1993, 57).

47. Edward M. Graham and David M. Marchick, *US National Security and Foreign Direct Investment* (Washington, DC: Institute for International Economics, 2006), 21–22; Edward M. Graham and Paul Krugman, *Foreign Direct Investment in the United States*, 3rd ed. (Washington, DC: Institute for International Economics, 1995), 14–15; Edward M. Graham, "Foreign Investment in the United States and US Interests," *Science* 254:5039 (1991), 1740–45, at 1740–41.

48. Jonathan Crystal, *Unwanted Company: Foreign Investment in American Industries* (Ithaca, NY: Cornell University Press, 2003), 1.

49. Graham, "Foreign Investment," 1741–42.

50. Crystal, *Unwanted Company*, 1.

51. Graham, "Foreign Investment," 1741, table 1.

52. See, e.g., Emmot, *Japanophobia*, 1–64.

53. See Norman J. Glickman and Douglas P. Woodward, *The New Competitors: How Foreign Investors Are Changing the US Economy* (New York: Basic Books, 1989), 13–19; Prestowitz, *Trading Places*. For another example of anti-Japan literature, see William R. Nester, *American Power, the New World Order and the Japanese Challenge* (Houndsmills: Macmillan, 1993).

54. William C. Rempel and Donna K. H. Walters, "Trade War: When Chips Were Down," *Los Angeles Times*, November 30, 1987, 1.

55. Jose E. Alvarez, "Political Protectionism and the United States International Investment Obligations in Conflicts: The Hazards of Exon-Florio," *Virginia Journal of International Law* 30:1 (1989), 1–187, at 61nn331 and 334.

56. Graham, "Foreign Investment," 1741.

57. For a very thorough, in-depth discussion of this concept and US defense industrial base policies in the 1980s and early 1990s, see Pages, *Responding to Defense Dependence*.

58. The following arguments are derived from *Final Report of the Defense Science Board 1988 Summer Study on the Defense and Industrial Base*, vols. 1 and 2, Washington, DC, 1988, http://www.dtic.mil/dtic/tr/fulltext/u2/a202469.pdf; http://www.dtic.mil/dtic/tr/fulltext/u2/a212698.pdf; *Report of the Defense Science Board Task Force on Defense Semiconductor Dependency*, Washington, DC, 1987, http://www.dtic.mil/dtic/tr/fulltext/u2/a178284.pdf (all accessed August 14, 2018).

59. *Final Report of the Defense Science Board 1988 Summer Study*, 1:1.

60. Alex Roland, *The Military-Industrial Complex* (Washington, DC: American Historical Association, 2001); Paul A. C. Koistinen, *State of War: The Political Economy of American Warfare, 1945–2011* (Lawrence: University Press of Kansas, 2012); James Ledbetter, *Unwarranted Influence: Dwight D. Eisenhower and the Military-Industrial Complex* (New Haven, CT: Yale University Press, 2011).

61. *Final Report of the Defense Science Board 1988 Summer Study*, vol. 1, appendix A, "Draft Presidential Directive: The National Industrial and Technological Base," A-1.

62. Paul N. Edwards, *The Closed World: Computers and the Politics of Discourse in Cold War America* (Cambridge, MA: MIT Press, 1996).

63. Kenneth Flamm, *Creating the Computer: Government, Industry, and High Technology* (Washington, DC, 1988).

64. On Japan, see Marie Anchordoguy, *Computers Inc.: Japan's Challenge to IBM* (Cambridge, MA: Harvard University Press, 1989).

65. Jackson, *Committee on Foreign Investment*, 5.

66. *Report of the Defense Science Board Task Force on Defense Semiconductor Dependency*, 1.

67. *Report of the Defense Science Board Task Force on Defense Semiconductor Dependency*, cover letter, Charles A. Fowler, Memorandum for the Secretary of Defense, February 9, 1987.

68. *Report of the Defense Science Board Task Force on Defense Semiconductor Dependency*, 2–3, 5–9, 60, quote at 9.

69. *Report of the Defense Science Board Task Force on Defense Semiconductor Dependency*, 4.

70. *Report of the Defense Science Board Task Force on Defense Semiconductor Dependency*, 2.

71. *Report of the Defense Science Board Task Force on Defense Semiconductor Dependency*, 66.

72. See *Report of the Defense Science Board Task Force on Defense Semiconductor Dependency*, 3.

73. *Report of the Defense Science Board Task Force on Defense Semiconductor Dependency*, 66.

74. See *Report of the Defense Science Board Task Force on Defense Semiconductor Dependency*, 2, 10.

75. *Report of the Defense Science Board Task Force on Defense Semiconductor Dependency*, 3.

76. *Report of the Defense Science Board Task Force on Defense Semiconductor Dependency*, 4.

77. *Report of the Defense Science Board Task Force on Defense Semiconductor Dependency*, 9.

78. Donald MacKenzie and Graham Spinardi, "Tacit Knowledge, Weapons Design, and the Uninvention of Nuclear Weapons," *American Journal of Sociology* 101:1 (1995), 44–99; Michael Polanyi, *The Tacit Dimension* (Garden City, NY: Doubleday, 1966).

79. *Report of the Defense Science Board Task Force on Defense Semiconductor Dependency*, 10–11.

80. Page, *Responding to Defense Dependence*, 89–107; Larry D. Browning and

Judy C. Shetler, *Sematech: Saving the US Semiconductor Industry* (College Station: Texas A&M University Press, 2000).

81. Page, *Responding to Defense Dependence*, 80–81, 83, 103 (quote). For the broader context of these consortiums, see Office of Technology Assessment, *Making Things Better: Competing in Manufacturing* (Washington, DC: Government Printing Office, 1990).

82. Testimony, C. Scott Kulicke, in *Competitiveness of the US Semiconductor Industry: Hearing before the House Subcommittee on Commerce, Consumer Protection, and Competitiveness of the Committee on Energy and Commerce*, June 9, 1987 (Washington, DC: Government Printing Office, 1988), 58–59.

83. Exon-Florio Amendment, sec. 721 (d) (2); Testimony of Allan I. Mendelowitz, General Accounting Office in *Foreign Acquisitions of US Owned Companies: Hearing before the Subcommittee on International Finance and Monetary Policy of the Committee on Banking, Housing, and Urban Affairs, US Senate*, June 4, 1992 (Washington, DC: Government Printing Office, 1992), 64.

84. Testimony of Olin L. Wethington, Treasury Assistant Secretary for International Affairs, in *Foreign Acquisitions of US Owned Companies: Hearing*, 48, 51.

85. Statement, Mendelowitz (GAO), "National Security Review of Two Foreign Acquisitions in the Semiconductor Sector," before the Subcommittee Commerce, Consumer Protection and Competitiveness of the Committee on Energy and Commerce, House of Representatives June 13, 1990, 1, https://www.gao.gov/assets/110/103326.pdf (accessed March 24, 2019).

86. Statement, Mendelowitz (GAO), "The President's Decision to Order a Chinese Company's Divestiture of a Recently Acquired US Aircraft Part Manufacturer," before the Subcommittee Commerce, Consumer Protection and Competitiveness of the Committee on Energy and Commerce, House of Representatives, March 19, 1990, at 7, https://archive.gao.gov/t2pbat11/140886.pdf (accessed March 24, 2019).

87. Department of the Treasury, Office of International Investment, "31 CFR Part 800: Regulations Pertaining Mergers, Acquisitions, and Takeovers by Foreign Persons," *Federal Register* 56:225 (November 21, 1991), 58774–88, at 58785, sec. 800.402.

88. Statement of William N. Rudman, Deputy Undersecretary of Defense, Trade and Security Policy, in *Foreign Acquisition of Semi-gas Systems: Hearing before the Senate Subcommittee on Science, Technology, and Space of the Committee on Commerce, Science, and Transportation*, October 10, 1990 (Washington, DC: Government Printing Office, 1990), 24.

89. Testimony of Frederick W. Volcansek, Acting Assistant Secretary for Trade Development, Department of Commerce, in *Foreign Acquisitions of US Owned Companies: Hearing*, 58–61, quotes at 60–61. See also Testimony of Chester Paul Beach Jr., Acting General Counsel, Department of Defense, in *Foreign Acquisitions*, 54–57.

90. On CFIUS's present profile, see James K. Jackson, *The Committee on Foreign Investment in the United States (CFIUS)*, Congressional Research Service,

January 16, 2018, 18–19; Graham and Marchick, *US National Security*, 53–56; Brandt J. C. Pasco, "United States National Security Reviews of Foreign Direct Investment: From Classified Programmes to Critical Infrastructure, This Is What the Committee on Foreign Investment in the United States Cares About," *ICSID Review* 29:2 (2014), 350–71.

91. Beverly Crawford, "Changing Export Controls in an Interdependent World: Lessons from the Toshiba Case for the 1990s," in *Export Controls in Transition: Perspectives, Problems, and Prospects*, ed. Gary K. Bertsch and Steven Elliott-Gower (Durham, NC: Duke University Press, 1992), 249–90. Stephen D. Kelly, "Curbing Illegal Transfers of Foreign-Developed Critical High Technology from Cocom Nations to the Soviet Union: An Analysis of the Toshiba-Kongsberg Incident," *Boston College International and Comparative Law Review* 12:1 (1989), 181–223.

92. Raymond Vernon, Debora L. Spar, and Glenn Tobin, *Iron Triangles and Revolving Doors: Cases of US Foreign Economic Policymaking* (New York: Praeger, 1991), 58–59. Chapter 3 of this book is the best concise overview over the FSX controversy. For a more detailed, albeit less balanced, account, see Jeff Shear, *The Keys to the Kingdom: The FS-X Deal and the Selling of America's Future to Japan* (New York: Doubleday, 1994).

93. Defense Science Board, "Keeping Access to the Leading Edge" (June 1990), in *Foreign Investment in the United States, Hearings before the House Subcommittee on Commerce, Consumer, Protection, and Competitiveness of the Committee on Energy and Commerce*, June 13 and July 1990 (Washington, DC: Government Printing Office, 1991), 253–60, at 253, 255.

94. Shintaro Ishihara, *The Japan That Can Say No: Why Japan Will Be First among Equals* (New York: Simon and Schuster, 1991).

95. See, for example, Peter F. Drucker, "Our Irritable Friend," *New York Times*, January 13, 1991, 246; Ronald E. Goldsmith, "The Japan That Can Say 'No': The New US-Japan Relations Card," *Journal of the Academy of Marketing Science* 19:3 (1991), 269; Michael Lewis, "The Samurai behind the Bow," *Los Angeles Times*, January 20, 1991, 7.

96. Ishihara, *Japan That Can Say No*, 23; the DSB report is discussed at 21–22.

97. Ezra F. Vogel, "Foreword," in Ishihara, 7–10, at 9.

98. Ishihara, *Japan That Can Say No*, 21.

99. Ishihara, 43.

100. General Accounting Office, *International Trade: US Business Access to Certain Foreign State-of-the-Art Technology* (GAO/NSIAD-91-278) (Washington, DC: Government Printing Office, 1991).

101. DSB, "Keeping Access to the Leading Edge," 254. Office of Technology Assessment (OTA), *Arming Our Allies: Cooperation and Competition in Defense Technology* (Washington, DC: Government Printing Office, 1990), 36–37.

102. National Academy of Sciences (NAS), National Academy of Engineering, and Institute of Medicine, *US National Security Export Controls and Global Economic Competition* (Washington, DC: National Academy Press, 1987), 54.

103. National Academy of Sciences et al., 30.
104. National Academy of Sciences et al., 10–11.
105. National Academy of Sciences et al., 33 (quote), 150–66.
106. That is the gist of OTA, *Arming Our Allies*, 36: "The US strategy appears to be to reach agreement [within Cocom] on a small number of technologies that must be controlled, and then rigorously enforce that regime."
107. DSB, "Keeping Access to the Leading Edge," 257.
108. Paul Freedenberg, "The Commercial Perspective," in *Export Controls in Transition*, 37–58, at 50–56. For critical view of a former DoD strategic trade adviser, see Peter M. Leitner, *Decontrolling Strategic Technology, 1990–1992: Creating the Military Threats of the 21st Century* (Lanham, MD: University Press of America, 1995).
109. Henry R. Nau, "Conclusion: Export Controls in a Changing Strategic Context," in *Export Controls in Transition*, 317–35, at 317.
110. Eduardo Lachica, "Export for Computers are Softened," *Wall Street Journal*, April 24, 1992, A5; John E. Yang, "US Relaxes Restrictions on Technology," *Washington Post*, April 24, 1992, C2.
111. The French government had "a 58 percent equity stake in Thomson-CSF, as well as a majority stake in Credit Lyonnais, one of the banks underwriting the acquisition." Glenn J. McLoughlin (Congressional Research Service), "The LTV-Thomson-CSF Sale: Issues in National Security and Technology Transfer," June 8, 1992, 3. For a more detailed summary and discussion of the LTV-Thomson case, see Matthew D. Riven, "The Attempted Takeover of LTV by Thomson: Should the United States Regulate Inward Investment by Foreign State-Owned Enterprises?," *Emory International Law Review* 7 (1993), 759–91.
112. Statement, Linda Spencer (Economic Strategy Institute), in *Defense Department's Role in Reviewing Foreign Investment in US Defense Companies: Hearing before the House Investigations Subcommittee of the Committee on Armed Services*, August 12, 1992 (Washington, DC: Government Printing Office, 1993), 2–9, at 2, 7; McLoughlin, "LTV-Thomson-CSF Sale," 2.
113. Statement, Linda Spencer, 5–8.
114. Prepared Statement of Norman R. Augustine, in *Pending Transactions under the Exon-Florio Amendment, Hearing before the Senate Committee on Commerce, Science, and Transportation*, May 14, 1992 (Washington, DC: Government Printing Office, 1992), 51–63, at 61.
115. See, for example, Prepared Statement of Senator Robert Byrd on the Senate Floor, in *Pending Transactions under the Exon-Florio Amendment, Hearing*, 38–39.
116. SSAs stipulated that "the company's chairman of the board, all principal officers, and a majority of its board of directors must be cleared US citizens. A company cleared under an SSA must establish a permanent committee of the board of directors, known as the Defense Security Committee (DSC). Department of Defense-approved outside directors are selected to serve on the DSC. The

DSC is required to ensure that foreign nationals and representatives of the foreign interest are effectively precluded from access to classified and controlled unclassified information." Prepared Statement of Chester Paul Beach, Acting General Counsel, Department of Defense, in *Sale of LTV Missile and Aircraft Division, Hearings before the Investigations Subcommittee and the Defense Policy Panel of the House Committee on Armed Services*, May 14 and June 25, 1992 (Washington, DC: Government Printing Office, 1993), 8–10 at 9.

117. McLoughlin, "LTV-Thomson-CSF Sale," 2. The DoD's Defense Technology Security Administration (DTSA) claimed that the DIA report was much more ambiguous, even contradictory, and also stated it was not aware that France was involved in sales violating export controls. Statement, George Menas, Director of Strategic Policy, DTSA, Department of Defense, in *Defense Department's Role in Reviewing Foreign Investment in US Defense Companies: Hearing*, 65–67.

118. Testimony of Susan J. Tolchin, Professor of Public Administration, George Washington University, at *Foreign Direct Investment, the Exon-Florio Acquisition Review Process, and H.R. 2624, the Technology Preservation Act of 1991, to Amend the 1988 Exon-Florio Provision, Hearings before the House Subcommittee on Economic Stabilization of the Committee on Banking, Finance and Urban Affairs*, March 31 and April 2, 1992 (Washington, DC: Government Printing Office, 1992), 107–22, at 117. Statement, Linda Spencer, 6.

119. Martin Tolchin and Susan J. Tolchin, *Selling Our Security: The Erosion of America's Assets* (New York: Alfred A. Knopf, 1992), 69.

120. Opening Statement, Senator J. James Exon, in *Pending Transactions under the Exon-Florio Amendment, Hearing*, 2.

121. Prepared Statement of Senator Robert Byrd on the Senate Floor, in *Pending Transactions under the Exon-Florio Amendment, Hearing*, 38.

122. Riven, "Attempted Takeover of LTV by Thomson," 763; Statement, Linda M. Spencer, 3.

123. PL 102-484, National Defense Reauthorization Act for the Fiscal Year 1993, sec. 837: Defense Production Act Amendments, in *National Defense Reauthorization Act for the Fiscal Year 1993: Conference Report to Accompany H.R. 5006* (Washington, DC: Government Printing Office, 1992), 155–56.

124. Matthew R. Byrne, "Protecting National Security and Promoting Foreign Investment: Maintaining the Exon-Florio Balance," *Ohio State Law Journal* 67 (2006), 849–910, at 868.

125. For similar arguments, see "Report of Defense Science Board Task Force on Foreign Ownership and Control of US Industry, June 1990," in *Defense Department's Role in Reviewing Foreign Investment in US Defense Companies, Hearing*, 11–27, at 23.

126. Hugo Meijer, *Trading with the Enemy: The Making of US Export Control Policy toward the People's Republic of China* (Oxford: Oxford University Press, 2016), 64.

Chapter 8

1. B. R. Inman and Daniel F. Burton, "Technology and Competitiveness: The New Policy Frontier," *Foreign Affairs* 69:2 (1990), 116–34, 133.

2. Mitchel B. Wallerstein, testimony, *US Export Control and Non-proliferation Policy and the Role and Responsibility of the Department of Defense, Hearing before the Committee on Armed Services*, US Senate, 105th Congress, July 9, 1998 (Washington, DC: Government Printing Office, 1998), 2.

3. Richard T. Cupitt, *Reluctant Champions: US Presidential Policy and Strategic Export Controls* (New York: Routledge, 2000), 119.

4. Barry Naughton, *The Chinese Economy: Transitions and Growth* (Cambridge, MA: MIT Press, 2007), 92–93.

5. Cited in Council on Competitiveness, *Economic Security: The Dollars and Sense of US Foreign Policy* (Washington, DC: Council on Competitiveness, 1994), 7.

6. *National Security Implications of Lowered Export Controls of Dual-Use Technologies and US Defense Capabilities, Hearing before the Senate Committee on Armed Services*, 104th Congress, May 11, 1995 (Washington, DC: Government Printing Office, 1996), 23.

7. For more detail, see Donald R. de Glopper, *China's Import of Foreign Technology, Survey and Chronology, 1 January–31 December 1984* (Washington, DC: Library of Congress, Federal Research Division, n.d.); Hans Heymann Jr. *China's Approach to Technology Acquisition: Part III—Summary Observations*, Rand Report Prepared for the Defense Advanced Research Projects Agency, R-1575-ARPA, February 1975 (Santa Monica, CA: Rand, 1975); Jahnavi Phalkey and Zuoyue Wang, "Planning for Science and Technology in China and India," *British Journal for the History of Science Themes* 1 (2016), 83–113; US Office of Technology Assessment, *Technology and East-West Trade* (Washington, DC: Government Printing Office, November 1979), chapter 11; Zuoyue Wang, "US-China Scientific Exchange, 1971–1989," in *Physicists in the Postwar Political Arena: Comparative Perspectives*, Collective conference papers, University of California, Berkeley, January 22–24, 1988, 244–70; Zuoyue Wang, "Transnational Science during the Cold War: The Case of Chinese-American Scientists," *Isis* 101 (2010), 367–77; Zuoyue Wang, "Physics in China in the Context of the Cold War," in Helmuth Trischler and Mark Walker, eds., *Physics and Politics: Research and Research Support in Twentieth Century Germany in International Perspective* (Stuttgart: Franz Steiner Verlag, 2010), 251–76; Zuoyue Wang, "The Cold War and the Reshaping of Transnational Science in China," in Naomi Oreskes and John Krige, eds., *Science and Technology in the Global Cold War* (Cambridge, MA: MIT Press, 2014), 343–69; Zuoyue Wang, "The Chinese Developmental State during the Cold War: The Making of the 1956 Twelve-Year Science and Technology Plan," *History and Technology* 31:3 (2015), 185–205.

8. Barry Naughton, *The Chinese Economy: Transitions and Growth* (Cambridge MA: MIT Press, 2007), 354.

9. Naughton, 69.
10. Wang, "US-China Scientific Exchange," 249.
11. Heymann, *China's Approach to Technology Acquisition*, 12.
12. As described in detail by Julian Gerwitz, *The Remaking of China: Myth, Modernization, and the Tumult of the 1980s* (Cambridge, MA: Harvard University Press, forthcoming).
13. Heymann, *China's Approach to Technology Acquisition*, tables 2 and 3.
14. Office of Technology Assessment, *Technology and East-West Trade* (Washington, DC: Government Printing Office, November 1979), 283.
15. Hugo Meijer, *Trading with the Enemy: The Making of US Export Control Policy toward the People's Republic of China* (Oxford: Oxford University Press, 2016), 51, 53.
16. Presidential Directive PD/NSC 43, *US-China Scientific and Technological Relationships*, November 3, 1978, https://www.jimmycarterlibrary.gov/assets/documents/directives/pd43.pdf (accessed May 10, 2021).
17. Odd Arne Westad, *Restless Empire: China and the World since 1750* (New York: Basic Books, 2012), 373.
18. OTA (US Congress, Office of Technology Assessment), *Technology Transfer to China* OTA-ISC-340 (Washington, DC: Government Printing Office, 1987), 3.
19. Robert G. Sutter, *US-China Relations: Perilous Past, Uncertain Present* (Lanham, MD: Rowan and Littlefield, 2018), 76.
20. Warren I. Cohen, *America's Response to China: A History of Sino-American Relations* (New York: Columbia University Press, 2010), 222.
21. Meijer, *Trading with the Enemy*, 57.
22. Meijer, 62.
23. Evan A. Feigenbaum, "Who's behind China's High-Technology 'Revolution'? How Bomb-Makers Remade Beijing's Priorities, Policies, and Institutions," *International Security* 24:1 (1999), 95–126, at 104.
24. Cited by Julian Gerwitz, *The Remaking of China*, MS, 203. Gerwitz has emphasized the exciting open-ended nature of the debate over the best path to follow to modernize China in the early 1980s.
25. OTA, *Technology Transfer to China*, 206.
26. Meijer, *Trading with the Enemy*, 84–85, 76–77.
27. Meijer, 85, 65.
28. Meijer, 69–71.
29. Meijer, 85.
30. Meijer, 89–99.
31. Cited by Meijer, 91.
32. Testimony of Eugene T. McAllister, Assistant Secretary, Bureau of Economic and Business Affairs, Department of State, at *Hearings before the Committee on Science, Space, and Technology: The Administration's Decision to License the Chinese Long March Launch Vehicle*, US House of Representatives,

102nd Session, September 23, 27 1988 (Washington, DC: Government Printing Office, 1988), 25–26. These were detailed the next day in Testimony of Karl D. Jackson, Deputy Assistant Secretary of Defense (Asian and Pacific Affairs), *Hearings before the Subcommittees on Arms Control, International Security, and Science, on Asian and Pacific Affairs, and on International Economic Policy and Trade of the Committee on Foreign Affairs: Proposed Sale and Launch of United States Satellites on Chinese Missiles*, House of Representatives, 100th Congress, September 28, 1988 (Washington, DC: Government Printing Office, 1989), 32–33.

33. Meijer, *Trading with the Enemy*, 89.
34. Heymann, *China's Approach to Technology Acquisition*, 13.
35. Cited by Mary Brown Bullock, "The Effects of Tiananmen on China's International Scientific and Educational Cooperation," *Problems of Reforms, Modernization, and Independence* 2:611–28, at 612.
36. Meijer, *Trading with the Enemy*, 85–88.
37. OTA, *Technology Transfer to China*, 104.
38. De Glopper, *China's Import of Foreign Technology*, 1, 2.
39. Heymann, *China's Approach to Technology Acquisition*, 26.
40. OTA, *Technology Transfer to China*, 41.
41. Meijer, *Trading with the Enemy*, 86–87. In fact this is taken directly from de Glopper, *China's Import of Foreign Technology*, vii, 1. In similar vein, a CIA report to Congress on the state of the Chinese economy in 1991 noted that "during the past decade China imported nearly $30 billion worth of technologically advanced machinery, shifting gradually from direct purchases of complete sets of equipment to technology-licensing arrangements and co-operative production ventures that transferred foreign technical and managerial know-how as well as hardware." CIA, *The Chinese Economy in 1990 and 1991: An Uncertain Recovery*, in *Global Economic and Technological Change, Hearings before the Subcommittee on Technology and National Security of the Congressional Joint Economic Committee*, 102nd Congress, May 16 and June 28, 1991 (Washington, DC: Government Printing Office, 1991), 289–356, 301.
42. David E. Edgerton, "The Contradiction of Techno-nationalism and Techno-globalism: A Historical Perspective," *New Global Studies* 1:1 (2007), 1–32.
43. Panel on the Impact of National Security Controls on International Technology Transfer, *Balancing the National Interest: US National Security Export Controls and Global Economic Competition* (Washington, DC: National Academy Press, 1987), 167. The panel was particularly critical of the increased weight of the DoD in reviewing license applications and insisted that the MCTL (Militarily Critical Technologies List) should be no more than a reference document for the DoD and should not be integrated into the US Commodity Control List that was used to shape the Cocom Control List.
44. Mastanduno, *Economic Containment*, 329.
45. Panel on the Impact of National Security Controls, *Balancing*, esp. chapter 8.

46. Panel on the Future Design and Implementation of US National Security Controls, *Finding Common Ground: US Export Controls in a Changed Global Environment* (Washington, DC: National Academy Press, 1991), 2.

47. Cupitt, *Reluctant Champions*, 120.

48. Mastanduno, *Economic Containment*, 333.

49. The following data is from *Export Controls and Non-proliferation Policy*, figure 4.1.

50. Panel on the Future Design, *Finding Common Ground*, 166.

51. Cupitt, *Reluctant Champions*, 142.

52. Cupitt, 148–49.

53. US Trade Representative Joseph Massey to Congress, in *Global Economic and Technological Change*, 357.

54. Westad, *Restless Empire*, 382.

55. Barry Naughton, Prepared Statement, in *Global Economic and Technological Change*, 448.

56. Meijer, *Trading with the Enemy*, 104–7.

57. *The Foreign Relations Authorization Act*, FY1990 and FY1991, H.R. 1487, PL 101-246.

58. GAO, Report to the Chairman, Committee on International Relations, House of Representatives, *Export Controls: Some Controls over Missile-Related Technology Exports to China Are Weak*, Report GAO/NSIAD-95-82, April 1995.

59. Richard Johnston, in *Global Economic and Technological Change*, 401.

60. Quoted in Meijer, *Trading with the Enemy*, 104.

61. Dianne E. Renack, *China: US Economic Sanctions*, CRS Report for Congress, CRS 96-272, October 1, 1997. See also *Chinese Proliferation of Weapons of Mass Destruction: Background and Analysis*, CRS Report for Congress, CRS 96-767, September 13, 1996; Shirley A. Kan, CRS Report for Congress 98-485; *China: Possible Missile Technology Transfers from US Satellite Export Policy—Actions and Chronology*, updated September 5, 2001; Shirley A. Kan, *China and Proliferation of Weapons of Mass Destruction and Missiles: Policy Issues*, Report for Congress, RL 31555, February 26, 2003.

62. Hearings, in *Global Economic and Technological Change*, 287.

63. Massey to Congress, in *Global Economic and Technological Change*, 357. See also Andrew C. Mertha, *The Politics of Piracy: Intellectual Property in Contemporary China* (Ithaca, NY: Cornell University Press, 2005).

64. Johnston, in *Global Economic and Technological Change*, 382.

65. CIA, *Chinese Economy in 1990 and 1991*, 321. Johnston did not try to quantify the role of intra-Western competition, notably with Japan and Western Europe, on the decline of US exports to China. Nor did he consider the impact of a sharp devaluation of the Chinese currency (the yuan or renminbi, RMB) over the decade on China's increasingly favorable balance of trade with the United States. The official exchange rate rose from 1.5 RMB/US$ in the late 1970s to

3.7 RMB/US$ in June 1986, to 4.7 RMB/US$ in December 1989, to 5.22 RMB/US$ in November 1990. See Nicholas R. Lardy, "Redefining US-China Economic Relations," in *Global Economic and Technological Change*, 431–39, at 435.

66. This authorized the president "to take all appropriate action, including retaliation, to obtain the removal of any act, policy, or practice of a foreign government that violates an international trade agreement or is unjustified, unreasonable, or discriminatory, and that burdens or restricts US commerce."

67. Massey, in *Global Economic and Technological Change*, 359.

68. Meijer, *Trading with the Enemy*, 84.

69. Johnston, *Global Economic and Political Change*, 387. Modes of engagement included joint equity ventures, contractual joint ventures, wholly owned subsidiaries, and joint development of offshore oil resources.

70. CIA, *Chinese Economy in 1990 and 1991*, 337–49.

71. Statement by Kent Wiedemann, in *Global Economic and Technological Change*, 428.

72. The White House, *Fact Sheet on Nonproliferation and Export Control Policy*, September 27, 1993, www.rertr.anl.gov/REFDOCS/PRES93NP.html (accessed November 6, 2019).

73. Cocom was abolished in March 1994 and replaced two years later by the Wassenaar Arrangement. It differed crucially from Cocom in that no member could veto another's transfer of controlled items.

74. As Michael Borrus and John Zysman put it in an influential collection of articles published in 1992, "the continual erosion of America's international economic position is a national security issue. We are past the point where America's security dominance can be exploited to impose more favorable terms of trade. Rather, we are confronted with precisely the reverse: how others can exploit terms of trade to impose dominance, how they can structure and play the international system through economic rules. In that world, the only secure America is a competitively able one. The United States must regain its competitive standing in trade, technology, and finance if it wants to be in a credible position to effectively manage the changing security system." Michael Borrus and John Zysman, "Industrial Competitiveness and American National Security," in Wayne Sandholtz, Michael Borrus, John Zysman, Ken Conca, Jay Stowsky, Steven Vogel, and Steve Weber, eds., *The Highest Stakes: The Economic Foundations of the Next Security System* (New York: Oxford University Press, 1992), 52.

75. Richard Van Atta and Michael Lippitz, *Transformation and Transition: DARPA's Role in Fostering an Emerging Revolution in Military Affairs*, vol. 1, *Overall Assessment*, IDA Paper P-3698 (Alexandra VA: Institute for Defense Analysis 2003), 9. For a brief history and assessment, see Jeffrey F. Collins and Andrew Futter, "Reflecting on the Revolution in Military Affairs: Implications for the Use of Force Today," *Russia in Global Affairs*, Valdai Papers, January 28, 2016. On the debate in the Soviet Union, see Dima P. Adamsky, "The Conceptual Battles of

the Central Front: The Air-Land Battle and the Soviet Military-Technical Revolution," in Leopoldi Nuti, ed., *The Crisis of Détente in Europe: From Helsinki to Gorbachev, 1975–1985* (Abingdon: Routledge, 2009), 150–62.

76. Frank Kendall, "Exploiting the Military Technical Revolution," *Strategic Review* 20:2 (1992), 23–30.

77. Clay Wilson, *Network Centric Warfare: Background and Oversight Issues for Congress*, CRS Report for Congress, 2014.

78. Van Atta and Lippitz, *Transformation and Transition*, 6.

79. Muhammad Sadiq and John C. McCain, eds., *The Gulf War Aftermath: An Environmental Tragedy* (Dordrecht: Kluwer, 1993), 264.

80. Wilson, *Network Centric*, CRS-4.

81. Michael J. Mazarr, "The Revolution in Military Affairs: A Framework for Defense Planning," Strategic Studies Institute, US Army War College, June 10, 1994, section 6.

82. Mazarr, section 7.

83. OSTP (Office of Science and Technology Policy), National Economic Council, National Security Council, *Second to None: Preserving America's Military Advantage Through Dual-Use Technology*, February 1995, 12.

84. OSTP, *Second to None*, 7.

85. OSTP, 1.

86. Department of Defense, *Critical Technologies Plan*, May 1991.

87. Cited in US Congress, Office of Technology Assessment, *Export Controls and Non-proliferation Policy*, OTA-ISS-596 (Washington, DC: Government Printing Office, May 1994), 30.

88. OSTP, Second to None, is the key source for what follows here.

89. Hearing before the Subcommittee on Technology, Environment, and Aviation of the Committee on Science, Space, and Technology, *America's Dual-Use Technology Future: Are We Prepared?* House of Representatives, Committee on Science, Space, and Technology, Subcommittee on Technology, Environment and Aviation, 103rd Congress, May 17, 1994 (Washington, DC: Government Printing Office, 1994).

90. Statement, Jones, Hearing before the Subcommittee on Technology, Environment, and Aviation of the Committee on Science, Space, and Technology, *America's Dual-Use Technology Future*, 41.

91. OSTP, *Second to None*, 1.

92. GAO, *Export Controls: Issues in Removing Militarily Sensitive Items from the Munitions List*, Report to the Chairman, Committee on Governmental Affairs, US Senate, GAO-NSAID-93-67, March 1993.

93. GAO, *Export Controls: Concerns over Stealth-Related Exports*, Report to the Chairman, Subcommittee on Acquisition and Technology, Committee on Armed Services, US Senate, GAO-NSAID-95-140, May 1995; GAO, *Export Controls: Sale of Telecommunications Equipment to China*, Report GAO/NSIAD-97-5,

November 1996; GAO; Report to Congressional Requestors, *Export Controls: Change in Export Licensing Jurisdiction for two Sensitive Dual-Use Items*, Report to the Chairman, Committee on National Security, House of Representatives; GAO/NSIAD-97-24, January 1997; GAO, *Export Controls: Information on the Decision to Revise High Performance Computer Controls*, Report to the Chairman, Subcommittee on International Security, Proliferation, and Federal Services, Committee on Governmental Affairs, US Senate, GAO/NSIAD-98-196, September 1998; GAO, *Export Controls: Change in Licensing Jurisdiction for Commercial Communications Satellites*; Statement of Katherine V. Shinasi before the Committee of Commerce, Science, and Transportation, US Senate, GAO/T-NSIAD-98-222, September 17, 1998; GAO, *Export Controls: Statutory Reporting Requirements for Computers Not Fully Addressed*, Report to Congressional Requesters. GAO/NSIAD-00-45, November 1999; GAO, *Export Controls: National Security Risks and Revisions to Controls on Computer Systems*; Testimony before the Committee on Armed Services, US Senate, GAO/T-NSIAD-00-139, March 23, 2000.

94. This is particularly well summarized in Report GAO/NSIAD-97-24 and Report GAO/NSAID-98-222.

95. Office of Technology Assessment, *Export Controls and Nonproliferation Policy*, Report OTA-ISS-596, May 1994, 10.

96. Testimony of David Cooper, General Accounting Office, *National Security Implications of Lowered Export Controls*, 1995, 36.

97. EO 12981 of December 5, 1995, *Federal Register* 60:236 (December 8, 1995), 62981–85.

98. Testimony, Gary Milhollin, Executive Director, Wisconsin Project for Nuclear Arms Control, *US Export Control and Nonproliferation Policy and the Role and Responsibility of the Department of Defense*, 29.

99. Testimony, Senator Thad Cochran (R-MS), *Proliferation and US Export Controls: Hearing before the Subcommittee on International Security, Proliferation, and Federal Services of the Committee on Governmental Affairs*, US Senate, 105th Congress, June 11, 1997 (Washington, DC: Government Printing Office, 1997), 1.

100. Testimony, William A. Reinsch, Undersecretary for Commerce, *Proliferation and US Export Controls*, 11.

101. Interview with Meijer in 2010, cited in *Trading with the Enemy*, 185.

102. Reinsch, testimony at Hearings before the Committee on Commerce, Science, and Transportation, September 17, 1998, 22.

103. Report GAO/NAIAD-00-45, 17 (emphasis added).

104. *Proliferation and US Export Controls*, 1997, 11; *US Export Control and Nonproliferation Policy*, 51–52.

105. The term is used extensively by Cupitt, *Reluctant Champions*, chapter 6.

106. The term is central in Meijer, *Trading with the Enemy*, and describes a coalition of actors opposed to trade liberalization with China, introduced at viii.

107. Cupitt, *Reluctant Champions*, has a table on p. 129.

108. We will avoid this term. It is misleading because the Run Faster Coalition was not, as Meijer suggests, a new feature of the 1990s. "Running faster" than one's rivals had always been a key tenet of US export control policy, translated into the need to maintain a technological gap between it and its friends and foes. Meijer further misleads us by characterizing the Run Faster Coalition as if they were against export controls altogether. They were not. They simply tilted the balance between enhanced exports and national security in favor of the former in the name of economic security. Their opponents, the control hawks, insisted that, for dual-use items, national security should trump commercial opportunity. In some specific cases, like supercomputers, the Run Faster Coalition saw no point in having an export control bar at all, unless, and as long as, the US was the sole supplier. They could take a very different position over the use of Chinese launchers for US-built satellites. There a key member of the Run Faster Coalition like Walter Reinsch could both insist that licensing remain under jurisdiction of the "probusiness" Department of Commerce and construct a complex set of control mechanisms to ensure that technical data that could enhance the performance of ballistic missiles did not leak to Chinese launch operators (see chapter 9). Meijer's arguments are developed in *Trading with the Enemy*; and Hugo Meijer, "Actors, Coalitions, and the Making of Foreign Security Policy: US Strategic Trade with the People's Republic of China," *International Relations of the Asia-Pacific* 15 (2015), 433–75.

Chapter 9

1. Hugo Meijer, *Trading with the Enemy: The Making of US Export Control Policy toward the People's Republic of China* (Oxford: Oxford University Press, 2016), 21.

2. Mark A. Stokes, *China's Strategic Modernization: Implications for the United States*, Carlisle, PA, Strategic Studies Institute, US Army War College, September 1999, 25, 26. See also James C. Mulvenon and Richard H. Yang, *The People's Liberation Army in the Information Age* (Santa Monica, CA: RAND, 1999), especially chapter 7, www.rand.org/pubs/conf_proceedings/CF145.html (accessed November 1, 2019). The conference was organized by the national directors of the Commission on Science, Technology, and Industry for National Defense (COSTIND).

3. All data from Meijer, *Trading with the Enemy*, 121–22.

4. Tai Ming Cheung, "Dragon on the Horizon: China's Defense Industrial Renaissance," *Journal of Strategic Studies* 32:1 (2009), 29–66, at 57.

5. Yasheng Huang, "The Role of Foreign-Invested Enterprises in the Chinses Economy: An Institutional Foundation Approach," in Shuxun Chen and Charles Wolf, eds., *China, the United States and the Global Economy* (Santa Monica, CA:

RAND, 2001), 149. See also Philip S. Golub, "Curbing China's Rise," *Le Monde Diplomatique*, October 2019, 2–4.

6. Meijer, *Trading with the Enemy*, 121–22.

7. Tai Ming Cheung, *Fortifying China: The Struggle to Build a Modern Defense Economy* (Ithaca, NY: Cornell University Press, 2009), 2

8. Statement by the Chairman, Sam Gejdenson (D-CT), *Need to Reform US Export Controls, Hearings and Markup on H.R. 2343, House Subcommittee on Economic Policy, Trade and the Environment, Committee on Foreign Affairs*, 103rd Congress, June 9, 1993, 84.

9. Statement of Gregory H. Hughes, *Need to Reform US Export Controls*, 89, 91–92.

10. *Bill H.R. 2912, Export Controls on Advanced Telecommunications: Hearing before the Subcommittee on Economic Policy, Trade and Environment, Committee on Foreign Affairs*, 103rd Congress, 1st Session September 22, 1993 (Washington, DC: Government Printing Office), 56.

11. Meijer, *Trading with the Enemy*, 192.

12. This policy was not uncontested. One company that received broadband telecommunications equipment was a US-Chinese joint venture called Huamei. The export license was not reviewed because Huamei was treated as a civil end user by both the company and the government even though, on further investigation, the GAO established that it was "partly controlled by several high-level members of the Chinese military." Granted the close relationships between civil and military officials in the Chinese system, it was not clear "how much military involvement in a commercial entity was required before it was considered a military end user." See GAO/NSIAD-97-5, 9.

13. See Meijer, *Trading with the Enemy*, 182–83, and section 5.2 in general.

14. HPCs have been treated extensively by Meijer, *Trading with the Enemy*, chapters 4 and 5.

15. Seymour Goodman, Peter Wolcott, and Grey Burkhart, *An Examination of High-Performance Computing Export Control Policy in the 1990s* (Los Alamitos, CA: IEEE Computer Society Press, 1996), ix.

16. For the origins of high-performance computer safeguards, see Mario Daniels, "Dangerous Calculations: The Origins of the US High Performance Computer Export Safeguards Regime (1968–1974)," in John Krige, ed., *Writing the Transnational History of Knowledge Flows in a Global Age* (University of Chicago Press, forthcoming).

17. Richard T. Cupitt, *Reluctant Champions: US Presidential Policy and Strategic Export Controls; Truman, Eisenhower, Bush, and Clinton* (New York: Routledge, 2000), 171.

18. James Treybig, testifying on behalf of thirteen CEOs of computer companies at a field meeting in Silicon Valley, recommended that Congress "raise the Distribution License (DL) limit to 2,000 CTP [MTOPS] from 195 CTP for friendly

countries," and "raise the DL limit to 195 CTP from 100 CTP for signatories of the Nonproliferation Treaty." *Export Control Reform in High-Technology: Hearing before the Committee on Science, Space and Technology*, 103rd Congress, 1st Session, August 13, 1993 (Washington, DC: Government Printing Office, 1993), 18. At this meeting the lobbyists for the computer industry also submitted a bill for Congress entitled the "Commercial Export Administration Act of 1993."

19. James A. Lewis, *Computer Exports and National Security in a Global Era* (Washington, DC: Center for Strategic and International Studies, 2001). See also Goodman et al., *Examination*.

20. Glenn J. McLoughlin and Ian F. Fergusson, *High Performance Computers and Export Control Policy: Issues for Congress*, Report RL31185 (Washington, DC: Congressional Research Service 2003), 9.

21. Tier 1 comprised Western Europe, Canada, Australia, New Zealand, Mexico, and Japan—there would be no controls at all on exports to these nations. Tier 2 countries included some former Eastern bloc states, South Korea, South America, and ASEAN countries. The threshold for securing a license to export to them was 10,000 MTOPS. A complete embargo was placed on Tier 4 countries, which included "rogue states" like Iran and Iraq. Meijer, *Trading with the Enemy*, 170.

22. Robert Johnson, "US Export Control Policy in the High Performance Computer Sector," *Nonproliferation Review*, Winter 1998, 44–59.

23. Testimony of Stephen Bryen, *Proliferation and US Export Controls: Hearing before the Subcommittee on National Security, Information and Federal Services of the Committee on Governmental Affairs*, US Senate 105th Congress, June 11th, 1997 (Washington, DC: Government Printing Office, 1997), 41.

24. Opening Statement, Senator Thad Cochran (R-MS), *Proliferation and US Export Controls*, 2.

25. Bryen, Testimony, *US Export Controls and Nonproliferation Policy and the Role and Responsibility of the Department of Defense: Hearing before Senate Committee on Armed Services*, 105th Congress, July 9, 1998 (Washington, DC: Government Printing Office, 1997), 11, 16.

26. Milhollin, Testimony, *US Export Controls and Nonproliferation Policy and the Role and Responsibility of the Department of Defense*, 28. To the satisfaction of these critics, on July 31, 1998, the Department of Justice announced that IBM East Europe/Asia Ltd. had received the maximum allowable fine of $8.5 million for violating seventeen counts of US export laws in its dealings with Russian weapons labs. GAO/NSIAD-98-166, 1n1.

27. Another GAO report released in November 1999 found more than adequate for weather forecasting, advanced aircraft design, and submarine design, but well below the threshold needed for 3-D modeling and shockwave simulation for nuclear weapons applications (forty-six thousand to seventy-six thousand MTOPS). GAO/NSIAD-00-45, November 1999.

28. GAO, Report to Congressional Requesters, *Export Controls: Statutory Reporting Requirements for Computers Not Fully Addressed*, GAO/NSIAD-00-45m, November 1999.

29. Glenn J. McLoughlin and Ian F. Fergusson, *High Performance Computers and Export Control Policy: Issues for Congress*, Report RL31185, February 10, 2003 (Washington, DC: Congressional Research Service, 2003), 13.

30. James A. Lewis, *Computer Exports and National Security in a Global Era* (Washington, DC: CSIS, 2001), 14.

31. US Senate, Select Committee on Intelligence, *Report on Impacts to US National Security of Advanced Satellite Technology Exports to the People's Republic of China (PRC) and Report on the PRC's Efforts to Influence US Policy* (Washington, DC: Government Printing Office, 1999), 2.

32. The detailed description of the changes made to the regulatory system that follows has been constructed using a combination of GAO reports and congressional hearings. Most accounts of these procedures (notably Meijer, *Trading with the Enemy*) do not respect the chronology. It is essential to do so to properly grasp the administration's defense of its decisions to license space technology exports to China.

33. GAO, *Report to Congressional Requesters*, GAO/NSIAD-97-24, table 1.

34. Robert D. Lamb, *Satellites, Security, and Scandal: Understanding the Politics of Export Control* (College Park: Center for International and Security Studies at Maryland, 2005), 30.

35. Wallerstein, Testimony, *National Security Implications of Lowered Export Controls on Dual-Use Technologies and US Defense Capabilities*, Hearings before the Senate Armed Services Committee, 104th Congress, May 11, 1995, 44.

36. The White House, EO 12870, Trade Promotion Coordinating Committee, September 30, 1993, https://govinfo.library.unt.edu/npr/library/direct/orders/tradepromotion.html (accessed November 4, 2020).

37. Cited in the May 1999 report by the Senate Select Committee on Intelligence, 14.

38. Cited in the May 1999 report by the Senate Select Committee on Intelligence, 15.

39. GAO/T-NSIAD-98-222, 6.

40. In 1995 Clinton renewed the Bilateral Agreement on Space Launch Services with China first signed by Reagan in 1988. The United States authorized a limited number of new geostationary orbit launches so long as their cost was no more than 15 percent below the equivalent launch by American companies. It was updated in 1997 to accommodate launches of Motorola's Iridium satellite network to Low-Earth Orbit. See Lamb, *Satellites*, 14. On Iridium, see Martin Collins, *A Satellite Telephone for the World: Iridium, Motorola, and the Making of the Global Age* (Baltimore: Johns Hopkins University Press, 2018).

41. GAO/T-NSIAD-98-222.

42. Wallerstein Testimony, Senate Hearings, *US Export Control and Nonproliferation Policy*, July 9, 1998, 62.

43. Report, Senate Select Committee on Intelligence, May 1999, 4.

44. Jan M. Lodal, Testimony, *United States Policy regarding the Export of Satellites to China*, Joint Hearings before the Committee on National Security Meeting with the Committee on International Relations, US Congress, 105th Session, June 17, 18, and 23, 1998, 109.

45. Lodal, 87.

46. Slocombe, Joint Hearings, *United States Policy regarding the Export of Satellites to China*, 164.

47. GAO, Report to the Chairman, Committee on International Relations, House of Representatives, *Export Controls: Some Controls over Missile-Related Technology Exports to China are Weak*, Report GAO/NSIAD-95-82, April 1995. At this stage of its formulation (which evolved according to circumstances), the MTCR divided controlled items into two categories, which distinguished (roughly) between whether the entity exported a complete missile (Category I) or a dual-use item that could be used for missiles or satellite launches (Category II). There was a strong presumption of denial for the export of Category I licenses. Category II licenses were assessed on a case-by-case basis, with a strong presumption of denial if the item would assist in the proliferation of WMDs, as judged by who its end user was.

48. Dianne E. Renack, *China: US Economic Sanctions*, Congressional Research Service Report for Congress, CRS 96-272, October 1, 1997. See also *Chinese Proliferation of Weapons of Mass Destruction: Background and Analysis*, CRS Report for Congress, CRS 96-767, September 13, 1996; Shirley A. Kan, CRS Report for Congress, *China: Possible Missile Technology Transfers from US Satellite Export Policy—Background and Chronology*, CRS 98-485, updated June 12, 1998, reproduced in *Joint Committee Hearings*; Shirley A. Kan, *China and Proliferation of Weapons of Mass Destruction and Missiles: Policy Issues*, Report for Congress, RL 31555, February 26, 2003.

49. Report, Senate Select Committee on Intelligence, 4.

50. Lamb, *Satellites*, 37.

51. For this summary, see Senate, Select Committee on Intelligence.

52. Jeff Gerth and Raymond Bonner, "Companies Are Investigated for Aid to China on Rockets," *New York Times*, April 4, 1998, A1, A3.

53. Gerth and Bonner, A1.

54. Testimony of David Tarbell, Director of the Defense Technology Security Administration, Joint Hearings, *United States Policy regarding the Export of Satellites to China*, 389. Tarbell claimed that Coates had written an internal memo "questioning whether Hughes was conducting an unauthorized launch failure investigation" in collaboration with engineers from the China Great Wall Industry Corporation. A program file containing all Coates's approvals for the Optus-B2 program, which he claimed to have kept at the Defense Technology Security

Administration, might have resolved the dispute: it could not be found. In his deposition to the Cox Committee Coates claimed that, on the contrary, he would never have authorized Hughes to share information with the PRC on how to improve the rocket fairing. Cox Report, vol. 2, 26–27.

55. Lamb, *Satellites*, 30–31.

56. Reinsch, Testimony, Joint Hearings, *United States Policy regarding the Export of Satellites to China*, 169.

57. Lodal, Joint Hearings, *United States Policy regarding the Export of Satellites to China*, 385.

58. NSA Damage Assessment, "Lost Encryption Board from Crash of Loral Satellite," February 14, 1996, in Joint Hearings, *United States Policy regarding the Export of Satellites to China*, 337.

59. Meijer, *Trading with the Enemy*, 209.

60. Cited by Kan, Joint Hearings, *United States Policy regarding the Export of Satellites to China*, 283, from which the surrounding evidence is also extracted.

61. This was possible because the Department of Commerce argued that a "military sensitive technology" (cf. table 6) lost that identity if it was built into the satellite, and so was not subject to MTCR Category II sanctions. Thus while State could continue to deny licenses to export satellites on the USML list, "Commerce could license satellites under its jurisdiction while sanctions were in place." Report, Senate Select Committee on Intelligence, 13. This situation was extended to all satellites once they were moved off the USML in 1996.

62. Milhollin, Testimony, *US Export Controls and Nonproliferation Policy and the Role and Responsibility of the Department of Defense*, 28, 31.

63. The text is reproduced in Joint Hearings, *United States Policy regarding the Export of Satellites to China*, June 1988, 88.

64. Opening Statement, Joint Hearings, *United States Policy regarding the Export of Satellites to China*, June 1998, 1.

65. Testimony of Peter M. Leitner, Senior Strategic Trade Adviser, Defense Technology Security Administration, *The Role of DTSA in Approving Critical Technology Exports: Hearing before the Committee on Governmental Affairs*, US Senate, 105th Congress, June 25, 1998 (Washington, DC: Government Printing Office, 1998), 3–4.

66. Testifying to Congress in June 1998 Sokolski said that during the Bush administration, "every inch of the way, the contractor had a shadow and it was a DOD monitor who was a rocket engineer. He followed and took notes. When the fellow misspoke, in the opinion of the monitor, he was reported on. More important than that, the superior who got the reports actually read them and picked up the phone and made complaints." Testimony, Joint Hearings, *United States Policy regarding the Export of Satellites to China*, 37, 53.

67. Sokolski, Written Response, Joint Hearings, *United States Policy regarding the Export of Satellites to China*, 396.

68. Sokolski, Testimony, Joint Hearings, *United States Policy regarding the Export of Satellites to China*, 10–14.

69. Sokolski, 14.

70. Sokolski, Testimony, *Transfer of Satellite Technology to China, Hearings before the Senate Committee on Commerce, Science, and Transportation*, September 17, 1998 (Washington, DC: Government Printing Office, 1999), 53

71. Sokolski, 52.

72. Freedenburg, Testimony, *Transfer of Satellite Technology to China*, 82.

73. Sokolski, Written testimony, Joint Hearings, *United States Policy regarding the Export of Satellites to China*, 202.

74. PL 105-261, October 17, 1998, section 1513, PL 105-261.

75. Lewis M. Franklin, "A Critique of the Cox Report Allegations of PRC Acquisition of Sensitive US Missile and Space Technology," in Michael M. May, Alastair Iain Johnson, W. K. H. Panofsky, Marco di Capua, and Lewis Franklin, *Cox Committee Report: An Assessment* (December 1999), https://cisac.fsi.stanford.edu/publications/cox_committee_report_the_an_assessment (accessed May 12, 2021), 81–99, at 93.

76. Kan, *China: Possible Missile Technology Transfers*, CRS-32.

77. Kan, CRS-32.

78. Opening statement, Hearings, *Transfer of Satellite Technology to China*, 1–4.

79. Lamb, *Satellites*, 42–43, on which most of this section is based.

80. Cited by Lamb, 42.

81. Gerth and Bonner, "Companies," A1.

82. Jeff Gerth, "US Business Role in Policy on China Is under Question," *New York Times*, April 13, 1998, A1, A8.

83. The investigative team at the *New York Times*, with special mention of Jeff Gerth, were awarded a Pulitzer Prize in 1999. The citation praised them "for a series of articles that disclosed the corporate sale of American technology to China, with US government approval despite national security risks, prompting investigations and significant changes in policy," https://www.pulitzer.org/winners/staff-45 (accessed December 2, 2019).

84. John Holum, Senior Adviser for Arms Control and International Security, Department of State, in *Oversight of Satellite Export Controls*, June 7, 2000, 4.

85. Lamb, *Satellites*, 48.

86. James Risen and Jeff Gerth, "Chinese Stole Nuclear Secrets from Los Alamos, US Officials Say," *New York Times*, March 6, 1999, http://www.nytimes.com/library/world/asia/030699china-nuke.html (accessed June 5, 2018).

87. "What Happened," www.WenHoLee.org (accessed June 10, 2018). For a survey of the history of US-China relations after World War II, see Zuoyue Wang, "US-China Scientific Exchange: A Case Study of State Sponsored Scientific Internationalism during the Cold War and Beyond," *Historical Studies in the Physical and Biological Sciences* 30:1 (1999), 249–77. On informal intelligence gathering

by US and Soviet scientists, see John Krige, "Atoms for Peace, Scientific Internationalism, and Informal Intelligence Gathering," in John Krige and Kai-Henrik Barth, eds., *Global Power Knowledge: Science, Technology and International Affairs*, Osiris 23 (Chicago: University of Chicago Press, 2006), 161–81.

88. Kan, *China*, gives considerable detail on this case.

89. Jeff Gerth and James Risen, "Reports Show Scientist Gave US Radar Secrets to Chinese," *New York Times*, May 10, 1999, A00001.

90. Matthew Purdy and James Sterngold, "The Prosecution Unravels: The Case of Wen Ho Lee," *New York Times*, February 5, 2002, A1; Dan Stober and Ian Hoffman, *A Convenient Spy: Wen Ho Lee and the Politics of Nuclear Espionage* (New York: Simon and Schuster, 2001).

91. Quoted by Joseph Cirincione, "Cox Report and Threat from China," Presentation to the CATO Institute, June 7, 1999, https://carnegieendowment.org/1999/06/07/cox-report-and-threat-from-china-pub-131 (accessed November 7, 2020).

92. Lewis R. Franklin, in May et al., *Cox Committee Report*, 81. For a critique, in turn, of their report, see Nicholas Rostow, "The Panofsky Critique and the Cox Committee Report: 50 Factual Errors in Four Essays," manuscript, circa 2000.

93. John M. Spratt Jr., "Keep the Facts of the Cox Report in Perspective," *Arms Control Today* 29:3 (1999), 24–25, 34, at 34.

94. See the unclassified findings of the two reports, "Intelligence Community Damage Assessment on Chinese Espionage," April 21, 1991, www.fas.org/sgp/news/dci042199.html (accessed June 9, 2018).

95. Cirincione, "Cox Report," 4.

96. Report, Senate Select Committee on Intelligence.

97. *House Select Committee on US National Security and Military/Commercial Concerns with the People's Republic of China, Submitted by Mr. Cox of California, Chairman* (Washington, DC: Government Printing Office, May 1999) (hereafter Cox Report). The claims made in the Cox Report need to be handled with care. One could treat it simply as part of a ruthless political campaign against Clinton as does Robert Lamb. One could dismiss it as so ridden with errors of fact and of interpretation as not to be worthy of serious consideration, as did the group at Stanford. We use it here as a primary source that contains much useful information pertinent to our question.

98. For analysis that includes Chinese responses, see John Krige, "Regulating the Transnational Flow of Intangible Knowledge on Space Launchers between the US and China in the Clinton Era," in John Krige, ed., *Writing the Transnational History of Knowledge Flows in a Global Age* (University of Chicago Press, forthcoming).

99. *Facts Speak Louder Than Words and Lies Will Collapse by Themselves*, response the government of the People's Republic of China, July 15, 1999, https://fas.org/sgp/news/1999/07/chinacox/index.html (accessed December 3, 2019).

100. Report, Senate Select Committee on Intelligence, 11.

101. Report, Senate Select Committee on Intelligence. 12.

102. Henry Sokolski, "US Satellites to China: Unseen Proliferation Concerns," *International Defense Review* 4 (1994), 23–26, at 23.

103. Sokolski, 24.

104. Cox Report, 2:76.

105. Cox Report, 2:77, 84, 84, respectively for the quotations (emphasis added).

106. Cox Report, 2:205 (emphasis in the original).

107. Cox Report, 2:93, 93, 85, respectively, for the quotations.

108. Cox Report, 2:84.

109. Kan, *China*, CRS 98-485, 6.

110. Cox Report, 2:213.

111. Cox Report, 2:212–13.

112. Report, Senate Select Committee on Intelligence, 6.

113. Cox Report, 2:98.

114. Report, Senate Select Committee on Intelligence, 11.

115. Report, Senate Select Committee on Intelligence, 12.

116. Testimony, *Oversight of Satellite Export Controls: Hearing before the Subcommittee on International Economic Policy, Export and Trade Promotion of the Committee on Foreign Relations, United States Senate*, 106th Congress, 2nd Session, June 7, 2000, 30.

117. Report, Senate Select Committee on Intelligence., 6.

118. The Cox Report cites a US trade representative as saying that some PRC bids were as low as half those proffered by Western bidders (2:174).

119. Meijer, *Trading with the Energy*, 208. See also Thomas S. Moorman, "The Explosion of Commercial Space and Its Implications for National Security," *Air and Spacepower Journal*, 1999, 6–20. In the event, the world market for commercial satellite launchings contracted after 2000. Globally there were thirty-five commercial satellite launches in 2000, sixteen in 2001, and twenty-four in 2002 of which four were launches from China. This was a mere 6.5 percent of government and commercial satellite launches that year. Kan, *China*, 37.

120. Reinsch, Testimony at Hearings before the Committee on Commerce, Science, and Transportation, September 17, 1998, 22.

121. Report, Senate Select Committee on Intelligence. 21.

122. Cox Report, vol. 2, chapter 7.

123. Cox Report, vol. 3, chapter 11.

124. "An Analysis of Export Control of US Technology—a DOD Perspective," in *A Report of the Defense Science Board Task Force on Export of US Technology, 4 February 1976* (Washington, DC: Office of the Director of Defense Research Engineering, 1976) (Bucy Report), 4.

125. Bucy Report, v.

126. Kan, *China*, CRS 98-485, 13.

127. *Public Law 105-261, October 17, 1998, Strom Thurmond National Defense Authorization Act for Fiscal Year 1999, Subtitle B—Satellite Export Controls*.

128. *Public Law 105-261*, section 1511.

129. *Public Law 105-261*. The full text of section 1514, (a) (B) (i) reads "technical discussions and activities, including the design, development, operation, maintenance, modification and repair of satellites, satellite components, missiles, other equipment, launch facilities, and launch vehicles."

130. *Public Law 105-261*, 22.

131. *Oversight of Satellite Export Controls: Hearing before the Subcommittee on International Economic Policy, Export and Trade Promotion of the Committee on Foreign Relations*, US Senate, 106th Congress, 2nd Session, June 7, 2000 (Washington, DC: Government Printing Office, 2000), 30–31.

132. Joan Johnson-Freese, "Alice in Licenseland: US Satellite Export Controls since 1990," *Research Policy* 16:3 (2000), 195–204, 200.

133. Lewis, *Preserving America's Strength*, 21.

134. John Krige, Angelina Long Callahan, and Ashok Maharaj, *NASA in the World: Fifty Years of International Collaboration in Space* (New York: Palgrave Macmillan, 2012), 276.

135. Daniel J. Sargent, "Pax Americana: Sketches for an Undiplomatic History," *Diplomatic History* 42:3 (2018), 357–76.

136. Department of Commerce, "Hydrocarbon Research, Inc., et al.: Consent Denial and Probation Order," *Federal Register* 27:44 (December 18, 1962), 12488.

137. "The Central Committee of the Communist Party of China Issued the 'Outline of National Innovation-Driven Development Strategy,'" Xinhua News Agency, Beijing, May 19, 2016, https://www.gov.cn/zhengce/2016-05/19/content_5074812.htm (accessed June 11, 2020). For a historical account of this quest for rejuvenation, see Elizabeth C. Economy, *The Third Revolution: Xi Jinping and the New Chinese State* (New York: Oxford University Press, 2019).

Chapter 10

1. Abrar al-Heeti, "US Hammers Huawei with 23 Indictments for Alleged Trade Secret Theft, Fraud," CNet, January 29, 2019, https://www.cnet.com/news/us-hammers-huawei-with-23-indictments-for-alleged-trade-secret-theft-fraud (accessed September 20, 2019).

2. Department of Commerce, Bureau of Industry and Security, "Addition of Certain Entities to the Entity List (Final Rule), Effective May 16, 2019," https://www.bis.doc.gov/index.php/all-articles/17-regulations/1555-addition-of-certain-entities-to-the-entity-list-final-rule-effective-may-16-2019 (accessed September 20, 2019); *Federal Register* 84:98 (May 21, 2019), 22961–68.

3. Angela Moon, "Google Suspends Some Business with Huawei after Trump Blacklist—Source," Reuters, May 19, 2019, https://www.reuters.com/article/us-huawei-tech-alphabet-exclusive/exclusive-google-suspends-some-business-with

-huawei-after-trump-blacklist-source-idUSKCN1SP0NB (accessed September 20, 2019).

4. "Lumentum Says Halting All Huawei Shipments, Cuts Quarterly Forecast," Reuters, May 20, 2019, https://www.reuters.com/article/us-huawei-suppliers/lumentum-says-halting-all-huawei-shipments-cuts-quarterly-forecast-idUSKCN1SQ1B5 (accessed July 25, 2019).

5. Ian King, Mark Bergen, and Ben Brody, "Top US Tech Companies Begin to Cut Off Vital Huawei Supplies," Bloomberg, May 19, 2019, https://www.bloomberg.com/news/articles/2019-05-19/google-to-end-some-huawei-business-ties-after-trump-crackdown?utm_source=twitter&utm_medium=social&cmpid=socialflow-twitter-business&utm_content=business&utm_campaign=socialflow-organic (accessed September 20, 2019).

6. Cheng Ting-Fang, Lauly Li, and Coco Liu, "Huawei Stockpiles 12 Months of Parts Ahead of US Ban," *Nikkei Asian Review*, May 17, 2019, https://asia.nikkei.com/Economy/Trade-war/Exclusive-Huawei-stockpiles-12-months-of-parts-ahead-of-US-ban (accessed September 21, 2019).

7. Julia Horowitz, "Huawei Phones Were Super Hot in Europe. Not Anymore," CNN Business, June 21, 2019, https://www.cnn.com/2019/06/21/tech/huawei-europe-smartphones/index.html (accessed September 20, 2019).

8. "Infineon Denies Report It Has Suspended Huawei Shipments," Reuters, May 20, 2019, https://www.reuters.com/article/us-huawei-tech-usa-infineon-denial/infineon-denies-report-that-it-has-suspended-huawei-shipments-idUSKCN1SQ118 (accessed September 20, 2019).

9. King, Bergen, and Brody, "Top US Tech Companies Begin to Cut Off Vital Huawei Supplies."

10. Julia Horowitz, "Huawei Phones Were Super Hot in Europe."

11. *Federal Register* 84:98 (May 21, 2019), 22961; Mark Landler, "Trump Abandons Iran Nuclear Deal He Long Scorned," *New York Times*, May 8, 2018, https://www.nytimes.com/2018/05/08/world/middleeast/trump-iran-nuclear-deal.html (accessed November 8, 2020).

12. Steven R. Weisman, "Sale of 3Com to Huawei Derailed by US Security Concerns," *New York Times*, February 21, 2008, https://www.nytimes.com/2008/02/21/business/worldbusiness/21iht-3com.1.10258216.html (accessed September 20, 2019).

13. House of Representatives, Permanent Select Committee on Intelligence, *Investigative Report on the US National Security Issues Posed by Chinese Telecommunications Companies Huawei and ZTE*, Report, October 2012, vi–vii.

14. House of Representatives, Permanent Select Committee on Intelligence, 3, 8, 29–32.

15. House of Representatives, Permanent Select Committee on Intelligence, 2.

16. Both laws enacted as part of PL 115-232, the John S. McCain National Defense Authorization Act for Fiscal Year 2019, https://www.govinfo.gov/app/details/PLAW-115publ232 (accessed October 31, 2020).

17. Ian F. Fergusson, Craig Elwell, and Jeanne Grimmett (Congressional Research Service), *Export Administration Act of 1979 Reauthorization*, Report, March 11, 2002, 3–4. The first EO to use IEEPA to extend the Export Control Act was signed by President Bill Clinton, EO 12923, June 30, 1994, https://fas.org/irp/offdocs/eo12923.htm (accessed September 21, 2019). For only a short time the export control authority did not have to rely on IEEPA. In November 2001, Congress extended the Export Control Act, but only until August 20, 2002. PL 106-508, November 13, 2001, https://www.congress.gov/106/plaws/publ508/PLAW-106publ508.pdf (accessed September 21, 2019).

18. For a quick overview of what the laws entail, see White and Case LLP, "Congress Finalizes CFIUS and Export Control Reform Legislation," July 2018, https://www.whitecase.com/publications/alert/congress-finalizes-cfius-and-export-control-reform-legislation (accessed September 22, 2019).

19. Congressional Research Service, *The US Export Control System and the Export Control Reform Initiative*, Report R41916, April 5, 2019, 9–22.

20. Skadden, Arps, Slate, Meagher, and Flom LLP, "Tightened Restrictions on Technology Transfer under the Export Control Reform Act," September 11, 2018, https://www.skadden.com/insights/publications/2018/09/tightened-restrictions-on-technology-transfer (accessed September 22, 2019).

21. 50 USC 4501, Foreign Investment Risk Review Modernization Act (FIRRMA), sec. 1719 (b) (1) and (b) (2) (H).

22. 50 USC 4501, Foreign Investment Risk Review Modernization Act, sec. 1706 (VI) (bb) (AA).

23. H.R. 5040, Export Control Reform Act of 2018, sec. 109. H.R. 5040 was signed into law, but the section quoted here was dropped for reasons unknown to us in the final version of act.

24. 50 USC 4501, Foreign Investment Risk Review Modernization Act, sec. 1703 (a) (6).

25. 50 USC 4501, sec. 1703 (a) (4) (B) (iii) (II) and sec. 1703 (a) (4) (D) (i) (I).

26. 50 USC 4501, sec. 1703 (a) (4) (D) (ii) (I).

27. "Smart Dust" is "a system of many tiny microelectromechanical systems (MEMS) such as sensors, robots, or other devices, that can detect, for example, light, temperature, vibration, magnetism, or chemicals. They are usually operated on a computer network wirelessly and are distributed over some area to perform tasks, usually sensing through radio-frequency identification." https://en.wikipedia.org/wiki/Smartdust (accessed September 29, 2019).

28. Department of Commerce, Bureau of Industry and Security, "Review of Controls for Certain Emerging Technologies," *Federal Register* 83:223 (November 19, 2018), 58201–2.

29. See, for example, Semiconductor Industry Association, *Comments on Advanced Notice of Proposed Rulemaking regarding Review of Controls for Certain Emerging Technologies*, January 10, 2019, https://www.semiconductors.org/wp-con

tent/uploads/2019/01/BIS-ANPRM-on-emerging-technology-jan-10.pdf (accessed March 23, 2020).

30. President Trump, "Executive Order [13859] on Maintaining American Leadership in Artificial Intelligence," February 11, 2019, https://www.whitehouse.gov/presidential-actions/executive-order-maintaining-american-leadership-artificial-intelligence/ (accessed September 22, 2019). The same day Trump signed a National Security Presidential Memorandum (NSPM) with the title "Protecting the United States Advantage in Artificial Intelligence and Related Critical Technologies." The memorandum appears to be classified, but the EO refers to it: "As directed by the NSPM, the Assistant to the President for National Security Affairs, in coordination with the OSTP Director and the recipients of the NSPM, shall organize the development of an action plan to protect the United States advantage in AI and AI technology critical to United States economic and national security interests against strategic competitors and adversarial nations."

31. Department of Defense, "Summary of the 2018 Department of Defense Artificial Intelligence Strategy: Harnessing AI to Advance Our Security and Prosperity," February 12, 2019, 5, https://media.defense.gov/2019/Feb/12/2002088963/-1/-1/1/summary-of-dod-ai-strategy.pdf (accessed September 22, 2019).

32. USC 50 4801, Export Control Reform Act of 2018, sec. 1752 (2) and (3).

33. "Export Control Act of 1979, as Amended," in Department of Commerce, Bureau of Industry and Security, *Principal Statutory Authority for the Export Administration Regulations*, January 2018, sections 2 and 3.

34. 50 USC 4501, Foreign Investment Risk Review Modernization Act, sec. 1702 (emphasis added).

35. For one of the few critical reflections of the impact of economic security thinking, see Ana Swanson and Paul Mozur, "Trump Mixes Economic and National Security, Plunging the US into Multiple Fights," *New York Times*, June 8, 2019, https://www.nytimes.com/2019/06/08/business/trump-economy-national-security.html (accessed March 23, 2020). Since "economic security" was rarely clearly defined, there was quite a bit of nuance in the (quite unsystematic) public discussions about the concept's scope, but it was never contested as a foundational framework. For glimpses of these debates, see, for example, *Perspectives on Reform of the CFIUS Review Process*, Hearing before the House Subcommittee on Digital Commerce and Consumer Protection of the Committee on Energy and Commerce, April 26, 2018 (Washington, DC: Government Printing Office, 2019).

36. President of the United States, *National Security Strategy of the United States of America*, December 2017, 17, https://www.whitehouse.gov/wp-content/uploads/2017/12/NSS-Final-12-18-2017-0905.pdf (accessed September 28, 2019).

37. President of the United States, 3.

38. President of the United States, 25, 2.

39. President of the United States, 2.

40. President of the United States, 17, 19.

41. EO 13797, "Establishment of Office of Trade and Manufacturing Policy," April 29, 2017, https://www.govinfo.gov/link/cpd/executiveorder/13797 (accessed September 28, 2019).

42. Peter Navarro, *The Coming China Wars: Where They Will Be Fought and How They Can Won* (Upper Saddle River, NJ: Financial Times Press, 2007); Peter Navarro and Greg Autry, *Death by China: Confronting the Dragon—a Global Call to Action* (Upper Saddle River, NJ: Prentice Hall, 2011).

43. White House Office of Trade and Manufacturing Policy, "How China's Economic Aggression Threatens the Technologies and Intellectual Property of the United States and the World," June 2018, 1, https://www.whitehouse.gov/wp-content/uploads/2018/06/FINAL-China-Technology-Report-6.18.18-PDF.pdf (accessed September 28, 2019).

44. White House Office of Trade and Manufacturing Policy, 20.

45. President of the United States, *National Security Strategy*, 27.

46. President of the United States, 55.

47. President of the United States, 26.

48. National Counterintelligence and Security Center, *Foreign Economic Espionage in Cyberspace*, July 24, 2018, 4–6, https://www.dni.gov/files/NCSC/documents/news/20180724-economic-espionage-pub.pdf (accessed September 28, 2019).

49. President of the United States, *National Security Strategy*, 28.

50. Senate Select Committee on Intelligence, *Open Hearing on Worldwide Threats*, February 13, 2018 (Washington, DC: Government Printing Office, 2018), 49–50.

51. David Dollar, Ryan Haas, and Jeffrey A. Bader, "Order Out of Chaos: Assessing US China Relations 2 Years into the Trump Presidency," Trump and Asia Watch, Brookings Institute, January 15, 2019, https://www.brookings.edu/blog/order-from-chaos/2019/01/15/assessing-u-s-china-relations-2-years-into-the-trump-presidency/ (accessed January 28, 2020).

52. Statement, Priestap, *Student Visa Integrity*, Senate Judiciary Committee, June 2018, 5.

53. Statement, Edie Bernice Johnson (D-TX), *Scholars or Spies? Foreign Plots Targeting America's Research and Development, Joint Hearing before the Subcommittee on Oversight and Subcommittee on Research and Technology, Committee on Science, Space and Technology*, 105th Congress, April 1, 2018, 17.

54. FBI, Counterintelligence Strategic Partnership Intelligence Note SPIN 15:006, "Preventing Loss of Academic Research," June 2015. See also FBI, Counterintelligence Strategic Partnership Intelligence Note SPIN 15:007, "Chinese Talent Programs," September 2015.

55. Permanent Subcommittee on Investigations, Senate Committee on Homeland Security and Governmental Affairs, *Threats to the US Research Enterprise: China's Talent Recruitment Plans*, November 18, 2019, 2, https://www.hsgac.senate.gov/imo/media/doc/2019-11-18%20PSI%20Staff%20Report%20-%20China's%20Talent%20Recruitment%20Plans.pdf (accessed January 24, 2020).

56. National Science Board, National Science Foundation, *Science and Engineering Indicators 2020: The State of US Science and Engineering*, NSB-2020-1, https://ncses.nsf.gov/pubs/nsb20201/ (accessed June 11, 2020).

57. National Science Board 2020, 12.

58. National Science Board 2020, 12.

59. Jenny J. Lee and John P. Haupt, "Winners and Losers in US-China Scientific Research Collaboration," *Higher Education*, December 3, 2019, https://doi.org/10.1007/s-10734-019-00464-7 (accessed January 20, 2020), concluding paragraph.

60. Richard P. Suttmeier, "State, Self-Organization, and Identity in the Building of Sino-US Cooperation in Science and Technology," *Asian Perspectives* 32:1 (2008), 5–31, at 22.

61. The Chinese authorities have encouraged scientists to publish in English-language journals. Margaret Cargill and Patrick O'Connor, "Developing Chinese Scientists' Skills for Publishing in English: Evaluating Collaborating-Colleague Workshops Based on Genre Analysis," *Journal of English for Academic Purposes* 5:3 (2006), 207–21.

62. Zuoyue Wang, "Chinese American Scientists and US-China Scientific Relations," in Peter Koehn and Xiao-Hung Yin, eds., *The Expanding Roles of Chinese-Americans in US-China Relations: Transnational Networks and Trans-Pacific Interactions* (New York: Routledge, 2002), 207–34. See also Suttmeier, "State."

63. By self-organization we mean building cooperative networks in which "the selection of a partner and the location of the research rely upon choices made by the researchers themselves rather than emerging through national institutional incentives or constraints." Caroline Wagner and Loet Leydsdorff, "Network Structure, Self-Organization and the Growth of International Collaboration in Science," *Research Policy* 34 (2005), 1608–18.

64. This is a statement by the JASON advisory group called on by the NSF to assess the threats China poses to the American science ecosystem: JASON Report JSR-19-21, "Fundamental Research Security," McLean, VA, Mitre Corporation, December 2019, https://www.nsf.gov/news/special_reports/jasonsecurity/JSR-19-2I FundamentalResearchSecurity_12062019FINAL.pdf (accessed January 27, 2020).

65. President of the United States, *National Security Strategy*, 22. STEM stands for the disciplines of science, technology, engineering, and mathematics.

66. *China's Threat to US Research/Innovation Leadership: Hearing before the House Permanent Select Committee on Intelligence*, 115th Congress, July 19, 2018 (Washington, DC: Government Printing Office, 2018).

67. *Student Visa Integrity: Protecting Educational Opportunity and National Security; Hearings before the Senate Committee on the Judiciary; Subcommittee on Border Security and Immigration*, 115th Congress, June 6, 2018 (Washington, DC: Government Printing Office, 2018); *China's Non-traditional Espionage against the United States: The Threat and Potential Policy Reponses*, Hearings before the Senate Judiciary Committee, 115th Congress, December 12, 2018 (Washington, DC: Government Printing Office, 2018).

68. Statement of E. W. Priestap, *Student Visa Integrity: Protecting Educational Opportunity and National Security*: Hearing before the Senate Committee on the Judiciary, Subcommittee on Border Security and Immigration, June 6, *2018*, 2, https://www.judiciary.senate.gov/imo/media/doc/06-06-18%20Priestap%20Testimony.pdf (accessed January 15, 2020).

69. US Senate, Permanent Subcommittee on Investigations, Committee on Homeland Security and Governmental Affairs, *Threats to the US Research Enterprise: China's Talent Recruitment Plans*, posted online on November 18, 2018, https://www.hsgac.senate.gov/imo/media/doc/2019-11-18%20PSI%20Staff%20Report%20-%20China's%20Talent%20Recruitment%20Plans.pdf (accessed January 15, 2020).

70. Annie Karni, "Trump Rants behind Closed Doors with CEOs," *Politico*, August 8, 2018, https://www.politico.com/story/2018/08/08/trump-executive-dinner-bedminster-china-766609 (accessed November 8, 2020).

71. Daniel Golden, *Spy Schools: How the CIA, FBI, and Foreign Intelligence Secretly Exploit America's Universities* (New York: Picador, 2017), 17.

72. Cited in Mitch Ambrose, "Universities under Pressure as Lawmakers Push Research Security," aip.org/fyi/2019/universities-under-pressure-lawmakers-push-research-security, July 12, 2019 (accessed January 20, 2020).

73. *Scholars or Spies?*, 3.

74. Statement, Abraham, *Scholars or Spies?*, 5.

75. Statement, Abraham, 5, 6.

76. Michael Wessel, Testimony, *Scholars or Spies?*, 34.

77. Department of Justice, *Chinese Professors among Six Defendants Charged with Economic Espionage and Theft of Trade Secrets for Benefit of People's Republic of China*, May 19, 2015.

78. Phil Muncaster, "Chinese Professor Jailed for Stealing US Trade Secrets," *Infosecurity Magazine*, September 2, 2020, available at https://www.infosecurity-magazine.com/news/chinese-professor-jailed-stealing/ (accessed November 5, 2020).

79. The core example cited by Michelle Van Cleave, former national counterintelligence executive, was also flawed.

80. Department of Justice, "Harvard University Professor and Two Chinese Nationals Charged in Three Separate China Related Cases," *Justice News*, January 28, 2020, https://www.justice.gov/opa/pr/harvard-university-professor-and-two-chinese-nationals-charged-three-separate-china-related (accessed February 8, 2020).

81. Mara Hvistendahl, *The Scientist and the Spy: A True Story of China, the FBI, and Industrial Espionage* (New York: Riverhead Books, 2020), 152.

82. Bethany Halford and Andrea L. Widener, "Harvard Chemist Charles Lieber Charged with Fraud," *Chemical and Engineering News*, January 28, 2020, https://cen.acs.org/research-integrity/misconduct/Harvard-chemist-Charles-Lieber-charged/98/i5 (accessed June 11, 2020).

83. Bill Chapell, "Acclaimed Harvard Scientist Is Arrested, Accused of Lying about Ties to China," NPR, January 20, 2020, https://www.npr.org/2020/01/28

/800442646/acclaimed-harvard-scientist-is-arrested-accused-of-lying-about-ties-to-china (accessed February 8, 2020).

84. Department of Justice, "Harvard University Professor."

85. Halford and Widener, "Harvard Chemist."

86. Cong Cao, "China's Brain Drain at the High End," *Asian Population Studies* 4:3 (2008), 331–45, at 334–35; Cong Cao, Jeroen Baas, Caroline S. Wagner, and Koen Jonkers, "Returning Scientists and the Emergence of China's Science System," *Science and Public Policy* (2019), 1–28, table 5.

87. Hepeng Jia. "China's Plan to Recruit Talented Researchers," *Nature, Career Guide*, January 17, 2018, https://www.nature.com/articles/d41586-018-00538-z (accessed November 8, 2020).

88. The Thousand Talents Plan, http://www.1000plan.org.cn/en/history.html (accessed January 28, 2020).

89. Hepeng Jia, "China's Plan to Recruit Talented Researchers."

90. Thousand Talents Plan, http://www.1000plan.org.cn/en/history.html (accessed January 20, 2020).

91. FBI, *Counterintelligence Strategic Partnership Intelligence Note (SPIN)*, SPIN15-007, September 3, 2015.

92. Staff Report of the Senate Subcommittee on Investigations of the Committee on Homeland Security and Governmental Affairs, *Threats to the US Research Enterprise: China's Talent Recruitment Programs*, released online on November 20, 2020, 28, https://www.hsgac.senate.gov/imo/media/doc/2019-11-18%20PSI%20Staff%20Report%20-%20China's%20Talent%20Recruitment%20Plans.pdf (accessed January 15, 2020).

93. Lawrence A. Tabak and M. Roy Wilson, "Foreign Influences on Research Integrity," slide 7, https://acd.od.nih.gov/documents/presentations/12132018Foreign Influences.pdf (accessed January 24, 2020).

94. Scott Kennedy, "Made in China 2025," June 1, 2015, Center for Strategic and International Studies, https://www.csis.org/analysis/made-china-2025 (accessed January 15, 2020).

95. Wayne Morrison, "The Made in China 2025 Initiative: Economic Implications for the United States," Congressional Research Service, April 12, 2019, https://fas.org/sgp/crs/row/IF10964.pdf (accessed January 15, 2020). They were next-generation information technology, high-end numerical control machinery and robotics, aerospace and aviation equipment, maritime engineering equipment and high-tech vessel manufacturing equipment, advanced rail equipment, energy-saving and new energy vehicles, electrical equipment, agricultural machinery and equipment, new materials, biopharmaceuticals, and high-performance medical devices.

96. Michael Brown and Pavneet Singh, "China's Technology Transfer Strategy," *Defense Innovation Unit Experimental*, January 2018, https://admin.govexec.com/media/diux_chinatechnologytransferstudy_jan_2018_(1).pdf (accessed January 15, 2018).

97. Kennedy, "Made in China 2025."

98. "The Central Committee of the Communist Party of China Issued the 'Outline of National Innovation-Driven Development Strategy,'" Xinhua News Agency, Beijing, May 19, 2016, https://www.gov.cn/zhengce/2016-05/19/content_50 74812.htm (accessed June 11, 2020). For a historical account of this quest for rejuvenation, see Elizabeth C. Economy, *The Third Revolution: Xi Jinping and the New Chinese State* (New York: Oxford University Press, 2019).

99. United States Census Bureau, https://www.census.gov/foreign-trade/bal ance/c5700.html (accessed January 15, 2020).

100. Office of the US Trade Representative, *Findings of the Investigation into China's Acts, Policies and Practices Related to Technology Transfer, Intellectual Property and Innovation under Section 301 of the Trade Act of 1974*, https://ustr.gov /sites/default/files/Section%20301%20FINAL.PDF (accessed January 15, 2020).

101. Office of the US Trade Representative, 17.

102. Morrison, "Made in China 2025 Initiative."

103. Alex Joske, *The China Defense Universities Tracker: Exploring the Military and Security Links of China's Universities*, Australian Strategic Policy Institute, International Cyber Policy Centre, undated. See also Alex Joske, *Picking Flowers, Making Honey: The Chinese Military's Collaboration with Foreign Universities*, Policy Brief Report No. 10/2018, Australian Strategic Policy Institute, International Cyber Policy Centre, 2018. A student at Boston University was accused of being an undercover PLA agent the day Harvard's Lieber was indicted. Department of Justice, "Harvard University Professor."

104. Jessica Wang, "A State of Rumor: Low Knowledge, Nuclear Fear and the Scientists as Security Risk," *Journal of Policy History* 28:3 (2016), 406–46.

105. Ana Swanson, "A New Red Scare Is Reshaping Washington," *New York Times*, July 20, 2019, https://www.nytimes.com/2019/07/20/us/politics/china-red -scare-washington.html (accessed May 3, 2021).

106. Peter Salovey, "Yale's Steadfast Commitment to Our International Students and Scholars," https://president.yale.edu/yale-s-steadfast-commitment-our -international-students-and-scholars, May 23, 2019 (accessed January 23, 2020).

107. Cited in Mitch Ambrose, "Universities under Pressure as Lawmakers Push Research Security," July 12, 2019, aip.org/fyi/2019/universities-under-pressure -lawmakers-push-research-security (accessed January 20, 2020).

108. Cited in Shan Lu et al., "Racial Profiling Harms Science," *Science* 363:6433 (March 22, 2019), 1290. See also Press Release, the Committee of 100, "Group of Leading Chinese-Americans Highlight Impact of US-China Tensions on American Science, Technology, Business and Education," September 28, 2029, https:// www.committee100.org (accessed January 24, 2020).

109. Council on Foreign Relations, "A Conversation with Christopher Wray," April 26, 2019, https://www.cfr.org/event/conversation-christopher-wray-0?utm _medium=social_owned&utm_term=conversation-christopher-wray&utm_content =042519&utm_campaign=event&utm_source=tw (accessed February 10, 2020).

110. The case is extensively covered in a variety of sources: by the FBI, *Chinese*

Talent Programs, SPIN: 15-007, September 2005, 4; by investigative journalist Daniel Golden in *Spy Schools*, chapter 1; in testimony, *Scholars or Spies?*, 53–67; by a major TV network, "How a Chinese Student Allegedly Stole Duke University Tech to Create a Billion-Dollar Empire," *NBC News*, July 24, 2018. See also Cynthia McFadden, Aliza Nadi, and Courtney McGee, "Education or Espionage? A Chinese Student Takes His Homework Home to China," *NBC News*, July 24, 2018, https://www.news.com/news/china/education-or-espionage-chinese-student-takes-his-homework-home-china-n893881 (accessed June 11, 2020); Kelly McLaughlin, "A Billionaire Known as 'China's Elon Musk' Is Suspected of Spying While He Was a Duke Student and Stealing a Professor's Invisibility Technology," *Business Insider*, July 24, 2018, https://www.businessinsider.com/chinese-billionaire-is-accused-of-stealing-research-from-a-duke-lab-2018-7 (accessed January 27, 2020).

111. McFadden, Nadi, and McGee, "Education or Espionage?"

112. Golden, *Spy Schools*, 32.

113. McFadden, Nadi, and McGee, "Education or Espionage?"

114. Golden, *Spy Schools*, 16, 8; *Scholars or Spies?*, 60.

115. Golden, *Spy Schools*, 13.

116. Testimony to the Judiciary Committee, 4.

117. Testimony to the Judiciary Committee, 4 (emphasis added).

118. FBI News, "Protecting Vital Assets: Pilfering of Corn Seeds Illustrates Intellectual Property Theft," December 19, 2016, https://www.fbi.gov/news/stories/sentencing-in-corn-seed-intellectual-property-theft-case (accessed January 15, 2020).

119. FBI, SPIN 15:007, September 2015, 3.

120. Statement before Senate Judiciary Committee on Students Visa Integrity, June 6, 2018.

121. Elizabeth Redden, "Professor Indicted for Alleged Undisclosed China Links," *Inside Higher Ed*, August 23, 2019, https://www.insidehighered.com/news/2019/08/23/kansas-professor-indicted-allegedly-failing-disclose-appointment-chinese-university (accessed May 2, 2021).

122. Lawrence A. Tabak and M. Roy Wilson, *Foreign Influences on Research Integrity*, 117th Meeting of the Advisory Committee to the [NIH] Director, December 13, 2018, 2021, https://acd.od.nih.gov/documents/presentations/12132018ForeignInfluences.pdf (accessed March 2, 2021), slide 6.

123. Staff Report of the Senate Subcommittee on Investigations, 50.

124. Francis Collins, Testimony, *Prioritizing Cures: Science and Stewardship at the National Institutes of Health*, US Senate Committee on Health, Education, Labor, and Pensions, August 23, 2018. The letter is available at https://www.insidehighered.com/sites/default/server_files/media/NIH%20Foreign%20Influence%20Letter%20to%20Grantees%2008-20-18.pdf (accessed February 13, 2020).

125. Francis S. Collins, *Statement on Protecting the Integrity of US Biomedical Research*, August 23, 2018, https://www.nih.gov/about-nih/who-we-are/nih-director

/statements/statement-protecting-integrity-us-biomedical-research (accessed February 8, 2020).

126. Circular letter from Francis S. Collins, August 20, 2018, https://www.in sidehighered.com/sites/default/server_files/media/NIH%20Foreign%20Influence %20Letter%20to%20Grantees%2008-20-18.pdf (accessed March 8, 2020). See also Collins, *Statement*.

127. NIH Advisory Committee to the Director, *ACD Working Group for Foreign Influences on Research Integrity*, December 2018, https://acd.od.nih.gov/documents /presentations/12132018ForeignInfluences_report.pdf (accessed March 10, 2020).

128. Jocelyn Kaiser and David Malakoff, "NIH Investigating Whether US Scientists Are Sharing Ideas with Foreign Governments," *ScienceMag*, August 27, 2018, https://www.sciencemag.org/news/2018/08/nih-investigating-whether-us-scien tists-are-sharing-ideas-foreign-governments (accessed May 2, 2021).

129. Jeffrey Mervis, "NIH Probe of Foreign Ties Has Led to Undisclosed Firings—and Refunds from Institutions," *ScienceMag*, June 26, 2019, https://www .sciencemag.org/news/2019/06/nih-probe-foreign-ties-has-led-undisclosed-firings -and-refunds-institutions (accessed March 2, 2021).

130. Peter Waldman, "The US Is Purging Chinese Cancer Researchers from Top Institutions," *Bloomberg Business Week*, June 13, 2019, https://www.bloom berg.com/news/features/2019-06-13/the-u-s-is-purging-chinese-americans-from -top-cancer-research (accessed May 3, 2021).

131. Jodi B. Black, *Protecting US Science from Undue Foreign Influence: The NIH Experience*, Presentation to the APLU Annual Conference, San Diego, November 11, 2019 (private communication), slide 18.

132. Cited in Waldman, "US Is Purging."

133. Black, *Protecting US Science*, slide 19.

134. Jeffrey Mervis, "Powerful US Senator Calls for Vetting NIH Grantees at Hearing on Foreign Influences," *ScienceMag*, June 6, 2019, https://www.science mag.org/news/2019/06/powerful-us-senator-calls-vetting-nih-grantees-hearing -foreign-influences (accessed March 2, 2021).

135. "Statement from NSF Director France A. Córdova," *NSF Response to the JASON Report: Fundamental Science and Security*, https://nsf.gov/news/spe cial_reports/jasonsecurity/NSF_response_JASON.pdf (accessed March 3, 2020).

136. *Fundamental Research Security*, JASON Report JSR-19-21 (McLean, VA, Mitre Corporation, December 2019), https://www.nsf.gov/news/special_reports /jasonsecurity/JSR-19-2IFundamentalResearchSecurity_12062019FINAL.pdf (accessed March 3, 2020).

137. JASON Report JSR-19-21, 8.

138. JASON Report JSR-19-21, 2.

139. JASON Report JSR-19-21, 16.

140. *NSF Response to the JASON Report "Fundamental Science and Security."*

141. The White House, *National Strategy for Critical and Emerging Technologies*,

October 2020, https://www.whitehouse.gov/wp-content/uploads/2020/10/National-Strategy-for-CET.pdf (accessed November 2, 2020).

142. Staff Report, Senate Permanent Subcommittee on Investigations, 12, items 9 and 11.

143. US Department of Energy, Order DOE O 486.1, June 7, 2019, www.directives.doe.gov. See also William Thomas, "DOE Bars Its Researchers from Participating in Rival Nations' Talent Programs," *Physics Today*, June 20, 2019, https://physicstoday.scitation.org/do/10.1063/PT.6.2.20190620a/full/ (accessed May 2, 2021). The restriction also applies to employees on detail or appointed under the terms of the Intergovernmental Personnel Act (IPA) by either the DoE or the NSF. See also Staff Report to the Senate Subcommittee on Investigations, 71–72.

144. Jeffrey Mervis, "NSF Hopes JASON Can Lead It through Treacherous Waters," *ScienceMag*, March 18, 2019, https://www.sciencemag.org/news/2019/03/nsf-hopes-jason-can-lead-it-through-treacherous-waters (accessed May 2, 2021).

145. Wessel, Testimony, *Scholars or Spies?*, 37.

146. President of the United States, *National Security Strategy*, 22.

147. Jeff Tollesro, "Chinese-American Scientists Uneasy amid Crackdown on Foreign Influence," *Nature* 570:7759 (2019), 13–14, doi:10.1038/d41586-019-01605-9 (accessed May 2, 2021).

148. Viggo Stacey, "US Dept. of State Begins Vetting Student Visa Applicants' Social Media," *Pie News*, June 25, 2019, https://thepienews.com/us-state-dept-begins-vetting-visa-applicants-social-media/ (accessed March 10, 2020). Two documentary film organizations have sued the Trump administration over the policy change, charging that demanding social media pseudonyms exposes people who live in authoritarian regimes to additional risks Charlie Savage, "Trump Administration Sued over Social Media Screening for Visa Applicants," *New York Times*, December 5, 2019, https://www.nytimes.com/2019/12/05/us/politics/visa-applications-social-media.html (accessed May 2, 2021).

149. See the document distributed by the AAU (Association of American Universities) and the APLU (Association of Public and Land-Grant Universities) entitled "Actions Taken by Universities to Address Growing Concerns about Security Threats and Undue Foreign Influence on Campus," updated April 22, 2019, https://www.aau.edu/sites/default/files/Blind-Links/Effective-Science-Security-Practices.pdf (accessed March 10, 2020).

150. For the following, see Maria T. Zuber, "New Review Process for 'Elevated Risk' International Proposals," https://orgchart.mit.edu/node/27/letters_to_community/new-review-process-elevated-risk-international-proposals (accessed March 10, 2020).

151. Newsletter Law360, "Tech Giants Warn H-1B Rules Will Push Business Abroad," November 3, 2020, https://www.law360.com/articles/1325023/tech-giants-warn-h-1b-rules-will-push-businesses-abroad (accessed November 7, 2020).

152. Deirdre Fernandes, "Nobel Prize Winners and Other Scientists Come to

Defense of Harvard's Charles Lieber," *Boston Globe*, March 1, 2021, https://www.bostonglobe.com/2021/03/01/metro/nobel-prize-winners-other-scientists-come-defense-harvard-professor-charles-lieber/ (accessed May 2, 2021).

153. Rebecca Trager, "MIT Nanotechnologist Arrested for Hiding His Ties to China," *Chemistry World*, January 18, 2021, https://www.chemistryworld.com/news/mit-nanotechnologist-arrested-for-hiding-his-ties-to-china/4013051.article (accessed May 3, 2021).

Index

Abraham, Ralph, 318
academia, 16–17, 83–84, 97; and active knowledge transfer to foreign nationals, 120, 134; FBI activity on campus, 312–13, 317, 321–22, 327; and increasing emphasis on applied research, 141, 142; and international scientific collaboration, 136, 147–49, 325; and IP theft, 313–14, 320; sharing of sensitive knowledge with the Soviets, 170–71; Sino-US academic cooperation, 315–18; targeted by Soviet visitors, 141
academic exchanges, 90–95, 315–18
academic freedom, 6, 13, 16, 83–84, 98, 154, 158
Act to Expedite the Strengthening of the National Defense of 1940, 20, 42–43
advanced materials, 23, 306
Advisory Committee on Export Policy (ACEP), 80–83
Afghanistan, 24, 135, 137, 145, 236–37, 243
aircraft and aerospace, 25, 41, 102, 105, 106, 110, 228. *See also* satellites
Alien Property Custodian, 40
Allen Report, 244
American Association for the Advancement of Science (AAAS), 145; Committee on Academic Freedom and Tenure, 178, 181; Committee on Scientific Freedom and Responsibility, 174
American Association of University Professors, 179
American Civil Liberties Union (ACLU), 163
American Council of Learned Societies, 94
American decline, 25, 29, 194, 195, 199, 201–3, 206, 207, 211, 214, 215, 217
American Physical Society, 144
American Society for Engineering Education, 83
American Vacuum Society, 143
Apple, 335
Apstar-2 satellite, 273, 282, 283
Arms Control and Disarmament Agency (ACDA), 274
Arms Export Control Act, 255
Armstrong, Bill, 139
Armstrong, C. Michael, 279
Army and Navy Patent Advisory Board, 47
artificial intelligence (AI), 306, 312
ARZAMAS-16, 263
Asian Tigers, 246
Association of American University Professors (AAUP), 152
Association of University Export Control Officers, 17
AT&T, 261
Atomic Energy Act, 9, 67, 79, 82, 96
atomic espionage, 59–60
Atoms for Peace, 90, 93
Augustine, Norman, 156, 188, 225, 274
Australia Group, 8, 245, 246, 264
Avago Technologies, 320

Bacher, Robert F., 55
Baird, Ian, 256
Baker, Lara, 170
basic and applied science, 148–49, 162, 177, 184, 185

Bayh-Dole Act, 142
Beckman Instruments Incorporated, 101
Berkner, Lloyd W., Berkner Report, 75–77, 79
Berman, Harold J., 103
Bingaman, Jeff, 248
Black, Jodi, 330
Blue Team, anti-Clinton lobby, 278
Boeing, 101, 117
Bok, Sissela, 178
Bonner, Raymond, 272, 278
Borrus, Michael, 203, 403n74
Brady, Lawrence, 139, 158
Bretton Woods, 71, 107
Brezhnev, Leonid, 108
Broadcomm, 299
Bromley, Allan, 145
Brown, Harold, 124, 127, 237; definition of critical technology and technical data, 124, 127
Bryen, Stephen, 179, 210, 240, 263, 264, 275
Bubble Memory Materials and Process Technology, 143
Bucy, J. Fred: allied trade with the Soviets, frustration with, 122; and critical technologies, 129; definition of technology, 102; détente, criticism of, 110; and exports, 126; lead time, emphasis on, 113. *See also* Bucy Report
Bucy Report: allied skepticism on know-how controls, 132; and centrality of know-how, 102–3; as complicating export control system technically and administratively, 133; concerns for knowledge loss in academic contexts, 120; critical of Cocom, 122; and economic warfare, 123; as exaggerating loss of technical data, 130; implementation of, 104, 123–27; and intra-Western trade, 121–23; and lead time, 103, 112, 114; as "local" response to global economic changes, 116, 133; main argument, 117–23; major breakthrough in export policy, 102, 103, 116, 117; and MCTL, 128; as overlooking reverse engineering, 130; as responding to DoD and Congressional hawks, 118; technical knowledge required for implementation, 120. *See also* Bucy, J. Fred
Bulletin of the Atomic Scientists, 78

Bundy, William, 93
Burton, Daniel F., 231
Bush, George H. W.: administration of, 30, 194, 225–26, 232, 237; Enhanced Proliferation Control Initiative, 245; lifting sanctions on China, 248, 271; nonproliferation controls, 245; total satellite launches in China, authorization of, 270
Bush, Vannevar, Bush Report, 61–62, 64
Byrd, Robert, 226
Byrd Amendment, 223, 226–27, 228

California Institute of Technology, 147, 181, 185
calutrons, 245
Canada, 43, 80, 82, 218
Carey, William, 158
Carlucci, Frank, 141, 179
Carter, Ashton B., 189, 332
Carter, Jimmy: inter-allied trade, refusal to impose tight controls on, 123; on limits to classification, 177; relationships with China, 235; and sanctions on Soviets in early 1980s, 136
Casey, William, 141
censorship, 39, 40–41, 48–49
Center for Strategic and International Studies (CSIS), 200, 264, 323
Central Intelligence Agency (CIA), 93, 94, 249, 287; and Farewell papers, 137–40; and foreign nationals, order to be disinvited from conferences, 143; reports of illegal Soviet technology acquisition, 135; threats to academic freedom, 156–59; US telecommunications exports to China, 261
Chalk, Rosemary, 174, 181
Challenger accident, 265
Chamberlain, Neville, 203
Chelyabinsk-70, 263
chemical industry, 40, 42, 43, 44, 53, 73, 81, 83, 92, 109–10, 130, 218, 321
China, 24, 25, 31, 32, 33; acquisition of intellectual property, 248; CFIUS and, 218–19; changing patterns of R and D, 315; constructive engagement in Clinton era, 257; defense budget, 260; export control policy in Reagan era, 238; FDI into, 260; as friendly non-allied partner of the US, 238; legal and illegal technology

acquisition, 26; missile proliferation, 239; most-favored-nation trade policy questioned, 247–49; non-proliferation system, abiding by, 249–50; sanctions and space cooperation with, 26, 247–49; scientific output, 316; STEM education abroad, promotion of, 242; threat to US economic security, 323; trade balance, 238, 247, 248, 323; US trade war with, 7, 26

China Academy of Launch Vehicle Technology (CALT), 284

China Aerospace Corporation (CASC), 265, 285

China differential, 235, 238

China Great Wall Industry Corporation (CGWIC), 265, 272

China National Aero-Technology Import and Export Corporation (CATIC), 218–19

China State Council, 323

Chinese Academy of Sciences (CAS), 263

Christiansen, Gene, 273

Christopher, Warren, 232

Chu, Judy, 325

Churchill, Winston, 60

Cirincione, John, 281

citizenship, 2

Clark, William, 175

classification. *See* secrecy (classification)

Clausewitz, Carl von, 200

Clinton, Bill: administration of, 25–26, 30, 33–34, 229; business interests vs. national security, 266; China, constructive engagement with, 257; China, liberal trade policies with, 261–63; economic security, emphasis on, 255, 256; export controls and nonproliferation, 250; lifting sanctions on satellite launches, 271; lobbied by business, 262, 264, 279; reform of controls on sensitive dual-use technologies, 254–57; total satellite launches in China, authorization of, 270

Coates, Al, 273

Cocom, 8, 9, 21, 72, 224, 250, 255; dispute over technical data control, 85–90; end of, 25–26, 72; meeting at Ministerial level in 1982, 136; and the MTCL, 132; and national sovereignty, 121; reduction in controls at end of Cold War, 244, 254; US requests for exceptions in, 122, 238

Cohen, William, 141

Collins, Francis, 328–29

Columbia University, 181

Comecon, 120

Committee of 100, 325

Committee on Foreign Investment in the United States (CFIUS), 9, 26, 206–7, 218, 220, 224, 226–28, 301, 304, 305, 309, 318; members, 218

Committee on Public Information, 40

Committee on Release of Scientific Information, 54

Commodity Control List (CCL), 8, 266, 268, 277, 290, 292; and MCTL, 128; reduced to foster trade after passage of EAA, 111

competitiveness, 16, 23, 25, 29, 33, 203, 207, 209, 211, 217, 223, 228, 231, 302, 308, 324, 334

Compton, Arthur H., 55

computers, 10, 11, 24, 94, 196, 210, 213, 216, 224, 228; high-performance computers (HPCs), 9, 25, 215, 229, 261–64; quantum computing, 306; trade liberalization with China under Clinton, 264; US requests for exceptions in Cocom, 122

Control Data Corporation, 131

control hawks, 257

controlled unclassified information (CUI), 13

Córdova, France, 331, 332

Coriden, Guy, 93

Cornell University, 147, 167, 182, 263

Corson, Dale, 167, 182

Corson panel: Corson Report, 32; critique by the AAAS, 175; critique by the Technology Transfer Intelligence Committee, 175; export controls in university research, objections to use of, 172; know-how, importance of, 170; knowledge loss to Soviet Union in academic contexts, 168–69; security in accomplishment vs. secrecy, 169. *See also* gray zone of research

coupling load analysis, 283

Cox, Christopher, 280

Cox Report, 34, 259, 280–89; critiques of findings, 281

Credit Lyonnais, 225
critical technology, 124, 125, 127, 253, 305–6; allies in Cocom, opposition to critical technology approach, 132
cryptography, cryptology, 48, 143
Cuba, 4
cultural nationalism, 317
Cultural Revolution, 234, 241
Currie, Malcolm, 101, 110, 115
customs, 9

Daladier, Édouard, 203
Davida, George, 150, 152
Davis, Ruth, 128, 129, 130
deemed export, 32, 167, 187–89, 292
Deemed Export Advisory Committee (DEAC), 188
Defense Intelligence Agency (DIA), 225–26, 243
Defense Production Act of 1950, 207, 218
Defense Science Board (DSB), 152, 156, 211, 212, 214–16, 221, 223
Defense Science Board Task Force on University Responsiveness to National Security Requirements, 153–56; use of research contract to control knowledge circulation, 155
Defense Technology Security Administration (DTSA), 219, 240, 263, 269, 272, 273, 275
DeLauer, Richard, 153, 166, 179–83
Democratic Party, business support and lobbying, 279
Deng Xiaoping, 7, 232, 237; on capitalist development in the US, 236; and self-reliance, 241; Sixteen Characters, 259
Department of Energy, 31, 218; restrictions on collaboration with China, 333
Department of Treasury, administration of sanctions, 8, 9, 31; and CFIUS, 218
détente, 24, 31; benefits to Soviet economy and military build-up, 110; criticism of, by Bucy, 32; functionalist justifications for US-Soviet collaboration, 109, 121; and Huntington, 109–10; and trade liberalization, 101, 104, 108, 109, 117, 121; and use of US technical consultants, 120
Dicks, Norm, 280
Dickson, David, 174

diversion from civilian to military use of technology, 114
Division of Research and Engineering, DoD, 120
DoD-University Forum, 166; Working Group on Export Controls, 166, 179, 180
Dong Feng missiles, 239
Dresser Industries, 104, 126
dual-use technologies, 2, 8, 25, 68, 113–14, 196, 213, 217, 237, 252–57, 265–66, 286, 289, 299, 303
Duke University, 326
Dulles, Allen, 94
DuPont, 93

EAA. *See* Export Administration Act of 1969 (EAA)
East European agents, 138
Eastland, James O., 67
ECA. *See* Export Control Act of 1949 (ECA)
economic espionage: industrial espionage, 28, 51, 53, 78, 225, 301, 311, 312, 313; in university laboratories, 327
Economic Espionage Act of 1996, 9
economic sanctions, 8, 9, 31, 41–42, 301
economic security, 24, 25, 33, 34, 134, 220, 228, 229, 231, 318, 323; concept, 195, 198, 207, 211, 307–13; and national security under Clinton, 256, 266; and proliferation, 260, 264; and Trump administration, 26, 300–303, 304, 307–13
economic warfare, 6, 14, 22, 30, 40, 43, 70, 72, 86, 111, 123, 126, 200
Edison, Thomas Alva, 39
Ehrlich, Thomas, 178–79
Eighth Five-Year Plan (PRC), 261
Eisenhower, Dwight D., administration of, 82, 177
Elachi, Charles, 292
electronics, 67, 86, 105, 218
emerging technologies, 167, 169, 181, 253
Emerson, Thomas, 161
end-use assurances, 114, 246
Enhanced Proliferation Control Initiative (EPCI), 245
Espionage Act of 1917, 9, 37–39, 55, 63, 67
Europe (Common Market), 105; European Union, 260
European Space Agency, 292

INDEX 433

Executive Order 12356 (Reagan), 177
Executive Order 12981 (Clinton), 255
Exon, J. James, 226
Exon-Florio Amendment, 195–98, 206, 207–8, 217–18, 226–27, 304. *See also* Foreign Direct Investment (FDI)
export, definition, 12, 49, 75, 81–82
Export Administration Act of 1969 (EAA), 24, 102, 104, 108; as encouraging trade, 111; as loophole to import dual-use technologies, 113; 1974 amendment, 117; 1977 amendment, 125; as nonproliferation instrument, 245; used to restrict foreign nationals attending international conferences, 143–44
Export Administration Act of 1979, 24, 218
Export Administration Act of 1994, 187
Export Administration Regulations (EAR), 8, 9, 12, 26, 32, 219
Export Control Act of 1949 (ECA), 7, 22, 24, 72–73, 76, 102, 111; extension of, 111; sunset provision, 21, 304; and technical data, 21, 102
Export Control Reform Act of 2018 (ECRA), 8, 17, 26, 303–5, 307–9, 311
export controls: defined by Congress, 259, 278; economic security and, 256; on free circulation of knowledge, 143; on papers at international conferences, 143, 168; precipitous decline under Clinton, 256; reforming of by Clinton, on dual-use technologies, 254–57; regime change from denial to approval with the Soviet bloc, 245; restricting foreign nationals on campus, 144–49; university research, restricting of, 173
extraterritorial reach of export controls, 23; critics of, 4, 121

face-to-face knowledge transfer, 121, 187, 292; prized by Soviet visitors to US academia, 141; risks to national security, 134; in satellite launch campaigns, 269, 284, 285, 289
Fairchild Semiconductor International, 117, 208, 210, 211, 214, 217, 221
Farewell papers, 137
Federal Acquisition Streamlining Act of 1994, 253
Federal Bureau of Investigation (FBI), 31, 51, 52, 56, 78, 313; and research cooperation with China, 314, 315, 318, 319, 324–26
Federal Employee Loyalty program, 96, 189
Federation of American Scientists, 77, 84
Ferguson, James, 160, 164
Feshbach, Herbert, 144
First Amendment, 6, 13, 16–17, 83, 136–37, 143, 151–52, 155, 156, 172; controls on technical speech, 137, 159–65
5G technology, 302, 306–7
Ford Motor Company, 51
foreign availability and export controls, 10, 131–32
Foreign Direct Investment (FDI), 6, 34, 206–10, 218, 224; from China, 218–19, 304, 305, 312; from Japan, 195, 208–10, 221, 224; by US multinationals in Europe, 105
Foreign Investment and National Security Act of 2007, 304
Foreign Investment Risk Review Modernization Act of 2018 (FIRRMA), 26, 303–5, 307–9, 311
foreign nationals, 153–54; loyalty and deemed exports, 188; visits to universities, restriction of, 144–49, 187–89
Four Modernizations, 234, 241
France, 1, 2, 28, 42, 53, 77, 79, 88, 89, 106, 218, 225, 227, 245, 249; Bucy approach, criticism of, 132; exports to Soviet Union, 109, 112
Franta, W. R., 146
Freedenburg, Paul, 276
free market: and containment, 72; free trade, 6, 14, 19, 23–24, 26, 62, 308; 1980s pushback against, 195–96, 197–98, 205, 209, 216, 226; Trump, critique of, 310
Frost, Ellen, 128, 129
FSX jet fighter, 221
Fujitsu, 106, 123, 208, 210, 214, 217, 227
Fundamental Research: Fundamental Research Exclusion (FRE), 32, 167, 183–89, 292; under threat, 331, 332, 333
Fusfeld, Herbert, 115
Futron Corporation, 288

Garson, John R., 103
General Accounting Office (GAO), 219, 255, 256, 264, 271
General Electric, 52, 117

George Washington University, 226
Gerjuoy, Edward, 162
German Reich, 20, 27, 39, 40, 42, 57; rocket technology, 53; technology transfer from after World War II, 50, 52–53, 58
Germany (East), 92, 193
Germany (West), 1, 25, 47, 89, 106, 218, 245, 249; critical of Bucy approach, 132; exports to Soviet Union, 109, 112; "national style" of export controls, 23; "Ostpolitik," 22
Gerth, Jeff, 272, 278, 280
globalization, 14, 106, 108, 110, 112, 113, 117, 188, 194, 196, 205–6, 208, 212, 213, 228, 295, 300, 310
Global Positioning Satellites (GPS), 252
Goldberger, Marvin, and export controls in university research, 147–48, 181
Golden, David, 319
Golden, James R., 200
Gololobov, Mikhail, travel in US, restriction of, 145, 148
Google, 299, 335
Gorbachev, Mikhail, 196
Gore, Al, 274
Gouzenko, Igor, 59–60
Grassley, Chuck, 331
Gray, Paul, and export controls in university research, 147–48, 181
gray zone of research, 3, 7, 13, 68–70, 167; abolished by DoD, 181–83; abolished in NSDD, 189, 184; and Corson panel vs. DoD task force, 173; creeping grayness, 182; criteria as defined in the Corson report, 172; critique of definition in Corson panel, 174
Great Leap Forward, 234, 241
Groves, Leslie, 54–55
Gulf War (1991), 232, 251, 252, 259
Gustafson, Thane, 116

Hansen, Larry, 131, 132
Harry Diamond Laboratory, 140
Harvard University, 178, 185, 321, 335
Herter, Christian, 94
Hewlett Packard, 117
Hitler, Adolf, 203
Hong Kong, 249, 334
Hoover, J. Edgar, 56
Huawei, 299–303, 305, 334

Huels AG, 218
Hughes, Gregory, 261
Hughes Aircraft, 239, 272, 273, 274, 279, 283, 289, 293–94
Huntington, Samuel P., 109; characterization of détente, 109; and Japan, 201, 203–4; and Soviet military adventurism, 110
Hydrocarbon Research Incorporated (HRI), 1–5, 95, 97, 103, 293; IBM, 94–95, 106, 114, 213, 293

IBM, 106, 114
ICBM, 265, 282, 286
Ichord, Richard, 127
Ikle, Fred, 179, 183
Immigration and Nationality Act of 1952, 77–78
Immigration and Naturalization Service, 77
Independent Review Committee, 272, 284, 285
industrial base, 24–25, 33, 65, 205, 210–12, 216, 220, 225, 308, 309, 311
industrial policy, 205, 217
Infineon, 300
Ingersoll-Rand, 137
Inman, Bobby, 151; and economic security, 231; knowledge loss, responsibility of university in, 170; national security, criticism of scientific community's attitude toward, 156–58; open dissemination of cryptography research, opposition to, 150
Institute of Electronic and Electrical Engineers (IEEE), 168
instrumentation industry, 102, 106
insurance. *See* satellites
intangible knowledge, 283, 284
Intel, 117, 299
intellectual property (IP), 12; acquired through Thousand Talents Program, 322; expansive concept of IP theft, 330; mitigating risk of IP loss, 328–34; Soviet disregard of, 52, 57; theft by China, 248, 314–15, 318, 320; theft exaggerated, 320–21. *See also* patents
intelligence community, 17, 31, 94, 131, 133, 139, 141, 150, 157, 159, 170, 175, 185, 261, 281, 295, 313, 315, 329, 334, 339n29
Intelsat 708 satellite, 270, 271, 284

Internal Security Act of 1950, 77–78
International Atomic Energy Agency (IAEA), 8, 245, 249
International Emergency Economic Powers Act of 1977, 218, 303
International Technology Transfer Panel (DoD), 180
International Traffic in Arms Regulations (ITAR), 8, 9, 12, 13, 26, 32, 168, 219, 292. *See also* United States Munitions List (USML)
Invention Secrecy Act, 9, 40, 47–48, 56, 67, 92
Iran, 60, 225, 239, 245, 274, 328, 329
Iraq, 232, 251
Ishihara, Shintaro, 221–22
Israel, 107, 260
Italy, 41, 245, 249

Japan, 20, 24, 25, 27, 29, 33, 47, 50, 105, 106, 107, 134, 193–230, 249, 262; exports to USSR in the 1970s, 112
JASON group, 331
Jeremiah, David, 281
Jet Propulsion Laboratory, Pasadena, 292
Johnston, Richard, 248
Jones, Anita, 254

Kama River Motor Vehicle Plant (KamAZ), Western equipment in, 137
Keith, Percival C., 2, 4, 103, 293
Kennan, George F., 60, 62
Kennecott Copper, 115
Kennedy, Donald, 166, 185; and export controls in university research, 147–48, 181
Kennedy, Paul, 107, 199, 202–3
Kennedy, Scott, 323
keystone technologies, 118, 124, 128, 129
Keyworth, George, 181, 183
KGB, 137, 139, 141, 158
Khrushchev, Nikita, 203
Kissinger, Henry, 108
know-how, 2, 3, 11, 31–32, 33, 50, 64, 67, 80, 81, 92, 98, 102, 104, 194, 217; China's quest for "the hen and not just the egg," 242, 243; Defense Science Board on, 216; discussion in Cocom, 87–89; and historians of science and technology, 15; importance emphasized by Corson panel, 170; and know-why, 276; origin of term, 57–58; portability from civil to military domain, 114, 282–86; shared in launch accident inquiries with the Chinese, 282–86; Soviet quest for processing know-how, 114; as "unembodied" technology, 242
knowledge, circulation, 103; tacit, 103. *See also* know-how
Korean War, 21, 65, 71, 74, 77, 79–81, 85–87, 96, 235
Kuwait, 232, 245

Lamb, Robert, 279–80
Lawrence, Ernest O., 55
lead time, 27–28, 103, 104, 111–17; first mentioned, 44; as perishable asset, 116
League of Nations, embargo against Italy, 41
Lee, Peter, 280–81
Leitner, Peter, 275
Lend-Lease program, 42, 51
Les Trentes Glorieuses, 106
Lewis, James, 264, 324
Libya, 225. *See also* rogue states
Lieber, Charles, 321, 335
linkage strategy, 22, 29
Lockheed Martin, 101, 117, 140, 188
Long March rocket, 239, 265–66, 270–73, 283, 285, 287
Loral Space Communications, 271, 272, 289, 293–94; and loss of launch debris, 273
Los Alamos weapons laboratory, 170, 280–81
Low, Francis, on foreign visitors to research universities, government restrictions on, 148
LTV Corporation, 224–27
Lumentum Holdings, 299
Luttwak, Edward N., 200

machine tools, 11, 23, 25, 43, 83, 90, 217, 224, 228, 281
Made in China 2025, 322, 323, 333
Maier, Charles, 107
MAMCO Manufacturing Incorporated, 218–19, 226
Manhattan Project, 47, 54–55, 61
Mansfield Amendment, 153
Mao Zedong, 6, 234
Marshall Plan, 64, 71, 72

Martin, Edith, 112, 167, 174, 180–84
Martin Marietta, 225, 274
Massachusetts Institute of Technology (MIT), 147, 148, 181, 185, 335; as source of technology transfer, 170
Massey, Joseph, 248
Mastanduno, Michael, 14, 102, 104, 122, 132
Mazarr, Michael, 252
McCain, John, 279
McCarran Act of 1950, 77–78
McCarran-Walter Act of 1952, 77–78
McCracken, Ed, 262
McDonnell Douglas, 117
McFarland, Ernest W., 67
McGrath, Peter, on foreign visitors to research universities, government restrictions on, 148
McMahon Bill, 55, 56, 66
MD Anderson Research Center, 329–30
Meijer, Hugo, 238, 257, 406n108
mercantilism, neo-mercantilism, 195, 205–6
Milhollin, Gary, 263–64, 274
Militarily Critical Technologies, targeted by Soviet Union, 138
Militarily Critical Technologies List (MCTL), 104, 127–34; and CCL, 128; Congressional aspirations for, 128; corporate frustration with, 131; first draft list by DoD, 129–30; Initial MCTL, 130; National Academies' position on, 133; relation to Bucy Report, 128
military-civil fusion, 324
Milton, John, 161
Missile Technology Control Regime (MTCR), 8, 9, 245, 270, 271, 274; and sanctions, 247, 250, 290, 291
mitigating risk of IP loss, 328–34
Mitterrand, François, 137
monitors at launch campaigns, 269–70, 273, 288–91
Monsanto, 218
Moore's law, 263
moral embargo, 20, 41, 44
Morita, Akio, 221
Morland, Howard, 160
mosaic theory, 3, 74
most-favored-nation status for China, 236; debates over post-Tiananmen protests, 247–49
Motorola, 117

Nacy, Loring, 93
National Academy of Sciences, 18, 25, 32, 146, 223
National Advisory Committee for Aeronautics, 39
National Aeronautics and Space Administration (NASA), 39, 235, 292
National Center for Manufacturing Sciences, 217
national identity of technology, 3–4, 23
National Innovation-Driven Development Strategy, 323
National Institutes of Health, 322, 328–29, 330
National Research Council, 39, 71, 76
National Science Board, 185, 315
National Science Foundation, 235, 331–33
national security, 5, 6, 7, 16, 32, 34, 64, 69, 73, 207, 213; and civil liberties, 137; concept of, 19, 21, 24, 25, 33, 68, 70–71, 193–94, 223, 335; and democracy, 69–70, 98, 133, 198–206; and economic security, 198–206, 223, 228, 309–10; and innovation system, 62; intellectual property as object of, 12
National Security Act of 1947, 71
National Security Agency, 143, 149–52, 273
National Security Council (NSC), 39, 175, 247, 277
National Security Decision Directive (NSDD) 84 (Reagan), 179
National Security Decision Directive (NSDD) 189 (Reagan), 184, 186
national security state, 1, 7, 18, 24, 27, 31, 32, 37, 60, 61, 62, 65, 70, 71, 85, 95, 98, 112, 163, 176, 181, 183, 205, 206, 292–94, 295, 335
National Security Strategy for Critical and Emerging Technologies, 332
National Security Study Directive (NSDD) 14-82, 175
NATO, and Cocom, 22, 72; and the Warsaw Pact, 244
Naughton, Barry, 234
Naval Consulting Board, 39
Navarro, Peter, 310–11
Nelson, Richard, 106
Netherlands, 42, 208
Neutrality Acts of 1935 and 1937, 42
New York University, 94

Nixon, Richard, 107, 108; evenhandedness with China, 235
nonproliferation entrepreneurs, 257, 263, 274
Nonproliferation Policy Education Center, 275
North Korea. *See* rogue states
Northwestern University, 140
Nuclear Non-proliferation Treaty (NPT), 8, 250
Nuclear Regulatory Commission, 31
Nuclear Suppliers Group, 8, 9, 245
nuclear weapons, 8, 27, 47

Obama, Barrack, administration of, 186, 304, 314, 315
OECD, 105, 245
Office of Censorship, 48–49
Office of Foreign Assets Control (OFAC), 9
Office of Science and Technology Policy, 175, 227
Office of Scientific Research and Development (OSRD), 54, 61
Office of Technology Assessment (OTA), 222, 235, 236, 242
Office of the US Trade Representative, 324
oil and gas technologies, 11, 23, 44, 48, 52–53, 83, 104
Omnibus Trade and Competitiveness Act of 1988, 207
One China policy, 236
Operation Desert Storm, 251–52. *See also* Gulf War (1991)
Operation Exodus, 147
Oppenheimer, Robert D., 55
Optus-B2 satellite, 273, 282
Organization of the Petroleum Exporting Countries (OPEC), 107, 207

Pajaro Dunes, 185
Pakistan, 274
passports, 38, 49, 76–79, 80
patents, 2, 46–47, 50, 53, 57, 59, 65, 92; and Invention Secrecy Act, 40, 47–48, 56; and Trading with the Enemy Act of 1917, 39–40, 48; and Union, 52, 57, 63; unpatentable knowledge, 57–58
People's Liberation Army, 260, 265
Perkin-Elmer, 117
Perle, Richard, 102, 104, 179–80, 183

Perry, William, 113, 125, 129, 237, 253; exports and direct military significance, 126
Petersen, Peter G., 109
Philips, 106
Poland, 10, 89; declaration of martial law, 24
Pratt and Whitney, 117
Presidential Proclamation No. 2465, 45
Press, Frank, 144, 145, 146, 167, 235
Prestowitz, Clyde, 209, 224
Priestap, E. W. (Bill), 314, 318, 327
proliferation, 225, 250, 309
Public Cryptography Study Group, 151–52, 157; and prepublication review, 151–52

Qi Yulu, travel in US restricted, 146–47, 148
Qualcomm, 299

radar technology, 48, 54, 56, 65, 67
Radio Corporation of America (RCA), 51, 62–63, 65
RAND Corporation, 116
Reagan, Ronald: administration of, 6, 33, 135, 136; and civil liberties, 137, 176; and Japan, 194, 210; relaxation of export controls on trade with China, 237–40; and research on export controls, 16; restrictions on the free circulation of knowledge, 143; on satellite launches on Chinese rockets, 239–40; on scientific communication and national security, 175; weapons sales to China, 239
realism, 14, 195, 197, 200–201, 204, 209, 228
reciprocity, 57, 58–59, 66, 70, 243, 310, 332
reclassification, 177
reexports, relicensing, 4, 13, 81, 87, 96, 121, 122, 144, 195, 224, 225, 300
Reif, Rafael, 325
Reinsch, William, 25, 256, 287
Relyea, Harold, 41
restricted data, 66–67
reverse engineering, 6, 10, 11, 53, 116, 130, 242, 311, 377n5
Revolution in Military Affairs, 233, 251–52, 259
Rhodes, Frank, and export controls in university research, 147–48
Rice, Condoleezza, 186, 332
Risen, James, 280
rogue states, 7, 250, 256, 328, 329

Roosevelt, Franklin D., 42, 44, 45, 61, 62, 205
Rubio, Marco, 313
Rumania, 1–3, 10, 103
Run Faster Coalition, 257, 406n108
Ruopeng Liu, 320, 326–27
Russia, 52, 64, 141, 260, 261, 262, 263, 264, 290, 291, 307, 310, 312, 313, 328, 329, 333, 334

Saddam Hussein, 232, 244, 245, 251, 259. *See also* rogue states
safeguards on satellite launching, 247, 269, 276, 288, 290
Sakharov, Andrei, 135
Salovey, Peter, 325
Samuel, Richard J., 204
sanctions on trade with China, 247
Sargent, Daniel, 107
satellites: changing regulatory responsibilities in the executive, 266–69; communications satellites, 23, 25, 34, 239; components regulated by the USML in early 1990s, 267; insurance against failed launch, 265; monitoring launch campaigns, 269–70; presidential waiving of sanctions on launch, 271; restrictions imposed by Congress on launching, 247; safeguards, 240; sanctions on launching from China, 248, 270. *See also* Technology Transfer Control Plan
Saudi Arabia, 239, 334
Saxon, David, and export controls in university research, 147–48
Schwartz, Bernard L., 279
scientific and engineering "manpower," 105, 153, 214, 316
scientific internationalism, 75, 98, 317
scud missiles, 232
Searls, Melvin, 249
Second Cold War, 135
secrecy (classification), 2, 7, 13, 28, 39, 47–48, 68, 108, 206, 219; of CFIUS deliberations, 228; declassification trend at end of World War II, 53–56; as defined by Espionage Act of 1917, 38; Executive Orders, 82, 177; of export control licensing decisions, 18. *See also* Espionage Act of 1917; Invention Secrecy Act
Sedition Act of 1918, 38

self-reliance, 241, 242, 243
Sematech, 216–17
semiconductor and solid-state industry, 23, 24, 102, 106, 113, 131, 132, 133, 134, 196, 210, 214–17, 218, 222, 300
Semiconductor Industry Association, 131
Senate Select Committee on Intelligence, 259, 280
sensitive but unclassified (SUB), 13
Sensitive Compartmentalized Information (SCI), 178
Servan-Schreiber, Jean-Jacques, 105, 106
Shanghai Communiqué, 235
Shattuck, John, 163
Shelby, Richard, 281
Shils, Edward, 78–79
short supply controls, 43
Silicon Graphics, 262; Power Challenge XL supercomputer, 263
Silkworm missiles, 239
Skyworks Solutions, 320
Smith, David, 327–27
Smyth, Henry DeWolf, 55
Smyth Report, 54–55
SNECMA, 123
Snepp v. United States, 163
socialist market economy, 259
Society of Photo-optical Instrumentation Engineers (SPIE), 168
Sokolski, Henry, 275, 276, 278, 287; intangible knowledge, loss of, 283; know-how and know-why, 276
Solarz, Stephen J., 193, 198, 199
Sony, 208, 221
Soviet Academy of Sciences, 94, 139; exchange programs with US, restrictions imposed by State Department on, 146
Soviet Union: academic exchanges with US, 90–95; acquisition of technological know-how, 51, 101, 120, 135, 136, 138, 139–40, 141, 142; atomic weapons, 21, 56; embargo against in 1930s, 42; exploitation of exchange programs, 141–42; investment in R and D, 110; and orchestrated exploitation of trade with the US, 135, 139–40; as target of embargo policy before World War II, 20; trade relations with, 14, 50–51, 57, 65–66, 112. *See also* Farewell papers
Spedding, Frank, 55

Spence, Floyd, 274
Spencer, Lydia, 224–25
sponsored research contract, 154; and First Amendment, 155; used to legitimate restricting knowledge flows, 155
Spratt, John, 281
Sputnik shock, 91
Stalin, Joseph, 59, 85, 90, 203
Standard Oil, 52
Stanford University, 147, 166, 181; Center for International Security and Cooperation, 281; as source of technology transfer, 170
Steelman Report, 64–65
Steering Committee on Technology Transfer (DoD), 179
Stokes, Mark, 260
Strange, Susan, 14
Strategic Defense Initiative (SDI), 24, 176
strategic trade theory, 205
Strom Thurmond National Defense Authorization Act for Fiscal Year 1999 (Public Law 105-261), 276, 278, 289–90; transfer of all satellites from CCL to USML, 290, 292
Suettinger, Robert, 247
Supercomputer Control Regime, 262
supercomputers, sales to Russian weapons laboratories, 262
Sutter, Robert, 236
Suttmeier, Richard, 317
Swindell-Dressler, 137

Tabak, Lawrence, 322, 330
tacit knowledge, 6, 57, 103, 115, 133, 134, 216, 285, 293–94. *See also* know-how
Taft, William, 183
Taiwan, 233, 236, 237; Taiwan straits, 263
technical data, 2, 4, 5, 10, 11–12, 44, 50, 56, 67, 70, 73, 74, 92–94, 219; academic protest against, 83–84; and censorship, 48, 102; as defined by Harold Brown, 124; definition and control of in World War II, 45–46; regulations and definitions in 1950s, 79–85; shared with Chinese in launch accident investigation, 282–86; volume of license applications, 106–7, 130
techno-industrial base, national, 24, 33, 123, 134, 233, 251, 252–54

technological gap, 28–29, 103, 105, 115, 117, 135, 233, 406n108
technology: as defined by Bucy, 102; as "the detail of how to do things," 118–119; evolutionary vs. revolutionary development, 119–20, 124–25; and know-how, 102
technology transfer, 104, 115–17, 288; active and passive mechanisms of, 103, 119, 289; diffusion, 115; and irreversible loss of knowledge, 116; and learning, 283, 285; restrictions on active modes, 125
Technology Transfer Ban Act, 127
Technology Transfer Control Plan, 269
Technology Transfer Intelligence Committee, 174
Technology Transfer Panel, 131
techno-nationalism, 28, 34, 205, 243
telecommunications equipment, export policies, 261
terrorism, 7, 13, 32, 186, 256, 304, 307, 309
Texas Instruments Incorporated, 32, 101, 106, 117
Thomson-CSF, 224–28
Thousand Talents Program, 34, 312–22, 323, 330
3Com, 301
Thurmond, Strom, 233, 275
Tiananmen Square protests, 23, 33, 233, 246; and US trade sanctions, 247, 270, 290, 291
Tianjin University, 320
Tibet, 233
Tolchin, Susan, 226
Tolmans, Richard C., 55
Toshiba-Kongsberg case, 221
total war, 6, 20, 21, 38, 45, 63, 68, 71, 212
trade, US trade policy, 11, 32, 206; East-West trade, 7, 22, 85–86, 90, 111–12, 122, 125; and national security, 7, 11, 72, 301
trade balance, US, 104, 106, 107, 202
Trade Promotion Coordinating Committee (TPCC), 268
Trading with the Enemy Act of 1917, 20, 39–40, 47
travel, freedom to travel, travel restrictions, 3, 5, 6, 9, 38, 41, 49, 51, 57, 66, 75–79, 83, 90, 91, 93, 96, 206, 333–34. *See also* visas; passports
Triplett, William, 278–79

Truman, Harry, administration of, 54, 55, 62, 64, 66, 73, 82, 86, 189, 203, 205
Trump, Donald J., administration of, 26, 34, 134, 230, 233, 265, 293, 295, 299, 300, 302–4, 319, 332, 333, 335
turnkey plants, 106, 120, 135

Umnov, N. V., travel in US restricted, 146
Unclassified Technological Information Committee, 70
UNESCO, 77
Unger, Stephen, 182
United Kingdom, Great Britain, 8, 22, 42, 53, 56, 89, 106, 132, 208, 209, 218, 245, 249
United Nations, 53–54
United Nations Atomic Energy Commission, 56
United States, exports to USSR, 109, 112; changing patterns of R&D expenditure, 65, 113, 166, 315; and Soviet exchange agreements, 109, 145
United States Munitions List (USML), 8, 266, 267, 268, 269, 270, 273, 277, 290, 292
United States v. Edler, 162, 173, 174
United States v. Progressive Inc., 160
Universal Instruments and Federal Tool Company, 140
University of California at Los Angeles, 147, 166, 183, 185
University of Illinois, 94–95
University of Minnesota, 148
University of Pennsylvania, 178
University of Wisconsin, 150
Urey, Harold C., 55
US Censorship Board, 39
US Council of National Defense, 39
use of research equipment, criteria for, 187–88

Vance, Cyrus, 145
Vanderwerker, Francis H., 47
Varian Associates, 131
Very High Speed Integrated Circuits (VHSIC) program, 148–49, 172; attempts to exclude foreign nationals from, 148, 168

Vietnam War, 107, 236
visas, 7, 9, 12, 24, 34, 38, 76–79, 318; restrictions for students from China, 333–34
voluntary information control, 40–41, 73–75, 76, 79, 80, 83, 87, 151–52, 155–58
Vossen, John, 143, 144, 145

Wallace, Henry A., 59
Wallerstein, Mitchel, 257, 269
Wang, Jessica, 189
Wang, Zuoyue, 234, 317
Warner, Mark, 319
Warsaw Pact, 244
War Trade Board, 40
Wassenaar Arrangement, 9, 25, 250, 403n73
Weapons of Mass Destruction (WMD), 7, 8, 227, 232, 245, 264, 309
Weinberger, Caspar, 180, 239
Wen Ho Lee, 280–81
Wessel, Michael, 320, 333
Westad, Odd Arne, 236
West Point, 200
Wherry, Kenneth S., 67
Wilson, David, 161, 166, 181
Wisconsin Project for Nuclear Arms Control, 263, 274
Wray, Christopher, 313, 318, 325
Wright, Gavin, 106
Wuhan University of Technology, 321

Xerox, 117
Xiaoxing Xi, 324–25
Xi Jinping, 31, 233, 324
Xilinx, 299

Yale University, 325
Young, John, 189
Young, Leo, 182

Zangger Committee, 8
Zhao Ziyang, 237
Zhenbao Island, 236
Zhou Enlai, 234, 235
ZTE, 301, 305, 334
Zuber, Maria, 334
Zysman, John, 203, 403n74

www.ingramcontent.com/pod-product-compliance
Lightning Source LLC
Chambersburg PA
CBHW071952290426
44109CB00018B/1994